# Closed Captioning

Johns Hopkins Studies in the History of Technology

Merritt Roe Smith, Series Editor

# Closed Captioning

Subtitling, Stenography,

and the Digital Convergence of Text

with Television

## Gregory J. Downey

The Johns Hopkins University Press    Baltimore

© 2008 The Johns Hopkins University Press
All rights reserved. Published 2008
Printed in the United States of America on acid-free paper
9 8 7 6 5 4 3 2 1

The Johns Hopkins University Press
2715 North Charles Street
Baltimore, Maryland 21218-4363
www.press.jhu.edu

Library of Congress Cataloging-in-Publication Data

Downey, Gregory John.
  Closed captioning : subtitling, stenography, and the digital conver-
gence of text with television / Gregory J. Downey.
     p. cm.—(Johns Hopkins studies in the history of technology)
  Includes bibliographical references and index.
  ISBN-13: 978-0-8018-8710-9 (hardcover : alk. paper)
  ISBN-10: 0-8018-8710-0 (hardcover : alk. paper)
  1. Speech-to-text systems. I. Title.
  TK7882.S65D69 2008
  384.55'6—dc22          2007020389

A catalog record for this book is available from the British Library.

*Special discounts are available for bulk purchases of this book. For
more information, please contact Special Sales at 410-516-6936 or
specialsales@press.jhu.edu.*

The Johns Hopkins University Press uses environmentally friendly
book materials, including recycled text paper that is composed of at
least 30 percent post-consumer waste, whenever possible. All of our
book papers are acid-free, and our jackets and covers are printed on
paper with recycled content.

# Contents

# Acknowledgments

This project was conceived and completed while I was an assistant professor at the University of Wisconsin–Madison, in a split appointment between departments in the College of Letters and Sciences: the School of Journalism and Mass Communication, and the School of Library and Information Studies. Both departments went out of their way, during tough budget times and tight labor situations, to grant this junior scholar the time and resources necessary to finish his work. Research travel and summer salary costs were paid with the assistance of the University of Wisconsin–Madison Graduate School and the Irwin A. Maier Faculty Development Fund. Primary and secondary textual sources for this story were culled from libraries all over the nation thanks to the first-rate staff of the UW-Madison Memorial Library. Other materials were found at the Wisconsin Historical Society (Madison), the Library of Congress (Washington, DC), the National Public Broadcasting Archives (College Park, MD), and the Gallaudet University Library (Washington, DC). My editor, Bob Brugger, and two anonymous reviewers at the Johns Hopkins University Press helped bring this project to fruition, and I thank them for their flexibility and professionalism.

Many parts of this research were first presented to my colleagues at UW-Madison, and besides the faculty and students of my home departments, I would like to thank members of the Department of Geography, the Holtz Center for Science and Technology Studies, and the Department of History of Science for their insight and advice. Faculty and students at several other schools also gave me valuable feedback, especially at Cornell University, Drexel University, Indiana University, and the Massachusetts Institute of Technology. Fellow historians Tom Haigh and Pablo Bozcowski organized productive conference sessions at the Society for the History of Technology (2004) and the Society for the Social Studies of Science (2003), allowing me to introduce my work to these groups. Some of this research was published as "Constructing

'Computer Compatible' Stenographers: The Transition to Realtime Transcription in Courtroom Reporting" in *Technology and Culture* (January 2006), and I appreciate the efforts of editor John Staudenmaier, assistant editor Suzanne Moon, managing editor Dave Lucsko, and the anonymous reviewers who helped me mold that manuscript. Finally, thanks to my colleague James Baughman for reading through the whole manuscript.

Although they did not participate directly in the production of this book, I continue to owe an intellectual debt to David Harvey, Bill Leslie, and Erica Schoenberger, who mentored me through my doctoral study at the Johns Hopkins University in the late 1990s. I'm sure they will be delighted that I found a category of information laborers even more obscure than the telegraph messenger boys of my dissertation.

Originally, this historical project on speech-to-text labor and technology was to have been paired with a present-day observation and interview project. But when I was first trained in the theories and methods of historical and geographical research, I was wisely counseled by my advisers to "follow the sources" rather than to stubbornly stick to any preconceived notions of how a particular project should unfold. The observations and arguments that follow thus rely much more on historical evidence and argument (what we can glean from primary and secondary textual material, set in the material contexts and cultural meanings of the times) than on ethnographic evidence and argument (what we can draw from observation and interview of present-day historical actors). However, I am indebted to some half-dozen members of the current captioning industry who spoke with me about their work practices, training programs, and employment strategies (I have left these informants anonymous in the text). Hopefully, the historical and geographical narrative presented here will inspire others to conduct further firsthand studies of the speech-to-text professions as they unfold.

While this book was being written, my second child, Suzanne Rose Yendrek Downey, entered this world, and I would like to thank her, along with my first child, Henry, and my wife, Julie, for their patience during each long hour that Papa spent writing on the computer. (At least you got to watch some extra TV while I scrutinized the closed captioning of PBS Kids, right?)

Finally, I would like to dedicate this book to my two grandmothers, Maebelle Downey and Melinda Krunfus, who—like many of the actors in the story that follows—balanced family life and work life through many years of twentieth-century technological change, always taking seriously their social roles, not only as parents and producers but also as purposeful citizens, in order to make the world a better place.

# Closed Captioning

# Introduction:

# Invisible Speech-to-Text Systems

Perhaps some day it may be that a reporter will be able to sit at a machine, dictate his turn, and have another machine grind out the finished transcript.

*John D. Rhodes, court reporter, 1921*

Perhaps, in time, an invention will be perfected that will enable the deaf to hear the "talkies," or an invention which will throw the words spoken directly under the screen as well as being spoken at the same time.

*Emil S. Ladner Jr., a deaf high school student, writing in the* American Annals of the Deaf, *1931*

If you were living in the United States on the morning of September 11, 2001, your household probably possessed at least one television set (98.2% of US households did) and had access to a wide variety of news and entertainment channels via analog or digital cable (68% of all US households with television did). In fact, on an average day that year, each household in the United States had at least one of their TVs turned on for a total of seven hours and forty minutes.[1] But September 11, 2001, was no average day. Four airplanes were taken over by terrorists intent on striking high-profile targets in New York City and Washington, DC; three of those planes crashed into their targets and one crashed en route. By noon, the nation faced a tragic death toll of thousands of civilian lives, in rural Pennsylvania, at the US military headquarters of the Pentagon in Washington, DC, and at the symbolic commercial center of New York City, where the Twin Towers of the World Trade Center were utterly destroyed.[2] Around the country, whether at home, in school, at the workplace, or in other public contexts ranging from corner taverns to health clubs, people in America watched in horror as these tragic events unfolded on television. But this time, the visual and auditory information medium that had united its viewing audience during so many tragedies and triumphs over the previous half-century offered something new: words. In homes with small children, parents could turn down the sound and read the latest headlines without fear of their preschoolers listening in. In noisy train terminals or quiet hospitals, crowds could watch words scroll up the screen, describing events that they could see but not hear. Most importantly, viewers who were deaf or hard of hearing (D/HOH) could receive civil-emergency information, follow the debate over the newly branded "War on Terror," and mourn with the rest of the nation in realtime—a simple right of media participation that had been denied to them for most of television's history.

The technology that allowed these words to appear on television was closed captioning, and although it had been invented three decades before 9/11, this first great tragedy of the new millennium was in many ways its public debut. Only since 1993 had all new televisions sold in the United States come ready-equipped with the closed-captioning "decoder" chips inside, and only since 1996 had the government mandated that networks and stations caption all of their programming. The gradual phase-in of captioning quotas mandated by the Federal Communications Commission (FCC) was still under way in 2001, with several years to go before all broadcast content would be covered. This slow transition was demanded by the broadcast industry because closed captioning required more than just technology—it required labor as well. On the morning of September 11, as news anchors broke from their TelePrompTers and struggled for words to describe what they were witnessing, dozens of realtime "stenocaptioners"—combining the latest in high-tech communications with century-old stenographic tools and techniques—worked across the nation through the long day (and night) to translate these voices of authority into text.[3]

In Des Moines, Iowa—over a thousand miles from Ground Zero—Holli Miller, a thirty-two-year-old stenocaptioner for the nonprofit, Washington, DC–based National Captioning Institute (NCI), was logged in remotely to Fox affiliate WNYW in New York City, watching the morning news broadcast over satellite television and simultaneously typing the captions for that broadcast using a high-speed phone line, a personal computer, and a special "stenotype" keyboard. She had already been working for three and a half hours—nearly a full day's shift in the stressful world of realtime captioning—when the terrorist attacks began. However, she kept on working for eight more hours that day because "she feared that if she stopped, it might be impossible for anyone else to get a working phone line in Manhattan and replace her." In Santa Rosa, California, Gale Muehlbauer called her for-profit captioning company as soon as she heard of the attacks, to let them know she was available for remote captioning. Then she began preparing for the work: "making notes of spellings, trying to decipher if the correct spelling is *al-Qaida* or *al-Qaeda*, . . . writing names of reporters from the major networks, anyone whose face might show up on the screen during any minute of many days ahead." After a half-hour of entering such stenotype-to-English translations into her personal computer, she got the call to go on the air, in rotating six-hour shifts over the next six days. Emergency work for these captioners was much harder than their normal routines. According to the *New York Times,* "Networks suspended commercials, as did local stations, robbing captioners of their breaks and opportunities to use the bathroom." Phone lines were in short supply, so normally competing captioning firms had to share open connections for fear of losing their data links. And inevitably, captioning errors increased as fatigue set in, as an unfamiliar vocabulary spilled across the airwaves, and as captioning trainees expecting to cut their teeth on the Home Shopping Network were instead thrust into governmental press conferences and international debates. But the emotional intensity was perhaps the hardest challenge. Reported one live captioner, "After 16 years working in the Los Angeles Superior Court reporting horrific criminal cases, including death penalty cases, I thought I had the ability to steel my emotions and do my job. But nothing in my past prepared me for this."[4]

The tragedy of 9/11 brought the closed-captioning labor force a momentary visibility it had always lacked. In the year that followed, popular articles and television advertisements began to promote live captioning as "the career of the future." Prospective stenocaptioners could work at home and set their own hours; they could learn the trade in two short years of classwork at a technical college, taking courses over the Internet; and they could earn salaries ranging from $35,000 to $100,000 annually.[5] Live captioning was described as a wide-open field, with a demand for ten times more realtime stenographers than the few hundred who were currently employed. The government was even willing to subsidize captioner-training costs, granting millions of dollars per year to captioning schools. And as apparent both from the individuals portrayed in the television advertisements, and from the roster of stenocaptioners who worked throughout the first week after 9/11, over 90 percent of these live captioners were women.[6] Here was a high-tech career that promised to cross various "digital divides" of location, education, class, and gender, which the federal government had long feared might harm US competitiveness in the "new economy."[7] More than that, it was a career where one could "do well" (making lots of money) by "doing good" (bringing broadcast accessibility to marginalized populations). Why had so few people ever heard of it?

Closed captioning, like many technological systems designed for processing and moving information over time and space, suffers from a visibility paradox.[8] The very design of captioned text to be "closed" instead of "open"—that is, ever-present in the broadcast signal but invisible on the screen until the feature is turned on by the user—was meant to keep it out of sight of hearing television viewers. The very ability to perform such captioning remotely hides this work in decentralized home offices around the nation, at all hours of the day and night. And the task of live captioning itself—rendering a just-in-time, close-to-verbatim transcription of the audio track of a program—is supposed to proceed unobtrusively, becoming apparent only when an error is made or a caption is garbled. But this "sociotechnical" information system—involving both people and machines—hides itself in both time and space in other, subtler ways as well. The same publicity granted to a few "online" realtime captioners in times of crisis like 9/11 overshadows the day-to-day work of many more "offline" post-production captioners—those who work on prerecorded, rather than live, television shows. These captioners are subject to daily turnaround time pressures no less burdensome than those of their realtime-writing counterparts, all the while being expected to sustain higher degrees of accuracy and art in their caption writing, placement, and timing. The invisibly encoded captioning text, whether put there by online or offline laborers, is a digital trace hidden inside an analog television signal, information that can lurk within a videotaped program for years until that program is digitized, "repurposed," and revalued—with the closed captioning then serving as a crucial text-searchable time index to the tape's content. And media attention to television captioning minimizes the vast and diffuse daily captioning work of other realtime stenographic writers who labor not in home offices or broadcast booths but in public courtrooms, private lawyers' offices, and university lecture halls around the nation.

This book illuminates the sociotechnical system of television closed captioning by tracing both its history and its geography in parallel with two other intertwined "speech-to-text" systems, courtroom stenography or "court reporting" (both state-mandated and profit-driven) and film dialogue titling or "subtitling" (for both entertainment and education).[9] Along the way, other technological and social systems for translating speech to text—tape recorders and secretarial transcribers in the modern business office, sign-language interpreters and student note takers in the mainstreamed classroom, and artificial-intelligence computer systems and human language translators in the cold war research laboratory—play key roles as well. The common thread in all these stories is that as the entertainment, educational, and administrative realms of American society exploded with information in new auditory and visual "multimedia" forms in the twentieth century, producing a rapid and reliable textual representation of this information was increasingly necessary. But at different times and in different places, different texts were produced for different audiences and for different political, economic, and social purposes. Sometimes the audience was public and the purpose was juridical: in the courtroom, a transcription of every word that was uttered by judge, witness, and counsel was crucial to ensure the legal right of appeal. Sometimes the audience was private and the purpose was educational: in the classroom, a condensation of complicated spoken lectures was necessary for students unable to understand auditory communication. And sometimes, although the audience varied over time and space, the purpose was to reap risk-free profit from recycled entertainment content: first in the global cinema and later on the global television screen, translating a story for many different language markets helped to sell that story over a media-saturated globe.

Over the course of a century, the practices and players in these diverse areas of information manipulation sometimes connected with each other and sometimes ignored each other. In the end, however, technological processes of networked digital computing, political-economic processes of regulation and privatization, and social processes of inclusion and globalization all helped these various speech-to-text practices to converge, until today the boundaries between court reporting, film subtitling, educational interpretation, and closed captioning are all but broken. Only by tracing the histories that fed this technological, corporate, and labor convergence can we understand and evaluate these information practices today.

## Tracing Speech-to-Text Technologies over Time and Space

The exploration of these diverse but related speech-to-text technologies and practices must cross many scholarly disciplines—from disability studies, communication studies, and educational research (in terms of audiences) to legal studies, information studies, and science and technology studies (in terms of artifacts). As a result, in this book I take an interdisciplinary approach, offering a narrative that is both historical (tracing change over time) and geographical (paying attention to difference across space) and focused on technologies (tools, machines, and the systems of action in which they are put to use to achieve certain results) and on laborers (individuals and groups of varying power and position who produce and reproduce those results).[10] Tracing a history does not simply mean cataloging "what happened when" but revealing and explaining the moments when diverse social actors had the willingness and the power to change the conditions of their lives—or to resist such changes. Similarly, charting a geography does not simply mean mapping "what happened where" but describing and analyzing the temporal and spatial limits of social action—and the ways new technologies redefine those limits.

Two particular subfields of history and geography, known as "history of technology" and "human geography," are most useful in studying the construction of closed captioning. History of technology deals with the ways humans imagine, create, and use both conceptual and material tools in pursuit of personal and social goals.[11] Technologies, whether they are "things" or "ways of doing things," do not exist in isolation; they are embedded in material contexts with other technologies, working together and "scaling up" into systems, networks, and even internetworks. These wider-scale technologies may even come to represent such important preconditions for social action that they take on the status of "infrastructures," as with systems of power distribution, networks of transportation, or internetworks of communication. In this way, technologies are socially embedded within the fabric of our systems of production and reproduction.[12] Such technologies are never simply mechanical in nature; instead, they inevitably involve human decision making in a process of "sociotechnical change."[13] Human involvement means that technologies are both socially produced (created, used, and destroyed for particular political, economic, and social purposes) and socially constructed (imbued with meanings and messages that make sense only with reference to a particular cultural moment).[14]

Historians of technology have not had much to say about the production and construction of closed captioning, however; only recently have they begun to explore technologies of accessibility and/or disability (such as the history of the wheelchair).[15] But the history of technology has long dealt with information and communication technologies, or ICTs—from the earliest days of the telegraph to today's World Wide Web—that are fundamental to the speech-to-text industry.[16] Recent history reveals a pattern of "digital convergence" in ICTs, with formerly distinct modes of information like audio and text, formerly distinct consumer products like telephones and cameras, and formerly distinct industries like broadcast and telecommunications, all combining in new ways through the transformative power of digital computers.[17] But convergence is impossible without technical and social standards—the "instruments that encode knowledge and order labor relations" in sociotechnical systems.[18] The production of the algorithms, rules, and regulations embodied in standards—a task mediated not just by engineers but also by consumers and workers—is a crucial part of innovation and competition in ICT development.

Technology consumers have received renewed attention from historians of technology, but technology workers have not.[19] In the past, new technology has been alternately defined as the tool of the skilled laborer, the engine behind the productive laborer, or the automation that should replace the intransigent laborer.[20] And nearly every historical claim about an ICT "revolution" has involved some sort of broad shift to "knowledge work," explicitly valorizing a certain set of (supposedly new) mental labors while relegating a parallel set of physical labors to the dustbin of history.[21] Yet the notion of "information labor" as a historical unit of analysis is lacking any secondary synthesis or coherent body of theory.[22] Focusing on the social and spatial divisions of labor within ICT—analyzing how humans are both embedded within and alienated from these technological structures—reveals that not only physical connections and legitimized standards but also ongoing efforts by workers "on the boundaries" between different technologies and organizations are necessary to keep information moving through any network.[23]

The second academic subfield used in this book, human geography, encompasses all the ways that humans imagine, inhabit, and change both the natural and built environments in which they live.[24] Such human-environment relations might be studied within individual sites and places, over extended landscapes and spaces, or between various hierarchical scales and network topologies. Geographers understand all of these spatial units as inevitably intertwined with temporal concepts—in measures like speed and distance or concepts like closeness and care; to speak of "space" is inevitably to speak of "time."[25] Geographical units of space and time are paradoxically both absolute, because they can be applied to all historical times and places, and relative, as each different historical culture will interpret space and time differently, a process we might call the "social construction" of space and time.[26] After all, the meanings of "far" and "fast" changed dramatically once steam, electricity, and petroleum were brought to bear on the technologies of transportation and communication.[27] No matter what level of technology is being used, the various spatiotemporal contexts in which humans exist are not simply given by nature but must be both produced and reproduced through the mustering of human innovation, surplus capital, and ongoing labor.[28] As such processes are repeated over time, a perpetually reproduced landscape of "uneven geographic development"—in both material and social senses—is the expected outcome.[29]

Human geography has only recently mustered such concepts to the study of disability and accessibility in society—often under transportation and mobility studies.[30] However, human geography has long involved the study of ICTs, especially the connections between social processes in both "physical" space (usually urban areas) and "virtual" space (known as cyberspace).[31] Initial studies of cyberspace geography were focused on the bodily experience of interfacing with computer networks and the social experience of consuming networked information technology in cities.[32] Newer research demonstrates how social activity in cyberspace helps restructure patterns of social activity in urban space, such as in community formation or job seeking.[33] The simple-substitution model in which physical process are transported wholly to virtual realms has given way to an understanding that "materially constructed urban places and telecommunications networks stand in a state of *recursive interaction,* shaping *each other* in complex ways that have a history running back to the days of the origin of the telegraph and telephone."[34] The details of this dialectic are still being debated, however. Manuel Castells has argued that "a variable geometry of production and consumption, labor and capital, management and information," called the "space of flows," would help rearrange the hierarchy of power within the older "space of places" represented by firms, cities, and regions. Spatial actors trying to succeed in the "informational" economy would need not only unique characteristics in physical space but also adequate connections to the new space of flows.[35] Yet the degree to which either the space of flows or the space of places might dominate social action still seems both historically and geographically contingent.

Combining history of technology and human geography can help pin down some of these contingencies. Such a study demands the use of a variety of sources—some qualitative (writings that must be contextualized and interpreted) and some quantitative (numbers that can be counted up and mapped). The bulk of this narrative rests on textual sources, both archival and published, drawn from a century of writings by the practitioners and analysts of four professions: film subtitling, deaf education, television captioning, and court reporting. Some of these texts are rooted in academic fields such as translation studies, film criticism, disability studies and communication studies. Others had their origin in government reports, curricular experiments, and legislative debates. Press reports from national and local newspapers and professional articles from trade magazines provide key sources as well. Finally, primary and archival material found at the Library of Congress (Washington, DC), the Gallaudet University Archives (Washington, DC), and the National Public Broadcasting Archive (College Park, MD) helps bring to life the behind-the-scenes discussions and deliberations between key actors in the construction of closed captioning. Interpreting this historical data with a geographical eye illuminates how people, practices, and products are not only produced and reproduced *within* space but actually serve to produce and reproduce different *kinds* of space itself. The visual space available within the television screen, the physical space contained inside the public courtroom, and the organizational space among the institutions of deaf education all have historical significance, even though none of these may be conveniently mapped on paper or filed away in the archives.

We humans make technologies, it is true, but those technologies also make us. Our technologies both enable and constrain our individual and collective agency; they suggest new possible futures even as they circumscribe our imaginations. In the same way, the environments we humans build paradoxically shape our ability to act on (or with) our environment. Each round of investment in local, national, or global infrastructure, in an ongoing attempt to overcome the "friction of distance" in human activity, becomes a new obstacle to further demands for spatial and temporal change. Historians of technology describe an ongoing series of "technological fixes," attempts by human actors to bring knowledge, technique, and technology to bear on the human condition—fixes that nevertheless often seem "autonomous," devoid of human agency in a capitalist mode of production built upon constant competitive revolution.[36] In the same way, human geographers describe a long history of "spatial fixes," attempts to produce, define, or control space and time—often for purposes of disposing of surplus accumulation in order to maintain that same crisis-prone capitalist mode of production.[37] In the long history of US speech-to-text practices, analyzing how both kinds of fixes work will add to our understanding of how new communication technology affects lived human geography—and vice versa.

## A Century of Intertwined Technologies and Labors

This book is divided into two broad sections. Part one—encompassing chapters 1, 2, and 3—provides background on the history, geography, technology, and labor of three distinct speech-to-text practices—film subtitling, television captioning, and court reporting—up to about 1980. Chapter 1 describes how the birth of the "talkies" in the late 1920s necessitated the subsequent birth of a film-subtitling industry, a set of laborers and practices who evolved along parallel but independent tracks in the United States and Western Europe. Caught in the middle were deaf and hard-of-hearing (D/HOH) audiences, who took it upon themselves to grow a film subtitling organization of their own, which eventually gained official state subsidy and support. Chapter 2 shows how these early film-subtitling efforts of the 1950s and 1960s were translated into a wider push to caption broadcast television in the 1970s. The resulting public-private partnership that emerged in 1980 was a compromise solution that attempted to provide a small measure of broadcast justice to a previously excluded minority audience, but only to the degree that the profits of the broadcast industry (and the goodwill of the mass market of hearing viewers on which these profits rested) would not be threatened. Finally, chapter 3 details the growth of the court-reporting profession through three types of stenographic technology and practice: pen-based, keyboard-based, and computer-based. This speech-to-text profession was largely insulated from film subtitling and television captioning until the 1980s, but the divisions of labor that court reporters developed and the technological threats they faced would later have a profound impact on the development of audiovisual titling.

Part two—chapters 4, 5, and 6—describes how court reporting, educational interpretation, and closed captioning became intertwined from the 1980s onward, as a result of a series of opportunities and threats coming from the arenas of technological innovation, state regulation, and market restructuring. Chapter 4 describes how the tools and techniques of computer-aided stenography acquired realtime speed and compact portability through successive rounds of "context jumping," first into broadcast captioning, then into classroom interpreting, and finally back into courtroom transcription. Through this transition to realtime performance, previous standards for, and meanings of, both court reporting and captioning were challenged and redefined. Chapter 5 explains that, even with its new realtime capability, closed captioning remained under threat throughout the 1980s, until the federal government finally decided to re-regulate the system by mandating both the inclusion of chip-sized captioning decoders in every television and the captioning of virtually all broadcast content, new and old. However, this victory for D/HOH interests was only accomplished through the rhetorical widening of the captioning audience to encompass hearing viewers. Chapter 6 details the consequences of captioning's new technological underpinnings and regulatory legitimacy for the wider speech-to-text industry, both in terms of the nonprofit captioning pioneers and the for-profit court-reporting agencies who increasingly competed with them. More and more individual captioners and court reporters attempting to forge careers in these industries found themselves using their expensive tools and training within an uneven and fragmented virtual and physical geography, performing subcontracted telework within an invisible infrastructure of satellite television, leased telephone lines, and high-speed Internet connections.

The conclusion builds on the story presented in parts one and two to suggest some ways that the speech-to-text professions of film subtitling, television captioning, and court reporting have been valued, devalued, and revalued over their intertwined histories. Today the text that accompanies recorded audio and video is not only used as an accessibility measure for D/HOH audiences but as metadata for indexing and retrieving high-value audiovisual "assets" digitized within globally accessible databases. At the same time, the language-based labor practices that make up subtitling and captioning are being newly recognized as translations rather than transcodings, especially by a new cohort of scholars keen on discovering how textual transformations both embody and obscure power relations between different groups. These new understandings of the value of text and the value of those who produce text can be understood as key components of the speech-to-text profession—a contested profession that continues to face upheavals as broadcast television shifts to an all-digital format.

The analytical focus of this book rests less on the initial producers or end consumers of speech-to-text technologies and more on the paid and unpaid laborers through whom those technologies come to be produced and consumed. However, such boundaries often break down in practice, as both the innovators and consumers of speech-to-text systems have at times blurred together with the laborers who made those systems work. To understand the construction of closed captioning, we must realized that what appears today to be a coherent and computerized system of translating audio to text, used across different domains of cinematic subtitling, television captioning, and live-speech display, is actually the result of a nearly century-long history involving the widely differing positionalities, social goals, and skilled labors of foreign and domestic film translators, deaf and hard-of-hearing educators and activists, and public and private court reporters. At key moments these different groups connected and misconnected, learned from each other and ignored each other, petitioned the state as political actors and innovated in the market as entrepreneurial capitalists. And these moments were all structured by spatial relationships between sites of disability education and sites of technological production, sites of television markets and sites of film markets, sites of court-reporter power and sites of court-reporter irrelevance. Just as the key speech-to-text actors produced a history, they simultaneously produced a geography as well.

Finally, a personal note. I approach this research as a hearing, English-speaking person—and as a fan who consumes subtitled foreign film, a teacher who uses online video databases indexed through closed captioning, and a parent of small children balancing their love of TV with their acquisition of reading. I have attempted to treat all of the various cultural, language, and disability communities discussed in this book with respect, especially when many members of those communities disagree stridently over whether they *should* be classified as a cultural, linguistic, or disability community. My analytical focus in this book has required me to simplify and bracket these debates; this does not, however, mean that I intend to minimize or dismiss them. I see cultural, language, and disability communities as historically contingent and socially constructed categories—groupings subject to constant change and reinterpretation over time, both from within and from without. Such change and reinterpretation is also structured by the various technological and spatial fixes affecting such social groups, whether those fixes are made in the name of accessibility or profitability, public interest or personal choice. I hope that the story I tell can help map out a territory for mutual understanding between different interest groups within the speech-to-text industries and enable them to more effectively articulate their needs, desires, and rights to the larger society in which they live.

# Part One

# Turning Speech into Text
# in Three Different Contexts

# Chapter 1

# Subtitling Film for the Cinema Audience

Nothing is simple when it comes to subtitles; every turn of phrase, every punctuation mark, every decision the translator makes holds implications for the viewing experience of foreign spectators. However, despite the rich complexity of the subtitler's task and its singular role in mediating the foreign in cinema, it has been virtually ignored.

*Abe Nornes, film subtitler, 1999*

In 1930, a short three years after the popular introduction of "talking" motion pictures in the United States with the release of *The Jazz Singer* starring Al Jolson, a teenaged violin student in New York City named Herman Weinberg was working at the Fifth Avenue Playhouse to adapt the full-orchestral scores of foreign films to the smaller chamber orchestras of the US cinema. He soon found himself out of a job—not only had sound film enabled spoken dialogue, rather than pantomimed and printed on "intertitles" spliced between scenes, but it also allowed music to be recorded, rather than played live in the theater by musicians. Weinberg was luckier than the musicians he had scored for, however; because he knew German, the language of most imported films at that time, he was able to shift from adapting music to adapting dialogue, writing new intertitles for imported films that remained incomprehensible to English speakers without a textual translation. Yet intertitling was for Weinberg a clumsy solution to the language-transfer problem, "not only silly but annoying," and he longed for a better way to link sound, image, and text.[1]

Weinberg's technological fix soon came in the form of the Movieola, a self-contained personal movie-viewer used for editing film. As he recalled, "It was like a miniature projection room. You could start and stop the film at will" and "measure not only the length of every scene but that of every line of dialogue." The Movieola allowed Weinberg to create textual "subtitles" that would flow naturally with every scene instead of breaking into the action every few moments. Weinberg soon found a photographic laboratory that could create a "title negative" out of these printed subtitles and, running them together with the original film, construct a new copy for exhibition. Armed with these new film-editing and production resources, and situated as he was in New York City—home to the greatest demand for film translation in the United States—Weinberg began to develop the rules of this new discipline of subtitling. "At the beginning I was very cautious and superimposed hardly more than 25 or 30 titles to a ten-minute reel . . . Then I'd go into the theatre during a showing to watch the audiences' faces, to see how they reacted to the titles . . . This emboldened me to insert more titles, when warranted, of course, and bit by bit more and more of the original dialogue got translated until at the end of my work in this field I was putting in anywhere from 100 to 150 titles a reel."[2] Although studios would sometimes bring other subtitle translators to New York to work on films in less widely known languages such as Japanese, Weinberg continued to do the bulk of US subtitling work for decades. By the time he retired some fifty years later, he and his small company had subtitled some four hundred films in a dozen languages.[3]

What is surprising about Weinberg's place in the history of subtitling is not the serendipitous combination of marketing need, production innovation, and labor practice in the New York film economy of the 1930s but the lack of explicit connections between his work and that of the other subtitlers who emerged around the same time—both the European professionals who subtitled American films for global cinema audiences and the American educators who began to subtitle donated Hollywood films for deaf and hard-of-hearing (D/HOH) students. Subtitling was a practice carefully designed by its practitioners to meet the cultural needs of language translation, the economic needs of commodity distribution, and the artistic needs of cinematic integrity. However, this practice was often compromised by both technical and political-economic constraints of time and space. Professional subtitling developed more or less independently in the "art houses" of Weinberg's New York City and in the distribution centers of the "subtitling countries" in Europe, where the practice competed with the more expensive but less intrusive technique of postproduction revoicing, or "dubbing."

Shunned by some critics as intrusively distracting and lauded by others as preserving a film's authenticity, subtitled film in the United States remained a commodity largely consumed by the urban elite. However, for D/HOH audiences, both the memories of the silent era and the experiences of watching subtitled imports in New York City brought this labor practice out of the art house and into the schools for the deaf, where do-it-yourself efforts by educators and activists to bring back the era of accessible, titled cinema led to a federally funded program to subtitle—or "caption"—films for deaf adults and children all over the nation.

## From Text to Talkies

As film historian Harry M. Geduld noted, "The silent film was not silent." From the start of the cinema at the turn of the century, films were accompanied by live music—"mood music," according to film scholar Norman King, "making the images more atmospheric, intensifying the emotional impact, heightening the drama, underscoring the climaxes, providing a sense of complicity." After about 1908, pit-orchestra drummers or special-effects experts might also set off "traps" or sound effects along with the action of a film. In some cases a film would be promoted as a "talking picture play," in which, according to Geduld, "troupes . . . would tour theater circuits with several films for which they had rehearsed dialogue for all the roles."[4]

Nevertheless, most silent films did include text supplements that helped to convey the main action of the film. This text need not have been correlated in time and space with the film itself—ushers could simply hand out "short program descriptions that outlined the plot and explained any complex relationships between characters."[5] In the first decade of the twentieth century, the "magic lantern" slide projector, used together with a film projector, allowed a skilled projectionist to actually pause the moving film periodically to display words on the screen. By about 1909, however, such overlays were replaced with the intertitles that so vexed Weinberg: printed cards that were photographed and integrated with the film itself. Although they complicated the film-production process and interrupted the action of a scene, intertitles required less projection skill and equipment than the magic-lantern technique, and quickly became a staple of silent film.[6] By 1926, French artist Marcel Duchamp had created a work, *Anémic Cinéma*, "almost entirely composed of inter-titles, in which verbal expressions printed on revolving discs move in exactly the same spirals as the graphic images they interrupt."[7]

One audience delighted by this marriage of text to film was the deaf and hard-of-hearing community. One writer in the *Silent Worker* proclaimed, "Unlike the spoken drama [in the theater], the deaf can enjoy moving pictures just as much as the hearing do."[8] This would be remembered as a golden age by a generation of deaf filmgoers. One recalled in 1957, "My mother took me every week to see the then current 'silents.'"[9] Contemporary teachers of the deaf agreed that their students stood to gain the most from this new medium. In 1910, the *Silent Worker* declared, "Every one who knows the deaf child knows how dear the moving picture is to its heart, and, if properly selected, how full of educational value it is."[10] Another educator commented in 1911, "At least a half dozen of our sister schools now have moving picture machines . . . Every school for the deaf should possess one."[11] Through the 1910s to the 1920s, film was seen as a unique and special tool for deaf education, on par with musical education for the blind.[12] Deaf educators sometimes feared the negative effects of film on their students—"the motion pictures become to [them] teachers whose attraction and potency are in direct proportion to [their] isolation."[13] But most agreed that taking the control of the new medium in residential schools was the best strategy. "I control the pictures," argued one, who screened mainly films that were "to my mind educational in character and yet attractive-fairy stories."[14]

Many of these educators argued that it was not the intertitles but the visual nature of silent films—including pantomime and even sign language—that made them so attractive to D/HOH audiences.[15] "The occasional interlarding of printed information," according to a 1919 opinion in the *American Annals of the Deaf*, "tempts to some exertion, but not enough to count for much."[16] One writer predicted hopefully that the acceptance of nonverbal communication in the movies would reduce prejudice against the deaf: "Having arrived at this stage, that there is an art of gesture and signs, there is but a short step to its inevitable corollary, namely, that the practice of that art in every-day life would not only be tolerated but welcomed."[17] Advocates for the deaf pointed out that the film business itself was providing new occupational opportunities for their community. After all, Charlie Chaplin's Hollywood-based studio employed the "deaf-mute actor, Granville Redmond," in roles "working side-by-side with the million-dollar comedian."[18] However, the "art of gesture and signs" used by hearing silent-film actors also worried some deaf educators: "One bad feature [of moving pictures] is the highly wrought, silly interpretive pantomime of gesticulation which is thought necessary to make one understand the meaning of the pictures. If children with their great imitative instincts fall into this dull, vulgar, overwrought style of tinhorn dramatics, our [deaf] children will grow up as puppets and posers."[19] Both the educational potential of film and the fears of its corrupting influence led the National Association of the Deaf (NAD) to film its own experts in the act of performing sign language, not only to preserve manual communication methods for future generations but also to help bring consistency to the many regional dialects of signing that had developed their own vocabulary and techniques.[20] Signed George Veditz in one of these films, "As long as we have deaf people on earth, we will have signs. And as long as we have our films, we can preserve signs in their old purity."[21] The association ended up producing over a dozen of these films from 1912 through the end of World War I.[22]

The size of the D/HOH audience for text-enhanced silent films paled in comparison, however, to the size of the immigrant and international audience for such films. Whether at home or abroad, an English-text film could quickly be edited for a non-English-reading audience simply by cutting out the old intertitles and splicing in new ones.[23] Such translations served a contradictory purpose, according to film scholars Richard Maltby and Ruth Vasey: on the one hand, intertitles "could be used to 'fix' the meaning of a scene"; on the other hand, intertitles "could remain vague and euphemistic" so that the same film could be interpreted by (or censored for) different audiences in different ways.[24] As early as 1918, the French proposed to export silent films to Spain, Italy, Switzerland, Egypt, and Greece by serially translating and splicing intertitles with each leg of distribution.[25] A decade later, intertitles "were routinely translated into thirty-six languages" for audiences around the world.[26] Such translations were crucial to US studios because, by this time, American films had already captured significant portions of the British and French markets.[27]

Into this mix of audiences, then, came sound films and recorded dialogue. But while inventors had sought to marry recorded sound with film since the 1870s, they had not necessarily thought that the silent cinema would be destroyed as a result.[28] As one contemporary remembered in 1955, "a very large business in synchronized sound seemed assured (even without any use of the system for dialogue) in furnishing sound effects, background music, and providing voice for lectures, speeches and travelogue commentary."[29] The speed and scope of the industry transition to sound surprised nearly everyone. At the start of 1928, there were twenty thousand movie theaters in the United States, but only 157 were wired to show any kind of synchronized-sound film.[30] Just a year later, more than half of the 335 films produced by Hollywood were talkies.[31] As a result, noted cinema scholar Henry Jenkins, "some two thousand theaters closed their doors in the first few months of 1929."[32] We can never know exactly what those early sound experiences were like; as film historian Alexander Walker observed, "Nothing can now reproduce the acoustics in a cinema of that time, or the fidelity of a Vitaphone amplifier, or even an audience's shock, delight or embarrassment at hearing, actually *hearing* those Hollywood divinities uttering mortal words."[33] But we do know that audiences were sometimes ambivalent about the rapid change. In one case early in 1929, as Jenkins relates, "a group of Syracuse exhibitors surveyed their audiences and found that only 50 percent of them preferred talking pictures to silent movies and that only 7 percent wanted to see silent films eliminated altogether."[34] Regardless of such sentiments, however, by about 1931, Hollywood had abandoned its production of silent films.[35]

If the rapid transition to sound cinema came as a shock to mainstream audiences and industry insiders, it was absolutely devastating to D/HOH and non-English-speaking moviegoers. One writer in the *Volta Review* echoed a common lament: "Is a new-born art to be superseded before it has reached full growth?"[36] Some decried the shift from action to dialogue and regarded this as a betrayal of film's innate visual purpose. As one deaf moviegoer at the time recalled, "The theatrical stage, of course, is not for the deaf, and we're not happy when we've paid at the box office and find the movie is practically a photographed stage-play" relying heavily on dialogue.[37] A commonly expressed notion among the deaf was that on the silent screen, as opposed to the spoken stage, "the drama, play or comedy is thrown on the screen in such perfect sequence that even without the aid of captions one has no occasion for depending upon the hearing in order to know what it is all about."[38] One deaf viewer felt, "At the best, dialogue retards and slows down action, one of the greatest charms of motion pictures."[39]

And a deaf high school student argued in the pages of the *Annals,* "The disappearance of the silent film has been a calamity to the deaf" because "the action on the screen has diminished as the talking has increased."[40]

The shift from silents to talkies gave deaf writers one of their first opportunities to articulate the social costs of media exclusion—especially in terms of what was, in 1930, "new media." The president of the Colorado Association of the Deaf reminded readers, "The deaf are shut out from so many things that make life worth living to the normally sensed—to mention only one, the radio and all it means."[41] Another writer pointed out, "Motion pictures, since their inception, have provided the only opportunity for the person with defective hearing to thoroughly enjoy a program with his family [or] friends upon an equal basis."[42] For film, which had been equitable toward the deaf for so long, to now be rendered opaque was not just a "calamity" but an insult: "The movies have been our own so long that this threatened deprivation fills me with resentment."[43] Deaf activists tried their best to convince the film studios not to abandon them, arguing their case in terms of audience sizes and lost profits. In some places, local D/HOH groups started petition drives.[44] One advocate cited "recent surveys disclosing that more than 3,000,000 public school children have defective hearing" and pointed to the lost audience of child patrons.[45] In April 1929 the Volta Bureau sent a letter to "seventeen leading producers" of talkies arguing that an audience of some 13 million D/HOH children and adults would be lost as a result of the transition to talkies. Even if "only ten per cent of this number have been attending the movies once a week," they argued, "at an average cost of thirty cents a ticket . . . they have been patronizing movie theaters to the amount of $391,862.31 a week," or roughly $20 million per year.[46] A few studios tried to placate these activists by claiming that whenever they produced a "talkie," "a silent version of the same picture will be produced at the same time with the same cast and under the same supervision as the talking pictures" and that such versions would be "shown in a certain percentage of theaters in all parts of the country."[47] But such dual-format productions, if they occurred at all, were short lived. Looking back, one deaf patron remarked, "The deaf are such a small minority that they don't mean much at the box office, and advice to producers from the deaf is academic, if not absurd."[48] Deaf leaders resigned themselves to obtaining either corporate promotional films or out-of-date silent films from sympathetic firms and distributors before the copies were all lost.[49]

The only victories that D/HOH audiences were able to achieve in cinematic accessibility actually served to divide their community rather than unite it. To understand this requires an appreciation of the diversity of deafness as both a biological and a social condition, dependent on a particular culture's definition of—and tools for—communication. Biologically, D/HOH individuals vary dramatically in the range of sounds they can hear out of one or both ears, with deaf persons usually understood to perceive a smaller range of auditory vibrations (in both frequency and amplitude) than hard-of-hearing persons can. Socially, however, the divide between deaf and hard of hearing rests on spoken-language understanding. If a person were able to perceive a wide enough range of sounds to understand spoken language, and/or if a person became deaf after having acquired spoken language, that person would likely be considered "hard of hearing"—most persons who gradually lose hearing as a result of aging would fall into this category. Hard-of-hearing persons might be more likely to hide their hearing difficulties and to rely on speech-reading or lip-reading techniques. Conversely, even persons born with a measurable level of hearing who, as children, acquired a signed language rather than spoken English, would likely be considered "deaf" in this communicative sense. The variety of differences in degree of deafness, age of onset, and acquisition of English language has, historically, resulted in an uneven demography, geography, and sociology of deafness in the United States. Even today, the prelingually deaf (those who became deaf before acquiring spoken language) are skewed toward the young while the postlingually hard of hearing (those whose hearing loss happened after they acquired spoken language) are skewed toward the elderly. In other words, the rate of hearing difficulty per thousand persons in the United States increases dramatically with age, meaning that the hard-of-hearing constituency tends to vastly outnumber the deaf constituency.[50]

Just how stark this divide was during the advent of the talkies is difficult to tell. In a 1931 review of its statistics on deafness, the US Census Bureau admitted that its numbers had "no high degree of accuracy," especially in the 1880–1890 censuses when, as historians later revealed, there was a policy "to pay enumerators a bonus for every deaf person they identified."[51] The official estimate of the incidence of prelingual deafness in 1930 was small—less than one in one thousand—but the number of postlingual hard-of-hearing persons was growing as the overall life span of the population increased.[52] Geographically, this meant that prelingually deaf children and adults tended to be concentrated in a small set of locations offering day schools, residential schools, or social clubs for the deaf, while postlingually hard-of-hearing adults were scattered across the landscape along with the rest of the elderly population. As a result, some D/HOH organizations were considered "manualist" in mission, advocating more for the deaf who used signed languages in communication—such as the NAD's efforts to preserve signed languages on film. Other D/HOH organizations were "oralist" in mission, advocating more for the hard of hearing who attempted to use spoken English in communication—such as the Alexander Graham Bell Society's efforts to teach lip reading and speech vocalization to children.[53]

This age, ability, locational, and philosophical divide in the D/HOH community had particular consequences as more and more theaters were wired for sound in the 1930s. Both manualists and oralists advocated for the return of intertitles to film, but only the oralists could settle for an intermediary solution: enhancing the sound in theaters through personal listening devices. At the 1929 annual meeting of the American Federation of Organizations for the Hard of Hearing, Bell Laboratories sponsored a film screening so attendees could "judge the merits of auditorium phones in theaters." Out of 218 responding attendees, some 86 percent claimed they understood "perfectly" or "fairly" using such devices.[54] One woman wrote in the oralist *Auditory Outlook,* "Since theatrical managers have begun actually to wire seats for us it has become increasingly possible for some of us at least to enjoy going to plays."[55] That same year, E. A. Myers and Sons advertised their Radioear system, where theater patrons plugged personal headphones into seats containing speaker wiring. Such systems were not only good business but also good citizenship the firm proclaimed: "The theater manager who overlooks the advantages which a Radioear installation brings to his theater overlooks more than a chance to help make life more interesting for many of the millions of hard of hearing people throughout the country."[56]

Some advocates for personal sound systems crossed the line into quackery, predicting that "stimulation given to the auditory nerve from repeated attendance at these [talking] pictures will result in partial or total cure of deaf persons."[57] Actually, increasing the sound levels for the hard of hearing could create an annoying vibratory distraction for the deaf. Even before recorded sound, one deaf writer complained, "To me the vibrations [of the music played along with silent films] are a continuous, growling thunder—or worse than that—which sickens me soul and body."[58] A writer in the manualist *American Annals of the Deaf* still maintained in 1931 that "the real solution . . . is to have a silent theatre in every city where there are a large number of deaf people. Show the best films in silent versions, the theatre would draw not only the deaf, but some hearing people."[59] But in the end, the diversity of deafness diluted any hope for a unified demand for cinematic justice. The *Volta Review* reported in 1929, "Some would retain captions, some would distribute synopses, some would install hearing devices . . . As yet no one plan has been found which will satisfy all."[60]

Even without any technological assistance, however, deaf audiences continued to consume the talkies in two main ways. Children were drawn to the movie houses regardless of the presence of sound, such that soon the deaf associations were lamenting not the passing of the silents but "the flying hands [signed language] and histrionic abilities of the small narrators [which] express in no unmistakable terms the slinking criminal, the rat-tat-tat of the machine-gun, and the death struggle of the victim."[61] A decade after the changeover to sound, another writer observed that "the motion picture is still a favorite entertainment of the deaf child."[62] By the 1950s the manualist NAD journal *Silent Worker* featured a regular "Films in Review" column covering major Hollywood releases, though often the reviews would mention a "preponderance of dialogue," which might hinder enjoyment. For many deaf adults, dialogue-laden talkies held little appeal. In a 1943 survey of "deaf families" in Indianapolis, while 40 percent attended movies at least once a month, some 34 percent never attended the movies at all, reporting as reasons: "They are uninteresting, too expensive, or too far away from their homes, but the chief reason given was that they could not interpret sound movies."[63] According to one writer, after 1930, "Only the foreign films were left for the real enjoyment of deaf movie-goers. The titles in English enabled viewers to follow the story from beginning to end. However, these foreign films were all too few and were shown only in small, out of the way theatres in large cities."[64] Thus, for a time at least, the interests of deaf audiences coincided with those of non-English-speaking audiences, both in the United States and around the globe, in supporting the new technology and technique of language-transfer subtitling.

## Inventing the Technology and Practice of Subtitling

Even though Hollywood ignored the loss of the domestic D/HOH audience occurring from the removal of intertitles, most filmmakers were well aware that the talkies threatened their global earnings for similar reasons. In the late 1920s and early 1930s, about a third of the US film industry's total revenue came from the foreign market—especially France and Germany, where the Film Europe movement was already in decline.[65] The cultural diversity of the European market—one engendered by the international commodification of film on one hand and the cultural specificity of the meanings audiences took from those films on the other—meant that not one but many successive translations of each film might be needed.[66] Though some prominent filmmakers claimed to ignore the problem—Louis B. Mayer allegedly "assumed that the popularity of American films would lead to the use of English as a universal language," according to film historian Kristin Thompson—for most studios, the increased production costs associated with sound cinema provided an incentive to sell sound-enhanced films more widely than ever before.[67] Expanding deeper into the European market provided a sort of spatial fix to complement sound's technological fix.

The customary process of creating and splicing intertitles was initially used to translate dialogue for sound films in less lucrative markets like Poland and Egypt. But although intertitling was inexpensive and well understood, its use in sound-enhanced cinema was especially jarring. Not only did the words interrupt the action of the film, but they also created a blank spot in the musical track (now physically attached to the film rather than played live). Especially for the major markets of France and Germany, some other more aesthetically pleasing but cost-effective solution had to be found.[68] Studios in the United States and abroad initially tried to solve the language-translation problem by shooting multiple versions of a film, one in each language, often with different actors.[69] (This solution echoed the alleged dual sound-and-silent versions that studios claimed to be shooting for benefit of D/HOH audiences.) While this method reused scripts and sound stages, it was nevertheless extremely costly, sometimes adding tens of thousands of dollars to production costs, a 30 percent premium for each new language version.[70] By about 1932 the multiple-language strategy disappeared, "too standardised to satisfy the cultural diversity of their target audience, but too expensively differentiated to be profitable."[71]

If altering the film-production process in order to translate dialogue was too expensive, perhaps the solution lay in altering the film-distribution process. Screenings of *The Jazz Singer* in France used an adjacent screen and slide projector—similar to the first magic-lantern titles—to display dialogue translations in parallel with the action of the main film. The trick worked in this case because Al Jolson spoke only a few lines of dialogue in the picture.[72] But for more dialogue-heavy films, a different sort of solution was needed. Innovators quickly concentrated on the production problems of (1) easily rendering typewritten subtitles directly on a blank reel of film, and (2) automatically merging the subtitle track with the original cinematic negative to create an easily distributed, subtitled print that could be screened without any additional equipment or labor. By the mid-1930s, patents from inventors in Norway and Budapest emerged, one using a thermal method of "very small letterpress type plates" on moistened film, and another using a chemical method combining a wax coating and a bleach bath to reveal letters on film.[73] With a material-production process to render the subtitles at hand, all the distributors needed now was a labor process to actually construct the subtitles.

In the United States, it was Herman Weinberg who first started using the Movieola film-editing console for subtitling European films into English. Similar work was being done in Europe for translating US films into French, German, or other languages. Sometimes the Hollywood studios would have workers in New York put together a list of English subtitles for US-exported films that could be translated piecemeal without the need for further timing, a practice known as "pivot subtitling."[74] But separated by an ocean, the two subtitling communities largely failed to trade secrets. Even so, the overall labor process devolved along similar lines in both places, and despite Weinberg's own description of the work as a "one-man job," it involved a substantial division of labor.[75] First, the team viewed the film all the way through. Next, they typed a transcript of all of the dialogue (scripts were rarely provided by the studio). Then a "spotter" went to work with the film and the transcript, performing the precise fragmentation of the flowing dialogue into time- or length-coded, bite-sized lines per scene, noting the exact moments when these bits should both appear and disappear from view. After that, the "translator" altered those individual lines of dialogue from source language to target language, from speech to text, and from verbatim statements to brief summaries, all the while preserving both meaning and emotion. Sometimes more than one translator worked on a single film, one fluent in the source language and one fluent in the target language. A final editing and proofing step, usually by the translator but sometimes by a fourth person working as editor (and perhaps censor), rounded out the labor process, and the film and subtitles were then handed off to production. Although today's subtitlers use computer workstations, time-coded videotape, and laser-etching film-production equipment, their basic labor process and division of labor—viewing, transcribing, spotting, translating, and editing—remains the same.[76] In the 1930s, such a process from start from finish might have taken a two-person team (taking on all four roles of transcriber, spotter, translator, and editor) about a month to complete (one week of transcription and spotting, two weeks of translation, and one week of editing) with another month or so for the production laboratory to refilm the final master.[77] Total cost in the early days of sound? About $2,500—roughly 75 percent of which probably comprised labor costs for spotting and translating. This would be less than one-tenth the cost of making a full second-language version of the film.[78]

The subtitling process might sound straightforward; however, like any labor process dealing with aesthetic considerations, the devil was in the details. Subtitle transcribers, spotters, translators, and editors fought a constant battle against the constraints of time and space within the legibility limits of celluloid film, the standards of projection speed, and the audience's capacity for attention.[79] Subtitlers on both sides of the Atlantic soon settled on roughly similar rules for their craft, although these standards were never articulated or debated until the 1970s and 1980s, when experts like Jan Ivarsson and Georg-Michael Luyken began writing about their work.[80] Usually basing their calculations on a standard, san-serif typeface like Helvetica, subtitlers kept each title within the limits of two lines of up to forty characters each (proportionally spaced) for 16mm and 35mm film. Estimating an average audience reading speed of between 150 and 180 words per minute, subtitlers learned to leave each individual two-line title on the screen for at least eight feet of film, or a little over five seconds. Now add to those limits of size and speed the constraints of synchronization. After subtitlers realized that, as Ivarsson put it, "many people need a moment, perhaps no more than a fraction of a second—a 'fixation pause'—to identify the speaker," they learned to make each subtitle appear "about a quarter of a second after the start of a dialogue." Similarly, pauses of up to a quarter-second were needed after each subtitle, or else, Ivarsson noted, "the eye often does not register the fact that a new subtitle has appeared." Subtitlers also found that, in Luyken's words, "if a subtitle is retained on the screen during a shot or scene change it will result in an effect known as 'overlapping' which is detrimental to the aesthetics and intelligibility of the subtitle . . . [because] confusion is caused when the eyes return to the beginning of the text as the change of shot or scene is perceived." Thus spotters had to pay careful attention to shot changes and other cinematography techniques.

What all this meant in terms of the translator's job, according to the British Film Institute when it investigated the industry in 1978, was that fully one-third of a film's original dialogue had to be discarded in the subtitling process.[81] Paradoxically, good subtitling rendered this loss of information invisible. Instead it was only those small but maddening errors that reminded viewers that subtitles were created by human labor: "a disproportion in duration between spoken utterance and written translation or the failure of subtitles to register obvious linguistic disturbances such as a lisp or a stutter."[82] Cinema critic Antje Ascheid recently pointed out, "As subtitling creates a double text out of an originally single text, it draws attention to its own mode of production, ruptures the ease with which character identification normally proceeds, and makes room for intellectual evaluation and analysis," which "potentially leads to a considerable loss of pleasure during this experience."[83] Given such a contradictory mix of obliviousness and scrutiny by audiences, it is no wonder that longtime subtitlers thought of their work more as an art than a science, a real test of creative authorship. Weinberg believed that the ideal subtitler had to be "a humorist, a poet, should know his language, and should have the patience to bear with the film."[84] Jan Ivarsson agreed: "Subtitling, when it is done to high standards of excellence, includes so many of the elements essential to art and above all demands so much skill, imagination and creative talent that it *is* indeed an art."[85]

## The Global Geography of Subtitling and Dubbing

Unfortunately for subtitlers, not everyone appreciated their art. A new translation technology to compete with subtitling soon emerged in the early 1930s, decidedly more expensive but taking greater advantage of the audio technology that made talking films possible in the first place. This second synchronous-translation solution represented a sort of midpoint between the fragmented translations of subtitling and the wholesale duplications of multiple versions. Technically referred to as "lip-synchronized revoicing," it became better known as "dubbing."[86] First employed in "looping," or re-recording a film's existing but muddy dialogue in the same language under superior studio conditions, this technique soon emerged as a language-translation solution as well.[87] Understanding the differing costs, turnaround times, labor processes, and aesthetic constraints of this alternative to subtitling helps to explain how an idiosyncratic and uneven global geography of subtitling countries versus dubbing countries emerged in the world of film and television—and why so many materials that could have been rendered in a textual form accessible to deaf and hard-of-hearing audiences around the globe were instead kept silent to them.

Studios began experimenting with dubbing around the same time that they were producing multiple-language films, but almost all sound had to be recorded simultaneously with the action, as proper sound-mixing technology was not yet available.[88] After 1930, the same Movieola used by the subtitling pioneers gained the ability to mix multiple sound tracks of voice, music, and effects.[89] By about 1936 the technology and technique, first developed in the United States, was available in Europe—where, considering the dominance of English-language film in circulation, it was most needed.[90] Most dubbing, even when ordered by US studios, took place in Europe, which helped keep costs down. In the early 1930s, "United Artists estimated the cost of dubbing a picture into Spanish as $3,500."[91] Cost breakdowns today peg about 30 percent of the dubbing cost to the labor of actors, 10 percent to the translators, and another 25 percent to the technical crew and director, resulting in about 65 percent of the total cost being driven by labor. Of course, even within these estimates, some films would be more expensive to dub than others—films with more dialogue or with well-known celebrities (who needed to have their voices dubbed by the same foreign talent in each film).[92] Sometimes, especially in Eastern European markets, films would undergo "half dubbing," a cheaper production technique where the original soundtrack was maintained at a barely audible level.[93] Even so, dubbing was more expensive than subtitling and was often "only economically feasible for the major language groups—Spanish, German and French—and for Italian, which constituted a special case due to the governmental restriction on the use of foreign languages in the cinema."[94]

Like subtitling, dubbing was considered by its practitioners as more of an art than a science. Matching mouth movements in time and space to translated dialogue involved just as much creative adaptation, though not as much wholesale discarding of dialogue, as creating text equivalents of speech through subtitling. However, unlike subtitling, which could in theory be performed by a single individual acting as transcriber, spotter, translator, and editor, the more complete division of labor in dubbing could pit the skills of the translator against those of the revoicing actors. As one analyst put it, "The film audience is often more concerned with lips than literature."[95] And dubbing writers and performers faced an even greater problem of invisibility than subtitlers did, as in dubbing, unlike subtitling, "the work is well done when no one [in the audience] is aware of it."[96] All this led one film critic to exclaim in the translation journal *Babel*, "Dubbing is a kind of cinematic netherworld filled with phantom actors who speak through the mouths of others and ghostly writers who have no literary souls of their own, either as creative authors or translators."[97]

By the mid-1930s, both dubbing and subtitling were available for film producers and distributors to choose when trying to market an American film to a non-English-language community.[98] But deciding which method to choose did not simply come down to cost. Sometimes the film itself suggests one method or another; a dialogue-heavy "art film" may be subtitled, whereas an action-heavy blockbuster might be dubbed.[99] Typically, however, the size and political-economic power of the target language community took precedence, dividing the US film export market roughly (and unevenly) into dubbing countries (larger, more powerful, and more homogenous language communities) and subtitling countries (smaller, less powerful, and more diverse language communities), a division that largely persists to this day. In Europe, in general, "the larger and more economically powerful the country, the more likely it is that dubbing will have assumed a position of dominance," such as in France, Italy, and Germany.[100] In some smaller and less affluent countries, for example Hungary, Slovakia, and the Czech Republic, dubbing became the preferred (and expected) mode of translation.[101] Cultural preservation efforts can lead to dubbing as well, especially in "nations without states" like Wales.[102] Generally, though, audiences in smaller and/or less affluent language communities have been trained to rely on less-expensive subtitling, especially Belgium, Cyprus, Denmark, Finland, Greece, the Netherlands, Norway, Portugal, Romania, and Sweden.[103] And in officially multilingual countries like Finland or Belgium, even "multilingual subtitling" is practiced, simultaneously displaying each subtitle in both languages (requiring either twice the screen space or half the words).[104]

Over time, the subtitling and dubbing industries have been pulled to different sites as well, with varying cost and quality depending on where the work is done. Film critics have long argued that "in a country where sub-titling is applied on a large scale, the quality of dubbing will be relatively low, and vice versa."[105] This is directly related to labor-market size and history; for example, Barcelona, Spain, reportedly had at least seventy-five dubbing studios in the early 1990s, and in most dubbing countries, "studios generally have access to a much larger pool of actors conversant with the requirements of dubbing."[106] Today, according to dubbing and subtitling expert Josephine Dries, "A dubbing company is dependent on the availability of actors for the recording of the voices and therefore needs to be situated in an environment with a good infrastructure, studios and a lot of actors [who speak the target language]." Subtitlers, by contrast, "are used to working at home with their own subtitling equipment. Material and communication is transported by post, telephone and modem." Thus "competition in the subtitling business crosses national borders much more easily than it does for dubbing."[107] All of these factors result in a wide variation in cost for dubbing versus subtitling across the globe. In the early 1990s, according to Dries, "The average American movie would be dubbed for about 24,000 ECU [European currency units]" in Spain and perhaps 48,000 ECU in Germany but only 16,000 ECU in Italy.[108] Similarly, the average cost for subtitling one hour of programming in Europe as described by Luyken in 1991 ranged from 750 ECU in subtitling countries to 1,500 ECU in dubbing countries.[109]

In Europe, then, because of this patchwork geography of dubbing versus subtitling countries, actual subtitling labor practices evolved in quite a different competitive environment than in the United States. By the mid-1970s, Hollywood productions in English still accounted for "more than 50 percent of world screen time," according to Guback.[110] A decade later, in Western Europe alone, some 2,500 English-language cinema films were imported into non-English-speaking countries each year.[111] Such films are not subtitled in the United States, where even today only a handful of high-profile subtitlers—such as Austrian-born and Manhattan-based Helen Eisenman, who is fluent in six languages and has subtitled more than three hundred films—has taken over from Weinberg in recent years.[112] Instead, a slew of private subtitling companies emerged in Europe to handle this demand, contracting studio work out to freelance subtitlers across the continent who were paid a piece wage by the subtitle, resulting in a "rather modest" income.[113] While the European dubbing industry grew its own labor unions and professional associations, subtitling was only able to develop scattered, unofficial "cooperatives" of freelancers, who in working for a single large distributor were at least able to crowd out some competition.[114] Such organizations are easily thwarted, though, by hiring from more isolated populations of workers, according to Ivarsson: "Very often, film and video companies employ students or immigrant housewives from the country of origin to do their subtitling after a crash course, with predictable results."[115] Freelance subtitlers, usually working from home, often have to make due with audio-only versions of the film or poorly prespotted scripts because studios fear video piracy and are unwilling to let even their subtitlers view the completed film before release. Turnaround times are often cut to the bone, with only eight to forty-eight hours (one to six working days) allotted for a one hour segment of content that might require the construction of some 750 subtitles.[116] According to one subtitler, such time and space constraints on labor provide "one reason why so many subtitles are preposterously wrong."[117]

## Consuming Cinematic Subtitles in the United States

What does this polarized geography of subtitling consumption and production mean for cinemagoers in the United States? On the one hand, the United States makes a perfect dubbing country because the huge consumer market guarantees that the added labor costs of dubbing will be recouped through the wider distribution of a revoiced film. On the other hand, an urban geography of immigration and taste, centered in New York City, has kept a market for undubbed, intertitled, and subtitled foreign features alive ever since the "little cinema" movement in the 1920s and 1930s began to appeal to ethnic enclaves and art house audiences.[118] At the time of the talkies, Gomery has noted, New York offered at least "twenty-five theatres to present feature films in a staggering array of languages."[119] Even though the coming of World War II halted the screening of most foreign films (due to censorship at home and shuttered studios abroad), by the mid-1950s many returning GIs had embraced the films of Europe, a continent they now knew firsthand.[120] Thomas Guback pointed out that in the 1960s, with the coming of the end of vertical integration (where studios not only produced films but owned the theaters that screened them), the competition from television (showing older films that used to enjoy some reissues in theaters), and the consequent lowering of output from Hollywood, "the gap between exhibition's demand and Hollywood's supply [was] being filled by foreign films."[121] In 1960, one critic wrote, "It is only recently, with the importation of large numbers of foreign films resulting from Hollywood's production decrease, that dubbing films have been accepted at all in America and Great Britain."[122] In that same year, a debate arose in New York City, instigated by *New York Times* film critic Bosley Crowther, over the aesthetics and economics of subtitling versus dubbing—a debate that demonstrates the meanings thought to be bound up in each method at the time.[123]

Crowther launched the debate when he admitted in his August 7 column that "while in Europe this spring" he had become convinced that "the convention of English subtitles on foreign-language films" must be abandoned in the United States and "replaced by the use of dubbed English dialogue." First, Crowther lamented the need of the subtitle-watcher "to have to spend a lot of precious time reading instead of looking at what is going on." Next he pointed out that dubbing quality had improved greatly over the last few decades, "with voices matched so finely and lips so deftly synchronized, that only the most expert observer can detect the lingual trickery." Finally he derided the practice of subtitling as obsolete, "an old device that was mainly contrived as a convenience to save the cost of dubbing foreign-language films when they had limited appeal." Thus, he declared, "Subtitles must go!" Crowther of course knew that this was a controversial suggestion, predicting "that small but valiant band of zealots who count on the foreign-language films to give them opportunities to hear pure Italian, German, Japanese and French may well descend upon us. And Herman G. Weinberg, that enterprising man who makes a good business of writing most of the subtitles, will probably picket *The New York Times*." Weinberg may indeed have taken offense because, rather than praising his pioneering skill in rendering subtitles under severe constraints of space and time, Crowther called his subtitling "a thoroughly inartistic thing."[124]

Reaction to Crowther's pronouncement was vocal and swift, and two weeks later he was back defending his claims again, offering up two further arguments in favor of dubbing. First, he pointed out to those who might not realize, most "pure" original films were actually dubbed through "looping," even if there was no process of language translation: "How many people realize that in most of the foreign films we see—from Italy, especially—the voices as we hear them were not recorded when the action was being photographed?" The dialogue purists actually had no "pure" film release to complain about losing in the first place. Second, Crowther introduced a new, more democratic argument: "More Americans would see foreign pictures if they were effectively dubbed; more foreign films would be shown in this country and more of the ideas and aspects of other nations would be exposed." In other words, even though dubbing cost more than subtitling, it would be worth the time and effort because dubbed films would find significantly larger paying audiences.[125]

The subtitling-versus-dubbing debate spilled out of the *Times* and into "film culture journals and little magazines" throughout the early years of the 1960s.[126] Six years after his first article on the subject, Crowther revisited the "still smoldering controversy" and reasserted his argument that "the dialogue in most pictures—that is, the dialogue you hear—is as frankly controlled and manufactured, arbitrarily determined and arranged, after the shooting as is the music or the sound effects."[127] In 1973 the city Department of Consumer Affairs "ordered that exhibitors publicize if a film was being shown in dubbed or subtitled form."[128] Even Herman Weinberg was quoted as approving of the "new trend" of dubbing foreign films for smaller cities, drive-in theaters, television, and the "great mass of movie goers who are not used to see foreign films in foreign languages." But he also pointed out that because dubbing still cost about ten times as much as subtitling, it would likely only be used for foreign-language films that expected a wide distribution in the first place, rather than to democratize more obscure, controversial, and inaccessible foreign films.[129] And Weinberg noted that subtitling a film which was subject to censorship could, with care, paradoxically work to preserve the integrity of the dialogue (in the soundtrack) precisely by appearing to undermine that integrity (in the text). Even though US subtitlers were often "forced into the un-happy function of censor to the ribald seams of Rabelaisian candor which run through many French and Italian scripts," they could at least preserve the film's original spoken dialogue even while censoring their own translated subtitles.[130] Weinberg later recalled that "when nothing else could be done" he would "agree to eliminate the [sub]ti-tle to save the dialogue and the scene that went with it" because "the censors would let the scene stay in, with its racy dialogue, if it wasn't translated—probably feeling that those who understood the racy orig-inal were already too far gone to save."[131]

Although he never mentioned censorship, Crowther conceded one situation in which subtitling was superior to dubbing, recognizing "the obvious reasons why students, deaf people and such prefer for-eign films with subtitles rather than with dubbed dialogue" (likely in-dicating that he received mail from these interest groups).[132] But what Crowther did not acknowledge was that the deaf community, while small in numbers compared to the population as a whole, could actu-ally be a significant constituency in local urban sites of deaf education and culture. One writer in the *Deaf American* pointed to the Apollo Theater on West 42nd Street in New York City as pulling in "the world's largest regular audience of deaf movie goers" with its subtitled foreign fare in the 1960s, regularly attracting "between seven and eight hundred deaf and hard of hearing movie fans each week, or roughly one-tenth of the theater's total weekly attendance."[133] Even educators of the deaf screened subtitled foreign films in the classroom when they could find appropriate titles.[134] But the supply of subtitled cinema was insufficient to meet the demands of the deaf either in the theater or in the classroom.[135] In fact, the limited availability of subti-tled films for schools meant that most movies used in deaf education had long been "free films that are put out by industrial concerns for school use."[136] As a result, the biggest revolution in American subti-tling in the postwar years would come neither from the urban film critics nor from Weinberg's cottage industry but from a group immune to the lures of dubbing: deaf educators and activists.

## Captioned Films for the Deaf

In 1929, before the coming of the talkies, a University of Minnesota psychology professor had argued before the Convention of American Instructors of the Deaf that "the failure of schools for the deaf to take

full advantage of the moving-picture educational film is comparable
with the failure of the schools for the blind to take full advantage of
the wonderful opportunities afforded by the invention of the phono-
graph."[137] Calls for such "visual education" for the deaf continued
through the 1930s and 1940s, but as silent (intertitled) films became
increasingly scarce, educators grew more aware that film was "the most
expensive of all visual aids."[138] Through the 1940s, custom-made films
for the deaf were pioneered for particular pedagogical applications, such
as teaching lip-reading to returning GIs who had lost their hearing after being exposed to bomb blasts.[139] But without printed translations,
the use of commercial films in the deaf classroom became increasingly
frustrating. In one case from 1935, instructors resorted to starting and
stopping the film so that students could lip-read what their teachers
might have to say about a particular scene. These same deaf educators
had proposed, "If such a center as the Volta Bureau, for instance, could
be designated as a distributor for a group of the outstanding industrial
films and have them retitled, if possible at the expense of the industries
which they represent, a great service would be rendered."[140] The idea
would return a decade later, when Sylvia Sanders, a graduate student at
Gallaudet College for the deaf and hard of hearing, suggested that "a
special motion picture library from which different schools for the deaf
could draw [might] be established," which could then "urge that pictures be retitled and captioned to contain new words and build vocabulary without being beyond the comprehension of children."[141] The
uneven geography of deafness and deaf education made such a system
necessary. But was it feasible?

Deaf educators found their answer in a pilot program for the blind
that had been developed between the 1930s and the 1940s for a similar
purpose: to make specially adapted information media widely accessible to a geographically distributed, demographic-minority audience.[142]
For the blind, however, the media in question was not film but print.
The official story of the Talking Books program was ably told by historian Marilyn Majeska. As early as 1904, "free mailing privileges" had
been granted to libraries for sending books in braille to blind individuals. As a result of lobbying by the American Federation of the Blind
(AFB), this initial effort was expanded in 1931 with the Pratt-Smoot
Act, entrusting the Library of Congress with the task of producing and
distributing braille books across the country to eighteen other regional
depositories. But just as producing sign-language films did not com-

pletely serve the wide variety of D/HOH viewers (especially the elderly, hard-of-hearing subgroup), producing braille books was insufficient to meet the demand of the wide variety of blind readers (again, especially the elderly, for whom "learning braille may be difficult because of lessening fingertip sensitivity"). According to the AFB, in 1928 only 20 percent of the US blind population could even read braille books. Thus in 1933 the Pratt-Smoot act was amended to include "books in recorded form," and the federation agreed to fund the production and distribution of some 1,200 record players if Congress would fund the production and distribution of the books-on-record themselves.[143]

The most striking fact about the Talking Books project, though, was not the legislative victory but the technological innovation involved. At the time, almost all consumer recordings were 78-rpm discs capable of holding five minutes of sound per side; however, recording the narration of a sixty-thousand-word book required 33-rpm discs capable of holding at least twelve and a half minutes of sound per side (even then, the book would take eight discs to record). As Majeska put it, "In effect, blind persons got the long-playing record fourteen years before it was available commercially." Even the material technology of the records had to be redesigned for the space and time demands of distribution, making them "durable, tough, and flexible enough to ship by mail as well as sturdy enough to last through fifty to fifty-seven playings." Through the Great Depression of the 1930s, the Talking Books project was able to produce both the record players and the records in staggering numbers, using Works Progress Administration funds to muster unemployed white-collar workers in Manhattan to the task. By 1937, the project had distributed some 16,700 containers of talking books representing 145 unique titles, as well as 16,200 automatic players. Considering the economic context and the technological hurdles, it was an unprecedented government-funded spatial and technological fix to a media justice problem.[144]

The dramatic success of the Talking Books project inspired deaf educators and activists in New York City to experiment with new display technologies and distribution schemes for subtitling film. Ross Hamilton, a Columbia University graduate student working at the Lexington School for the Deaf, experimented with using two synchronized cameras to project film and captions on the same screen (reminiscent of early experiments in titling sound films for foreign audiences).[145] Emerson Romero, a former silent-film actor (and first cousin to the more famous

Cesar Romero) who was by then "working for an airplane manufacturing company on Long Island," used the old method of splicing intertitles into four feature films to create his own "national Film Library for
the Deaf," which he circulated to local deaf schools and clubs (including the Lexington School where Hamilton worked).[146] At the time,
Romero wrote that "if the deaf want to have anything done for their
benefit they should not ask the hearing people to do it but should go
out and do it themselves."[147]

In the immediate postwar period, many educators of the deaf
shared this idea and, shunned by Hollywood ever since the end of the
silents fifteen years earlier, began to experiment with subtitling films,
which they usually referred to as "captioning." But their individual efforts took place within an institutional geography of deaf education—
some oralist and some manualist, some day schools and some
residential programs—that defined one's place in D/HOH culture to a
striking degree. Even today, "when you meet a deaf person, his school
is a primary mode of self-identification; it's usually told after his name
but before his job."[148] Linked by this geography of educational affilia-
tion, the individual efforts of deaf educators and activists to "do it
themselves" eventually grew into a collective and federally funded
program, an effort that laid the groundwork for the captioning of television in the 1970s.

This collective effort first began around 1947 when the Conference
of Executives of American Schools for the Deaf, inspired by the work
of Hamilton and Romero, started to study film captioning.[149] Realizing
that neither Hamilton's dual-projection nor Romero's intertitling would
be acceptable solutions, Edmund Boatner, superintendent of the American School for the Deaf in West Hartford, Connecticut, charged his
vocational-education teacher, J. Pierre Rakow, with the task of "studying and learning the techniques of film captioning."[150] Rakow, who
was himself deaf, first "approached one of the major motion picture
companies" to do the subtitling but was "told that the cost of such an
operation would be prohibitive."[151] Undeterred, Rakow contacted
Titra Film Laboratories in New York (a subsidiary of the Paris-based
Titra Film, likely the same photographic laboratory that Weinberg
used) and secured an affordable production process for his subtitles.[152] Finally, despite the fact that film piracy made producers "unwilling to sell
or lease prints of any of their better films," Rakow convinced one studio

to lease out the Abbott and Costello film *The Noose Hangs High* (1948) for the captioning experiment, writing all the titles himself.[153]

Encouraged by Rakow's success, in 1949 Boatner and fellow educator Clarence O'Connor, director of the Lexington School in New York, incorporated Captioned Films for the Deaf (CFD) in Connecticut at the American School and began to seek financial backing.[154] As it happened, the sister of film star Katherine Hepburn, Marion Hepburn Grant, was president of the Junior League of Hartford and immediately organized a variety show to raise funds for the new project. Not only did the women raise some $7,500, but two members personally volunteered to transcribe and caption the first film, the twenty-five-minute narrated color documentary *America the Beautiful.*[155] By December 1951, copies of several features were circulating in both Hartford and New York City, renting for fifteen dollars a showing (twenty dollars if the film was in color).[156]

Over the next decade, CFD would purchase and caption nearly thirty feature-length films, some of them entertainment features purchased from studios (at up to $500 each) and other educational films donated by corporations. Although the target audience was children in residential schools for the deaf, Boatner maintained that the intent was to caption not educational films but "wholesome feature films, which the deaf child can understand."[157] The decision to work with the products of Hollywood meant that CFD was able to streamline two key steps in the captioning labor process: transcription and spotting. Transcription was so onerous, Boatner recalled, that "we learned early in the game to take only films for which we could get the script." Even so, the two-inch-thick loose-leaf scripts "had to be greatly condensed, not only into captions but also into language suitable for deaf viewers." Spotting was even harder, Boatner wrote: "The problem of timing captions was so complex that it threatened to defeat us altogether."[158] In the end, CFD chose only those films destined by the studios for export to Europe, since for these products "the producer worked out captions in English before translating them." With both transcription and spotting handled by professional subtitlers beforehand, Boatner recalled, "a skilled captioner could now arrange suitable captions without even viewing the film."[159]

By the mid-1950s the CFD experiment had become so successful that something extraordinary began to happen: the wider D/HOH community—both manualist and oralist—came together to demand that the federal government fund the program so it could scale up to serve the entire nation.[160] As a later president of CFD put it, "Not only did the NAD [National Association for the Deaf], the NFSD [National Federation of Schools for the Deaf], state associations, and local clubs work together, but you were joined by the Office of Vocational Rehabilitation, the Conference of School Executives, the Convention of Instructors, the A. G. Bell Association, parent groups—in fact everyone who has an interest in the deaf joined hands to promote this film legislation."[161] Unlike the efforts to stem the loss of the silents in the 1930s, which split the D/HOH community into those who could use assistive listening devices in theaters and those who could not, captioned media was a goal that all sides could agree on.

The D/HOH community first suggested that a national CFD program might be housed at the Library of Congress, which had managed the Talking Books for the Blind program since the 1930s.[162] But resistance on the part of new the new Librarian of Congress, L. Quincy Mumford, meant that the program ended up under the Office of Education in the Department of Health, Education, and Welfare.[163] Public Law 85-905, passed in September 1958, authorized up to $250,000 a year for the project, but its first actual appropriation (which came almost a year later in August 1959) amounted to only $80,000, enough to hire a staff of three people.[164] Nevertheless, it was enough to hire an educator at the Gallaudet College for the deaf and hard of hearing, John Gough, as the new director.[165] That Gough was not deaf initially caused some worry, but by the following year he had hired CFD's first deaf manager, Malcolm "Mac" Norwood.[166] Educated at both the American School and at Gallaudet, Norwood had "built up a collection of captioned filmstrips" and "established a budget to encourage his teachers to rent old-time silent films with captions for use in their classes" at his previous job at the West Virginia School for the Deaf.[167] With its first round of funding, its first director, and its first legitimate liaison to the deaf community, the federal CFD program was ready to get to work captioning films.

This time, however, a vocational instructor and some volunteers from the local Junior League would not be enough to perform the captioning on the scale the new organization demanded. Because CFD was now located in Washington, DC, it seemed to make sense to perform the captioning at Gallaudet College, where "the student body would serve as a test group for the aptness of captions."[168] In December 1959 Rakow made a trip to Washington to train three deaf Gallaudet instructors, Stanley Benowitz, Robert Panara, and Leon Auerbach, in his captioning technique.[169] Mac Norwood even tried his own hand at captioning, writing the titles for a 1960 film of "World Series and Football highlights."[170] Rakow continued to consult for CFD through the early 1960s, captioning at least a dozen films through his new private company, Superior Films, while the Gallaudet captioners learned their craft.[171] "Thus to a considerable extent," commented the *Silent Worker* approvingly, "the program is not only for, but also by the deaf."[172]

The captioning labor process was still bound by the time and money constraints that Rakow originally faced, however, meaning that CFD could not afford to caption a film unless the studio provided them with "a producer's continuity which gives a brief description of each scene on the film, the length of the scene and the dialogue."[173] Just as with subtitled foreign films for English audiences, the dialogue in a captioned film for the D/HOH audience had to be drastically modified for word difficulty and reading speed—a foot of film per word, at a fourth-grade reading level, was the target.[174] The *Silent Worker* described the labor in detail for its readers (who were likely impatient to see the final films): "In addition to writing the dialogue, the captioner must look out for difficult words, find short words to replace long ones, and then use arithmetic to figure where the title appears on the film. He writes—Begin 97.8 (97 feet and 8 frames from the start of the movie) End 99.4 . . . Thus, a captioner must be both a mathematician and a writer. Not an easy job."[175] All these rules of practice were developed as needed—there were no scientific studies of caption viewing to consult, although *Silent Worker* noted, "An investigation of this kind might be worthwhile."[176] In the end, each feature-length film resulted in some sixteen double-spaced, typewritten pages of captions, and from start to finish the process took about six months if all went well.[177]

All did not usually go well, however. While Gallaudet College provided an eager and abundant labor supply, even finding enough Movieola caption-editing stations was a challenge at first.[178] But the real problem came in producing the final prints. For each older, ninety-minute, black-and-white feature (new releases were usually cost-prohibitive), CFD spent about $2,400 to obtain four prints, and each had to be etched with captions by Titra Film Laboratories at an additional $325 per print.[179] At Titra, however, the demands of a growing foreign-film import-and-export business were pushing CFD's less-lucrative business aside. Reported the *Silent Worker* in 1960, "the company, which has an exclusive process, is swamped with work on 16mm films intended for use by South American TV. Brazil is an especially big user."[180] The journal observed a year later that "Business firms that are having films processed often pay for an 'expediter' to help push their work through the 'lab,' thus getting special, favorable treatment."[181] By late 1960, the delay at the processing laboratory had reached up to six months, causing nearly half of the scheduled screenings of CFD films to be canceled.[182] Boatner later recalled, "Titra had to farm out some of our captioning to European companies, and a few were even captioned in Egypt."[183] Ironically, one of the tactics that CFD used to meet the demand for its new service during these years was to turn to foreign films that had already been subtitled in English: "While these films may not fill the bill entirely in terms of what deaf groups wish to see, they have the one advantage of being immediately available with captions at a time when the film cupboard is all but bare."[184] Even these films needed a bit of editing before being screened, however; in one instance, "the caption writer who prepared the film for general circulation threw in a few cuss words here and there. Since captioned films circulate to church groups and schools as well as to adult groups and clubs, it was necessary to clip out these 'shockers.'"[185]

Such questions of content and audience became increasingly important as the CFD program grew rapidly through the 1960s. Although the program had been started by educators from residential deaf schools for children, it was initially seen as also serving an estimated 260,000 persons in the adult deaf community through screenings at a nationwide network of deaf clubs.[186] In 1960, out of the 259 screenings of captioned films across the United States, over a third of the audience was made up of adults, and the films themselves represented "a slight leaning to adult pictures."[187] But as additional laws granted larger budget appropriations to CFD, the mission of the agency was nudged more and more to education, not entertainment. Public Law 87-715, passed in fall 1962, increased the yearly appropriation limit from $250,000 to $1.5 million.[188] In contrast to the first CFD law, when Gough assured adult deaf activists that "the films will serve all of the deaf, not just some single groups such as school children," this time Gough reported, "Testimony in favor of the new law hammered at the educational and language retardation of the deaf and the promise of captioned films to help improve this situation."[189] Thus PL 87-715 expanded the definition of "film" to include the more school-friendly media of filmstrips, slides, and transparencies "to promote the academic, cultural, and vocational advancement of deaf persons."[190] A larger staff of eleven was authorized, including "a writer who will specialize in captioning and teaching guides."[191] And the agency used the labor of teachers in deaf schools as far away as New York and California to write captions for its first round of thirteen educational science films.[192] (A few decades later, a similar shift of focus from entertainment to education would be seen in federal funding of captioned television.)

The question of whether to focus on adults or children mattered, in part, because the circulation of the entertainment films to adult audiences was perhaps the greatest logistical challenge (and cost) CFD faced. For each feature, up to four or five prints might be made, and each print would be screened from twelve to fifteen times per year.[193] By 1961 some 670 groups had registered to receive captioned films, making "reservations 2 years in advance" to serve an average audience of eight thousand viewers each month.[194] Originally the Department of Agriculture had been contracted to ship the films, but by the time the catalog reached 150 prints (representing fifty unique films) the distribution center had to be shifted to the Indiana School for the Deaf at Indianapolis.[195] By the latter half of 1962, adults still made up only a third of the total audience for CFD, but that total audience had grown to 124,820 persons (a sixfold increase from 1960).[196] Despite this expansion in scale and scope, because of the leasing contract with the film studios, none of the costs of distribution and screening could be recovered by charging admission at the deaf clubs (although the *Deaf American* reported, "In some cases clubs of the deaf needing funds to pay projector rent have sold three-day memberships to non-members" to skirt this rule).[197]

By the time the CFD program was expanded a third time by Public Law 89-258 in late 1965—increasing the maximum authorization by $1.5 million dollars—a balance of sorts had been reached.[198] The program listed 125 educational films in circulation, compared with 212 entertainment features.[199] However, CFD was now captioning more educational films (51) than features (44) per year, and although children still outnumbered adults in the audience by a margin of two to one, the *Deaf American* reported, "Many of the educational titles originally captioned for use in schools for the deaf are now being made

available to the adult deaf through normal distribution channels."[200]
Two big changes in distribution aided the accessibility to the films for
both sides. For the adults watching entertainment features in deaf
clubs, films could now be ordered from one of three regional distribution centers housed at residential schools for the deaf in Colorado,
Indiana, and New York.[201] For the children watching educational films
in school, fifty-eight smaller distribution centers were organized (see
figure 1.1). Finally, CFD purchased some one thousand overhead projectors, one thousand filmstrip projectors, two thousand screens, and
fifty Technicolor projectors for "for release on loan" to deaf groups and
schools.[202] Over the next decade, "Four thousand classrooms and training centers were provided with basic media equipment."[203] Far from
merely captioning films (a labor and technical challenge in its own
right), Captioned Films for the Deaf had created an entire sociotechnical infrastructure for the distribution, screening, and return of visual
entertainment and education products targeted to one of the most
geographically dispersed and media-isolated interest groups in the nation (see figure 1.2).

One of the most significant results of this infrastructure was that
the national CFD program inspired and supported grassroots film-
captioning efforts in deaf education around the country. Recalled educator Robert Stepp Jr., "As a greater variety of new materials became
available, teachers became actively involved in the adoption process—
selecting, adapting, designing, and producing materials tailored to the
particular needs of their students and to their own teaching methods."[204] Some of these projects began to target adults. At the Maryland
School for the Deaf in 1964, instructors obtained a federal grant to produce thirty-one filmed and captioned lessons to train deaf adults to be
keypunch operators.[205] In general, the use of "visual education" for
deaf children was bolstered; in a 1962 survey of "all of the schools and
classes listed in the American Annals of the Deaf" (with a 56 percent response rate representing nearly twenty thousand students), researchers
Jerome Schein and John Kubis found that "almost every school and
class owns at least one movie projector."[206] But a striking difference
emerged when segregated, deaf-only schools were compared with classes
for the deaf held within hearing schools. The deaf-only schools were
likely to find films from (in order of preference) "the Captioned Film
program, free films from business and trade associations, and commercial sales or rental agencies," but the integrated classes mostly found

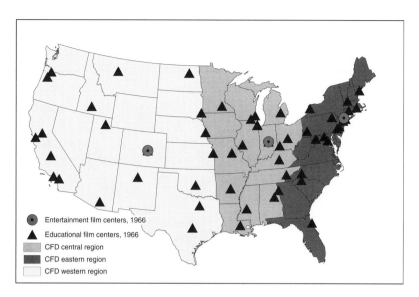

Fig. 1.1. Captioned Films for the Deaf distribution network, 1966. This nationwide network was created to distribute physical copies of captioned films across an uneven geography of deaf and hard-of-hearing day schools, residential schools, and adult social clubs. Source: Map created from data found in "Captioned films for the deaf," *AAD* 110 (1965), 323, and John A. Gough, "Captioned films for the deaf," *AAD* 111 (1966), 399–412

films from (in order) "their own school system," "the city or county public library," and "the free films from business and trade associations and those from a university film library." For deaf and hard-of-hearing students attending hearing institutions, "only a small proportion of the classes borrow films from the Captioned Film program."[207]

In response to such patterns, CFD, as well as the deaf and hard-of-hearing education and activist community at large, reached a turning point in the late 1960s. In 1966 the Department of Health, Education, and Welfare—which had housed CFD for nearly a decade—was restructured, and Captioned Films for the Deaf became the responsibility of the new Bureau of Education for the Handicapped (BEH). The new law that followed in 1968, PL 90-247, expanded the captioned films mission accordingly, "to cover all handicapped individuals," not just the deaf.[208] CFD was renamed Media Services and Captioned Films (MSCF) in part out of fears that the original mandate excluded the hard of hearing.[209] This was quite a change from the mid-1960s, when the National

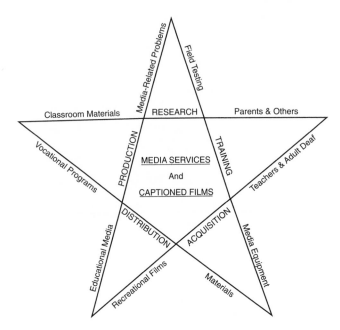

Fig. 1.2. Media Services and Captioned Films projects, 1968. Besides acquiring, captioning, and distributing films, MSCF developed classroom materials, provided teacher training, and conducted research into "visual education." Source: John A. Gough, "Report from Captioned Films for the Deaf," *AAD* 113 (1968), 1119

Association of the Deaf had been contracted to evaluate films for inclusion in the CFD catalog and when the law had mandated that whenever CFD films were screened, "the audience [be] composed predominately of deaf persons."[210] But just as the original program would not have been possible without the collective organizing efforts of groups from across the deaf spectrum—manualist and oralist alike—the next step in accessible media would depend on the deaf community linking with the hard-of-hearing community.

It was clear to almost everyone in the late 1960s that the next challenge was television. Soon after CFD had started, John Gough had spoken before the NAD in 1960 and proclaimed, "In providing you with a library of captioned films, the government is giving you an equal chance to use and enjoy this medium of communication which is shared by millions of your fellow Americans."[211] But when Gough retired in 1969, that promise of equality had not yet been fulfilled.[212] Neither the old CFD nor the new MSCF had the legal right or economic power to screen their captioned films on broadcast television.[213] But the demand for captioned television had grown ever since the first Symposium on Instructional Media for the Deaf, held at the University of Nebraska in 1964.[214] Two years after this conference, deaf educators had organized four Regional Media Centers for the Deaf (RMCDs) at the University of Massachusetts, the University of Tennessee, New Mexico State University, and the University of Nebraska.[215] Each center specialized in a particular form of media: transparencies at Amherst, films at Lincoln, "programmed instructional materials" at Las Cruces, and, most importantly, educational television at Knoxville.[216] It was not just the educators who were eager for captioned television but the deaf community at large; as early as 1965, *Deaf American* (at the urging of captioning pioneer Emerson Romero) had solicited readers to write in with their preferences for captioned television programs, hoping that "perhaps a heavy response from deaf video fans could have an effect upon those responsible for network programs."[217] By the end of the 1960s, recalled educators of the deaf William Jackson and Roger Perkins, "the Washington staff of the Captioned Films program had expanded and a television specialist was employed to coordinate efforts in this area."[218] The challenge of bringing captioned media to the living rooms of all deaf and hard-of-hearing Americans, outside the spatial and temporal limits of deaf clubs and school assemblies, would soon fall to these media pioneers, and to MSCF's new chief, Mac Norwood.[219]

The labor practice of subtitling had to be reinvented practically from scratch at least three times, for three different purposes, in three different social and economic geographies: (1) in the US film industry for translating foreign cinema into English for elite audiences; (2) in the European film industry for translating Hollywood cinema into a myriad of languages for local audiences in dozens of separate countries; and (3) in the deaf-education and activism community for translating spoken Hollywood cinema into readable media, accessible by mail to an uneven geography of deaf schools and social clubs around the nation. Each situation resulted in a different set of social relations of labor for the subtitlers, from Weinberg's cottage industry in New York City to the contingent freelance home workers of Europe's subtitling subcontractors, both of which contrast dramatically with the volunteer deaf educators and professionals of Captioned Films for the Deaf. But in each context similar practices of subtitling emerged. No matter what the purpose or who the sponsor, the methods of capturing, transforming, and reworking the time and space of speech in the cinema had to balance art and science, authorship and editing, visibility and invisibility. As one present-day subtitler poetically put it, "The translator must condense his translation in the physical space of the frame and the temporal length of the utterance. The reader cannot stop and dwell on an interesting line; as the reader scans the text, the machine instantly obliterates it."[220] Another wryly observed, "I suppose the way one ought to think of this enterprise is not with chagrin that so much gets lost, but with surprise that so much gets through."[221]

The global need for speech-to-text translation of filmed entertainment has only increased since the talkies first appeared over seventy-five years ago, especially since movie subtitles have moved from the silver screen to the television screen. According to one analyst of European television, "Up to half the material transmitted by many European stations is in English—American soaps, British comedy series, natural history programmes and so on—because these stations cannot, or will not, produce sufficient material in their native language to fill the transmission hours they deem vital."[222] Considerations of art and function demand that such materials be re-titled for home viewing: "Sitting in a darkened auditorium in the cinema and with comparatively little to distract one is a very different experience from watching television in one's domestic environment where nothing like the same concentration can be guaranteed and lighting conditions may make the deciphering of subtitles no easy task."[223] But the shortened turnaround time of television production means that subtitling quality suffers and subtitling working conditions worsen.[224] For all its limitations, though, film students today who consult their textbooks on subtitling largely find that this practice "is unanimously declared the victor over dubbing."[225] Debates over the purity of dubbing versus subtitling continue within elite culture, but in the age of the thirty-two-subtitle-track DVD, more seem to agree with subtitler Jan Ivarsson, who advises, "Act global, think local. People want to hear the languages of the rest of the world, but they want to be sure they have understood them in their own tongue, too. Subtitles seem to be the answer."[226]

Yet that answer may in many cases be irrelevant. In a world of global film where "most European programs are not profitable until foreign sales are made," the question of dubbing versus subtitling often reduces to a pure calculation of costs versus profits.[227] One critic even argued in the 1980s that in less-developed, more economically peripheral sites of smaller film revenue, "penetration of subtitled foreign films has indirectly led to the physical neglect of the sound systems in the theatres, exhibitors being guided by the spurious logic that spectators occupied in reading subtitles will not be overly concerned with the quality of sound."[228] In any case, from the point of view of Hollywood—still the most powerful and profitable site of production of global film content—both dubbing and subtitling are rendered unnecessary by "a mode of production characterised by action-oriented aesthetics" versus "walk and talk" movies, which historically "fared substantially less well in the non-English-speaking market than did action pictures."[229] In a transnational cinema of universally recognizable car chases, sex scenes, and corporate logos, who needs language?

For the practice of subtitling in the United States, as the 1960s came to a close the answer to this question was to be found neither in Hollywood nor in New York but in the uneven geography of the deaf education and activism community—a group that increasingly defined itself not as a disability community but as a language community. Experiments in finding automatic, cost-effective and decentralized methods of subtitling educational film continued through the 1970s.[230] However, it would not be until 1979 that Hollywood actually subtitled a first-run theatrically released film for deaf audiences in the United States—the Robert Markowitz picture *Voices* starring hearing actress Amy Irving as a deaf dancer. In fact, the studio only agreed to the subtitled screening after a vocal boycott by the deaf community.[231] Even the critically acclaimed Randa Haines film set in a school of deaf education, *Children of a Lesser God* (starring hearing actor William Hurt and deaf actress Marlee Matlin), was only subtitled in very limited release when it hit theaters in 1986.[232] Film critic Roger Ebert asked pointedly at the time, "How about a few silent scenes in which the signs are translated by subtitles, giving us [hearing audiences] something of the same experience that deaf people have (they see the signs, and then the subtitles, so to speak, are supplied by their intelligence)?"[233] But by and large, the US film industry was still unwilling to perform same-language subtitling for its domestic audience, whether to render signed speech intelligible to hearing viewers or to render vocalized speech intelligible to deaf viewers. Although the D/HOH community still relied on Media Services and Captioned

Films in order to experience cinema through the 1980s, the context of that cinematic experience changed dramatically when MSCF introduced open-captioned videocassettes to its subscribers, ushering in a third name change to Captioned Films/Videos and a new focus on the small screen.[234] As chapter 2 illustrates, the focus of D/HOH educational and activist efforts toward media justice through subtitling in the 1970s and 1980s had decisively moved away from the high culture of film and instead toward the mass market of television.

# Chapter 2

# Captioning Television for the Deaf Population

There is a common and very alarming misconception abroad that the long-awaited little black box (the decoder) will magically bring words to the home television screen. This, of course, is not the case. Those closed captions must first be encoded on the broadcast tape.

*Doris Caldwell, television captioner, 1979*

During the fall of 1970, Donald Torr, a professor at Gallaudet College in Washington, DC, was trying hard to get his students to watch more television. This was not an easy task, for Gallaudet was a school for the deaf, and television, as deaf educators well knew, "relies so heavily on an audio track that it is extremely difficult, or simply impossible, for deaf people to understand it."[1] Yet deaf children still watched television, just as they attended the movies; one writer reported in 1951, "Television tastes of deaf children seem to parallel their movie tastes. Action and sports are favorites."[2] Great things had been done in the 1960s for filmed entertainment, with the federally funded Captioned Films for the Deaf program adding subtitles to Hollywood features and farming them out to deaf schools and deaf clubs. Hoping to create something similar for television, Torr and his students began using Gallaudet's new video-production studio to "borrow" broadcasted shows by taping them off the air, captioning them with handmade titles, and redistributing them to student dormitories over the school's new closed-circuit cable system. As Torr recalled, "It appeared to me that many students were quite restricted in their thinking. For some the world seemed to be constrained to campus and home activities. Television would open a window on that world if we could caption it."[3] Their first effort, an episode of *The Wild Wild West*, was an immediate hit.[4] One student wrote, "Never before have I thoroughly enjoyed a T.V. program and been able to laugh through one! Thanks!" Another observed, "This type of programming just proves how badly we need more captioned language for T.V.—not only movies!" A third simply pleaded, "Please, keep up with it forever!"[5]

Such sentiments revealed what Torr and his fellow deaf educators already knew: the deaf and hard-of-hearing (D/HOH) community was starving for accessible television. Even at the beginning of television's ascendancy as the most ubiquitous mass medium in America, viewing was a popular activity for both adults and children at D/HOH clubs. In Philadelphia in 1949, "almost every afternoon, a group of youngsters [could] be found in the auditorium of the Friends of the Deaf Community Center . . . watching intently the news programs and laughing delightedly at the comedies"; adults took their turn in the evening, "and for them sports [were] the big attraction."[6] One writer argued in 1952, "To totally or partially deaf persons, television is a haven from the boredom of always reading books, playing games, or working in hobbies."[7] All through the 1950s, across the uneven geography of deaf education—from Nebraska and Oklahoma to Chicago and Washington, DC—instructors began experimenting with lip-reading, signing, and various forms of captioning on locally broadcast educational television (ETV) stations.[8] But in parallel with these experimental efforts, some writers in deaf journals pleaded for more consideration from commercial television producers: "more visual aids: headlines, captions, location signs, display of key clues, letters, recipes, maps, and characters' or participants' names; more gestures, less turning of backs and furtive action, and a program occasionally . . . tailored especially for the hard of hearing."[9] Such requests had never before been part of the formal agenda of deaf education; at the 1963 International Congress on Education of the Deaf, for example, there was a "total absence of any references to television" in the proceedings.[10] By 1970, however, these isolated pleas had become a torrent. One writer suggested that networks could "make plot summaries available ahead of time" for deaf viewers.[11] The Council of Organizations Serving the Deaf—a rare forum including both the manualist National Association for the Deaf (NAD) and the oralist Alexander Graham Bell Association—proposed, "For outstanding movies, flash the topic or partial dialogue as is now done for foreign films."[12] An author of an otherwise dry article on educational television ended with a surprising call to arms: "Let's start a clamoring demand for the rights of the deaf population that will bring about captioned public television."[13]

The social and technological system for producing and distributing captioned television to D/HOH viewers in America had its roots in the history of subtitled film but departed from that history in significant and necessary ways. Unlike Captioned Films for the Deaf, captioned television programs could not be parceled out piecemeal over the uneven geography of deaf schools and clubs; instead, some method of universal broadcast was needed. But the prospect of making television accessible to the deaf raised the fear of making television unpalatable to the hearing—putting the numerically small and geographically distributed D/HOH audience in direct economic competition with the mass market of television households containing hearing viewers. The technological fix that resulted from this dilemma—centrally hiding digital captions within the analog broadcast signal of a television program, only to be decoded and revealed through specially equipped consumer electronics—was thus also a spatial fix of sorts, as it took place across a decentralized landscape. It combined private household consumption and voluntary corporate participation in order to accommodate the cost of broadcast justice across a capitalist political economy. But such a fix was, by itself, not enough to make captioned television a reality. Both the state and the market were required to come together, at the urging of D/HOH activists, to craft an unprecedented public-private partnership for distributing and managing the significant and ongoing labor costs of captioning television programs. The closed-captioning system was intended to bring subtitling for the deaf out of the realm of volunteer labor and federal assistance and into the world of self-sustaining broadcast economics. But while the system achieved early success in terms of both technology and practice, it soon threatened to end up a political and economic failure.

## From the Captioned Classroom to the Captioned Living Room

By the late 1960s, deaf educators had witnessed enough success with captioned films—both with children in the classrooms of residential and day schools for the deaf, and with adults in deaf clubs across the nation—to know that captioning was a revolution not just in visual education but in media justice. A number of technological innovations that appeared at about the same time—especially closed-circuit cable television systems, helical-scan videotape recorders, and broadcast-quality character generators—motivated a new wave of experimentation in captioned television.[14] And the creation of four university-based Regional Media Centers for the Deaf (RMCDs) in 1966 meant that one center, the Southern RMCD in Knoxville, TN, could specialize as "a clearinghouse for ideas on utilization of television in programs for deaf children and adults."[15] The Southern RMCD trained a generation of media-savvy educators who "developed production and utilization skills which undoubtedly led to hundreds and hundreds of items of teacher-produced materials."[16] These projects ranged from the do-it-yourself classroom project at the Clarke School for the Deaf to caption documentaries in-house to the $100,000 closed-circuit television production and distribution installations of the Kansas School for the Deaf.[17] Yet by 1970, the year of the profession's first conference on Communicative Television and the Deaf Student, use of television at deaf schools that received materials from Media Services and Captioned Films was estimated at only 25 percent.[18]

Given such low numbers, the real aim of these educators was to create a system for the distribution of captioned television analogous to the system for distributing captioned film—a system that would depend on both captioning technology and captioning labor. One advocate imagined a world where "a simple, relatively inexpensive captioning process were devised whereby televised programs from the educational channel could be videotaped and suitable captions added by students and teachers in the school setting," eventually scaling up into a "Regional Captioning Center" with "professional staff."[19] Captioned Films director John Gough predicted that such a library could reach into deaf households, with "new techniques which can turn the home television set into a teaching machine through the use of self contained cartridges."[20] Education was always the focus of these initial ideas, as when the Southern RMCD experimented with storing captions on punched paper tape, or when the American School for the Deaf experimented with captioning videotaped episodes of *Sesame Street*.[21] But increasingly, educational labor was tied to broadcast justice. "Your efforts to train teachers of the deaf is an impossible task. I don't think this is where you must concentrate your energies," argued one speaker at the Communicative Television conference in 1970. "I think that every television set in the 98 percent of American homes [owning one] can and should be wired within the next few years so that it can be capable of handling materials for every deaf child."[22]

The shift from visual-education goals to media-equality goals in the captioning community mirrored a wider shift in the deaf community as a whole: long seen by outsiders as an "impairment," a "disability," or a "handicap," the fact that deaf persons could not hear was beginning to be reimagined not as medical deficiency but as cultural diversity.[23] One catalyst for this shift was the 1960 publication of William Stokoe's *Sign Language Structure,* which argued that American Sign Language (ASL) was "a full (though not written) language, with a logical internal grammar and the capacity to express anything."[24] This finding came as little surprise to generations of deaf persons who had grown up "speaking" in sign and, in the process, seeking out the company of others who did so as well; however, it was a dramatic shift from earlier oralist presumptions that signed languages such as ASL were inherently inferior to spoken and written English. As one deaf-culture activist later explained, "Although signed language has been suppressed in education for over a century in many lands, it could not be banished from the lives of Deaf people; most of them continued to take Deaf spouses and to use their manual language at home, with their children, and at social gatherings."[25] By 1972 linguist James Woodward had proposed a new terminology to capture these two different understandings: "using the lowercase *deaf* when referring to the audiological condition, and the uppercase *Deaf* when referring to the Deaf community and its members."[26]

This radical new understanding—a shift in both self-definition and research understanding that would continue to evolve over the 1980s—illustrates that the notion of disability is itself a historical and cultural construction. As scholar Gary Albrecht has argued, "Persons with disabilities are found in societies across all cultures and throughout history, yet their identification and adaptive patterns vary markedly by culture, political economy, and environment. Therefore any analysis of impairment, disability, and handicap in a society requires that persons with disabilities be studied in their cultural and political economic context."[27] Historian Douglas Baynton has illustrated that, with respect to deafness in the United States, "the meanings of 'hearing' and 'deaf' are not transparent . . . As with gender, age, race, and other such categories, physical difference is involved, but physical differences do not carry inherent meanings. They must be interpreted and cannot be apprehended apart from a culturally created web of meaning."[28] This web of meaning is woven out of not only social norms but also technological contexts and political-economic structures. For example, Albrecht argued that, particularly in the United States and Western Europe, "disability considerations are situated within a capitalist economic system in which individuals are judged to be disabled in terms of their economic contributions, potential, and value to society" such that "persons who are so certified as having disabilities become the 'raw material' for a series of businesses within a larger rehabilitation industry."[29] In the case of D/HOH populations, that "rehabilitation industry," as Susan Foster has shown, involves two overlapping models (which roughly correspond to the classic oralist and manualist positions in deaf education). The first, a medical or "personal deficit" model, highlights "the expert and active role of the diagnostic or therapeutic practitioner (i.e., physicians, counselors, speech therapists, audiologists, and teachers)" working to "restore function or minimize the effects of hearing loss through rehabilitation of the individual." The second, a social or "interaction" model, argues that "barriers experienced by deaf people in performing the activities of daily life—such as social interaction, work, shopping, attending school, or using public services" are really "a function of social, linguistic, and cultural differences between the majority and minority group" and thus demand "interventions . . . on the environment as well as the individual."[30]

Where any particular D/HOH individual or educator falls in this medical-versus-social divide depends to some degree on personal choice and to some degree on personal experience with deafness. For example, timing of onset is crucial. As demographer Jerome Schein has pointed out, prelingual deafness "presents a severe barrier to acquiring spoken language" and "is also associated with a tendency to affiliate with other deaf people."[31] Deaf children are rarely born to deaf parents, and thus, in Baynton's words, "they make up the only cultural group where cultural information and language has been predominantly passed down from child to child rather than from adult to child, and the only one in which the native language of the children is different from the language spoken by the parents."[32] By contrast, postlingual deafness—or "adult-onset" deafness—"usually does not interfere with speech and acquired language" according to Schein, "and persons deafened in adulthood typically do not seek out early deafened people for companionship or vice versa."[33] This means, as activist Harlan Lane has argued, "late deafening and moderate hearing loss tend to be associated with the disability construction of deafness."[34] But, Lane reminds us, "there are people who have very limited hearing or none at all but choose not to be part of the Deaf-World. Conversely, there are many Deaf people who hear well enough to use a telephone and speak well enough to be understood, but choose to live in the Deaf-World."[35] Sherman Wilcox and Phyllis Wilcox conceptualized the blurred boundaries of Deaf (capital-*D*) culture as an intersection of four factors: audiological condition ("degree of hearing loss"), social network ("the degree to which a person associates with Deaf people"), linguistic choice ("the extent to which the individual uses and supports the use of ASL"), and political activism ("the extent to which the person wields power in Deaf community affairs").[36]

That attention to the power dimensions of deafness is important, as captioning could become a tool for wielding power by both sides in the debate. On the one hand, captioning could be seen as the ultimate "mainstreaming" tool for the medical side, helping to teach deaf individuals better English skills and lip-reading techniques; on the other hand, captioning could be seen as the epitome of broadcast justice, a subtitling tool analogous to cinematic language translation for a cultural minority shut out of television participation simply because they communicate with their hands and eyes, not their mouths and ears. In other words, for both sides, universal captioned television could address what author Tom Humphries referred to for the first time in 1975 as "audism"—"paternalism and institutional discrimination toward Deaf people."[37]

The idea of overcoming audism by placing media-access technology in every home drew on a very particular success story within D/HOH activism: the telecommunications device for the deaf or TDD. As one deaf educator later described, "Until 1964 there was no real way a deaf person could communicate over a telephone without assistance from an interpreter."[38] But in that year, "an enterprising deaf ham radio operator" (and physicist) named Robert Weitbrecht figured out a way to attach the nearly obsolete teletype (TTY) keyboard that he used for radio communication to his telephone.[39] What he invented—a cradle for a standard telephone receiver that could translate electronic pulses back and forth into high-pitched sounds to travel along telephone wires—came to be known as the "acoustic coupler." Realizing the profit potential of his innovation, Weitbrecht soon formed a corporation to sell his device, which became a mainstay of home computing for decades.[40] At the same time, though, others were considering the potential for nonprofit applications of the new modem.

In an unusual gesture of cooperation across the medical and cultural divide, the oralist A. G. Bell Association and the manualist National Association for the Deaf pooled their efforts in the mid-1960s to create the Teletypewriters for the Deaf Distribution Committee and bring surplus AT&T teleprinters into a network that would be glued together with Weitbrecht modems.[41] The TTY network stalled at about only twenty-five nodes until 1968, when AT&T decided to donate eight thousand of its own obsolete terminals to the effort. Soon the D/HOH community had established a network of local groups to distribute these free terminals to interested households. Before the end of that year, the Teletypewriters for the Deaf Distribution Committee of Indianapolis, IN, was established as one of these local TTY refurbishing and distribution centers. By 1969 the Indiana group had gone national, changing its name to Teletypewriters for the Deaf, Inc.[42] Here, then, was a case in which the deaf community combined its innovative skills and its organizational resources—together with a hefty corporate donation—to bring a previously inaccessible communications medium into daily use. As one writer put it in the early 1970s, "For each of the media, except radio, deaf people recently have succeeded in promoting adjuncts which permit their use without the audio dependence. The . . . TTY makes the telephone available to deaf users. Captioned Films for the Deaf has restored motion pictures to deaf users."[43] If they could conquer all these, why not TV?

Following the example of TDD, the New York University (NYU) Deafness Research and Training Center soon offered a similar model of captioned television distribution: nationwide "community antenna television" (CATV), as the nascent cable TV system was then known. At the 1971 National Conference on Television for the Hearing Impaired, Jerome Schein enthusiastically argued that through CATV, "the number of possible channels increases to at least 82! The difference between 12 and 82 is the difference between a precious resource to be expended only for huge audiences and a relatively common one available to small groups."[44] For Schein, this meant that the low numbers and geographical dispersion of deaf populations no longer posed a barrier to accessible television: "With only one or two members per thousand population, deaf people cannot make a strong case for their needs to be served by the limited over-the-air television capacity. But when an area can have 82 channels [through CATV], it becomes reasonable to request that one channel be set aside for the deaf community."[45] As a standard of equity, Schein proposed that "time can be allotted so that the proportion of time assigned for deaf programming is approximately equal to the proportion of deaf people in the general population."[46] By 1972 Schein and the NYU center were pushing a CATV-based National Television Cooperative for Deaf Viewers where each member organization would produce one television show per year at a cost of $1,000 to $2,000.[47]

In the NYU proposal, however, it was not clear whether "television for the deaf" meant captioned television, as Malcolm Norwood's Media Services and Captioned Films (MSCF) desired, or "signed television"—that is, programs either "spoken" entirely in American Sign Language, or translated through a live ASL interpreter using some sort of split-screen effect. The ASL option was particularly attractive to those members of the deaf community who relied on manual communication, especially since, just as in the early days of silent film when the NAD had attempted to preserve on celluloid the art of signing, filming the practice could only bolster its acceptability. This cultural argument, coupled with the fact that adding an ASL translation was technically trivial, was powerful: "It requires but one or two persons and no other equipment except, perhaps, another camera. And we really do need to see more signing on TV. Visibility means awareness. We will not be successful in finding justice for the deaf until there is awareness."[48]

Yet seeking justice through televised ASL posed problems of its own. Most obviously, not all deaf or hard-of-hearing individuals knew ASL, especially those who had lost their hearing after first learning English. And it was not necessarily easy to translate between ASL and English because ASL was a separate language, not based on English except in the rare case of "finger-spelled" English words.[49] Even for those who knew ASL, "geographic variation in signs, corresponding to dialects in spoken language," meant that "a sign in use in North Dakota might be different from that used in Florida."[50] Also the small screen spaces allocated to interpreters either made their hands difficult to see or focused exclusively on their hands, ignoring ASL's use of facial expressions and body language to communicate meaning as well.[51] Yet

ASL did hold one dramatic advantage over captioning: it could be performed in realtime. ASL was thus the only solution available in the early 1970s that could handle live news. When the National Technical Institute for the Deaf (NTID) experimented with ways of translating the *CBS Evening News* for its students, it chose live ASL interpreting.[52] And when television station KRON in San Francisco decided to provide a five-minute news summary for deaf viewers each morning at 8:30 a.m., it chose two Gallaudet College graduates, Jane Wilk and Peter Wechsberg, to host "Newsign Four" in ASL.[53]

Beyond such a short news summary, however, a problem emerged: the labor of simultaneous signing, especially in a complex, 250-word-per-minute news program, was too much for all but the most skilled interpreters to handle. As Schein warned, "Under the stress of the moment, TV interpreters sometimes become poker-faced. That blank look disrupts communication more than a similarly uninflected voice, because the viewer cannot detect possible irony and digressions and other features of speech which can be heard."[54] One way to alleviate the speed stress was to switch from ASL to one of several different coding systems known as Manually Coded English (MCE), which could be transcoded (rather than translated) more directly.[55] Of course, this defeated the purpose of preserving ASL as the gold standard of deaf culture and communication. Captioning could not yet handle live news, it was true, but in almost all other respects it would offer a more widely accessible solution to a D/HOH audience that, because of its small size, internal diversity, and wide dispersal, could hardly afford to fragment itself further.

The push for text-based news alerts had, in fact, been one of the first broadcast battles that the D/HOH community took on with the Federal Communications Commission (FCC). Text had long been an occasional part of the news screen; as early as 1952, the debut of the *Today Show* on NBC featured "a continuing running ribbon of news framing the lower screen."[56] And starting in 1965, ABC made its "news bulletin" interruptions deaf-accessible using "specially designed cards so that the deaf and those with severe hearing losses can receive the important message visually."[57] But it was the dramatic Apollo moon landing in 1969 that inspired hope of wider news accessibility for the deaf. As the *Volta Review* argued, "When we recall the TV broadcasts of the Apollo mission with printed words of events streaking across the bottom of the TV screen the deaf were benefited and the hearing were not distracted. This facility might be extended to important daily news programs."[58] Shortly after this worldwide television event, in April 1970, Nancy Lipshultz of the Illinois Association of the Deaf filed a petition with the FCC requesting not only that emergency broadcasts be presented textually but also that "special programming such as subtitled movies, educational programs and whatever else will help to provide entertainment for the hearing impaired person" be required of television broadcasters. The FCC responded by recommending (but not requiring) that all television broadcasters air titled emergency broadcasts for the deaf.[59] The FCC also suggested that stations in large markets cooperatively rotate the presentation of captioned material for the deaf, in order that "it might be possible for them to do more in the way of visual presentation of value to the deaf than each station would be able to do (or justified in doing) continuously." But the FCC's faith in station cooperation was misplaced, as no such experiments came to pass, even after publications like *Volta Review* and *Deaf American* urged their readers to lobby their local broadcasters.[60] In fact, apart from the network policy of ABC, by December 1971, a Gallaudet survey of captioned newscasts found only half-a-dozen local stations across the nation that used text to alert deaf viewers of emergency news.[61]

## Deaf Captioning as a Mission for Public Television

Fortunately for the deaf and hard-of-hearing community, their growing demand for accessible television came at a politically opportune moment. As media historian James Baughman put it, "By the mid 1950s most of the older mass media accepted the obvious: TV had won the war for the largest share of time Americans spent consuming mass media."[62] But a decade later, in the 1960s, the still-new medium of television was under attack, as media scholar William Hoynes described: "Critics argued that commercial television's reliance on advertiser revenue and its need to attract a mass audience made it structurally incapable of serving the broader cultural, informational, and educational functions of a democratic mass communication system."[63] In a surprising example of legislative speed, Congress created the Corporation for Public Broadcasting (CPB) in 1967 as a direct result of the Carnegie commission's report on educational television issued earlier that same year.[64] By 1969, the CPB had created the Public Broadcasting System (PBS) "to select and distribute television programming among all public

television stations."[65] Neither CPB nor PBS created programming; that
task was left to a handful of affiliate stations in large urban areas which
took on the role as production centers, including WNET New York;
WGBH Boston; WETA Washington DC; WTTW Chicago; and KCET Los
Angeles.[66] Soon after PBS began broadcasting in 1970, MSCF's new
chief, Mac Norwood, contacted Phil Collyer at WGBH "to produce a
demonstration captioned program for use by PBS," choosing the Boston
affiliate "because of its reputation as an innovator."[67] By October of the
next year, Norwood had secured $274,000 from the Bureau of Educa-
tion for the Handicapped for a pilot project to caption the popular
cooking program *French Chef* with Julia Child. PBS agreed to broadcast
the captioned versions of eight episodes of these shows "in at least six
major cities" in the fall of 1972.[68] For CPB and PBS, it was a great example of a "market failure" that only publicly supported television could
remedy. But the official purpose of this pilot project, as Norwood later
described, would be to discover the "reaction of the normal viewing
public to open captions" and to estimate "the size of the audience captioned television can be expected to reach."[69]

These two project goals each affected the path captioning would
take in the 1970s. The fear that captioning for the deaf might depend
on the reaction of hearing viewers had first been suggested a year before,
in a study commissioned by Norwood's agency and performed by
Robert Root of the defense-contracting firm HRB-Singer.[70] The hearing
teenaged and adult members of 522 families were invited to view two
captioned Disney programs (a half-hour nature documentary and a full-
length film) over their local cable system. Some 229 persons from 124
families (24% of the sample) responded. After viewing the films, the respondents were asked for their reactions to the captions in a multiple-
choice survey: only 4 percent reported that "the captioning bothered
me a great deal" on the two programs, though another 26 percent reported that "the captioning bothered me a little." When asked, "What
would be your general reaction to captions on selected TV programs
(not all programs)?," 10 percent of respondents said "unfavorable," 43
percent said "neutral" and 47 percent said "favorable."[71] After the results were reported in the *American Annals of the Deaf*, they quickly
became common knowledge within the deaf-education community.
Many interpreted the results optimistically, as 70 percent of hearing
viewers reported that captions did *not* bother them. Over time, argued
one writer, the remaining hearing viewers would become accustomed to

such captions: "It's a matter of making the unfamiliar the customary."[72] But HRB-Singer concluded, more pessimistically, "considering the economics of network television, it may be presumed that the three major networks would be reluctant to present captioned material on programs of high general interest or, for that matter, for most programs carrying advertising."[73] This fear that hearing audiences would reject on-screen captioning was to be tested in the *French Chef* pilot project.

The second goal of the pilot, estimating the size of the audience for captioning, was also seen as crucial to building network support for accessible TV. The last nationwide census of the deaf had taken place in 1930, putting the incidence of deafness in the population at about one in two thousand.[74] If this rate held for 1970 population levels, it would mean a deaf population of only about 102,000 people out of all 203 million persons in the United States.[75] But the audience for captioning might include both deaf and hard-of-hearing individuals. In 1969 the US Department of Health, Education and Welfare (HEW) awarded a grant to the NAD to perform a National Census of the Deaf Population (NCDP) contracted out to Schein at the NYU Deafness Research and Training Center.[76] By late 1971, using preliminary numbers, Schein estimated that there were "nearly 2 million people who cannot appreciate most television programs," including "about 150 to 200 deaf persons per 100,000 population" (three or four times the 1930 estimate). "The rate was higher in rural areas, but actual numbers of hard-of-hearing and deaf people were greater in urban-suburban areas."[77] The figures for each subgroup also varied by state, yielding an uneven geography which would not necessarily match up with the most lucrative broadcasting markets of the television networks (see figure 2.1). Overall, the NCDP counted some 13.4 million "hearing-impaired" persons (having some loss of hearing in at least one ear) in the United States—or about sixty-six per thousand.[78] A related HEW study, sampling forty-four thousand households (134,000 individuals) in 1971, came to a similar figure of "13.2 million persons 3 years of age and over" with "an impairment of hearing in one or both ears."[79] Not surprisingly, advocates of captioning would cite the 13.2 million figure estimating the hard-of-hearing population, rather than the 2 million figure estimating only the television-deprived D/HOH population. The bigger the interest group, it was thought, the more seriously its claims for recognition would be taken—not only by the federal government but also by networks and their advertisers.

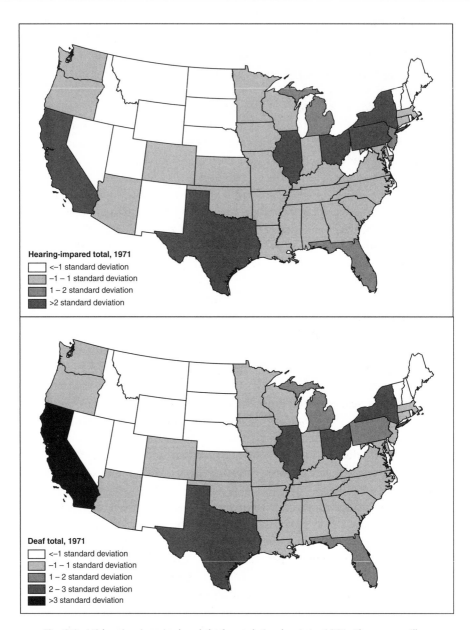

Fig. 2.1. US hearing-impaired and deaf population by state, 1971. These maps illustrate some of the complex geographies of deafness broadcast captioning advocates had to contend with in the early 1970s. The total numbers of hearing-impaired and deaf individuals per state both were tied to population size. However, the per-capita rates of hearing-impaired and deaf individuals per state each mapped out differently because hearing impairment increases statistically with age. The West and South

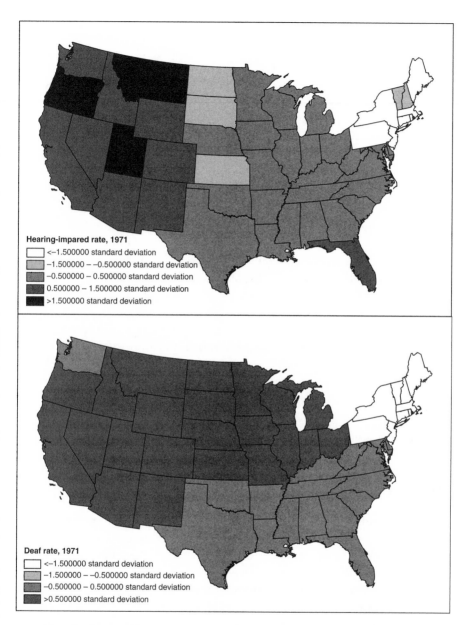

**Hearing-impared rate, 1971**

☐ <−1.500000 standard deviation
▨ −1.500000 − −0.500000 standard deviation
▦ −0.500000 − 0.500000 standard deviation
▩ 0.500000 − 1.500000 standard deviation
■ >1.500000 standard deviation

**Deaf rate, 1971**

☐ <−1.500000 standard deviation
▨ −1.500000 − −0.500000 standard deviation
▦ −0.500000 − 0.500000 standard deviation
■ >0.500000 standard deviation

showed proportionally higher rates of hearing impairment, while these same areas showed proportionally lower rates of deafness. Despite its large total hearing-
impaired and deaf populations, the Northeast showed low rates per capita in
both categories. Source: Maps created from data found in Jerome D. Schein and
Marcus T. Delk Jr., *The deaf population of the United States* (Silver Spring, MD: NAD,
1974), 26–28

Armed with the somewhat contradictory goals of serving its D/HOH audience while not offending its hearing audience, WGBH brought together a group of three employees in what would eventually become known as the Caption Center to figure out how to proceed with the *French Chef* pilot project.[80] There were two main challenges. First, just as in the film industry half a century before, they had to develop a system to automatically display and remove text captions overlaid on existing video at the proper times. Instead of a chemical process of etching characters into film, WGBH dealt with an electronic process of adding characters to videotape. Most of the components engineers needed were already at hand. Sharon Early, later a director of the Caption Center, recalled how "WGBH used a standard Vidifont character generator to create the captions, and 'married' the Vidifont to a computer (General Automation SPC-16) which could store on disk an unlimited number of captions."[81] Or, as fellow Caption Center alumnus Mardi Loeterman put it, they "built a captioning computer the size of a closet."[82] To control the caption timing and presentation, "the system was programmed to read timing beeps laid down on the videotape's cue track—SMPTE time code—and pop captions in and out as programmed."[83] This was not a simple plug-and-play operation, however, as "most of the parts were made by different companies and were not necessarily manufactured to work together."[84]

The second challenge was developing the labor skills and aesthetic techniques required to caption television for the deaf. Thanks to the experimentation within deaf education, the Caption Center did not have to start from scratch. According to Earley, Phil Collyer's team "enrolled themselves in a self-directed crash course in the science of reading and in deafness. They called on others who were producing captions, consulted with reading experts, deaf educators, deaf people themselves." The station soon developed a particular caption-editing philosophy. "WGBH's decision to edit for reading speed and language level was essentially an affirmation of its role as a communicator," explained Earley. "Our job was to make viewing a program enjoyable and meaningful to deaf viewers. When editing was necessary to insure comprehension or to prevent viewer frustration, we edited."[85] Just as with subtitling foreign-language film, captioning meant more than simply transcribing spoken dialogue into text; it meant translating meanings intended for one audience into meanings intelligible to another. What the Caption Center staff probably did not realize at this point, though, was that the debate over edited-versus-verbatim captions in TV accessibility was only just beginning in the D/HOH community.

Having overcome hurdles of both technology and labor, WGBH successfully aired eight captioned episodes of *The French Chef* from August 1972 to September 1972, though only over selected PBS stations (failing to cover the whole geography of deafness in the United States).[86] To the FCC, these were broadcasts like any other; no special approval was necessary. But how much of the deaf community saw the broadcast, and what did the hearing community think of it? Amazingly, even though "WGBH-TV had hoped to make a study to determine the success or failure of their efforts," they were unable to answer their own research questions: "they ran out of money before this could be accomplished."[87] PBS compiled anecdotal viewer comments as a substitute for actual audience research, and in this respect "the experiment was so well received by the deaf community that WGBH captioned fifty other programs during the next two years, all with funding from HEW's Bureau of Education for the Handicapped."[88] But a crucial opportunity had been lost, as there had been no new claims to address the argument that captioning television for a deaf minority would drive away the vast and lucrative hearing majority.

## Hiding Digital Data Inside an Analog Signal

Deaf educators and activists had already thought long and hard about this problem and had come up with a tentative solution: a "hidden" captioning system. Unlike in filmed cinematic subtitles, television captions could be created electronically and inserted into an existing broadcast signal. In the *French Chef* experiment, the captions were inserted at the production site before the episode was broadcast, but why couldn't they just as well be inserted at the user site, while the broadcast was being viewed? As early as 1968, deaf educator E. Jack Goforth argued for such a system, where "captions are stored on audiotape as digital impulses, similar to an audio note in music . . . To place the captions on the screen, the digital impulses must be transmitted to the home receiver which is fitted with the caption-producing circuit." Captions would be sent out separately from the normal TV signal, on a multiplexed FM radio channel. Goforth estimated that a custom-built TV/FM caption decoder for the home would cost around $2,000 at first but argued, "Mass production should bring the price down considerably."[89] Other experiments along similar lines emerged in the 1960s as well, according to communication researchers Sandra Danielson and David Howe: "One method involved the use of two separate transmitters and two television sets, one each for sending and receiving captions, and another pair for sending and receiving the regular picture . . . Another technique used a telephone connected to an electronic device inside a television set."[90] But none of these efforts gained widespread notice or funding.

The innovation that finally allowed captions to be hidden within the television signal came from an unlikely source, the National Bureau of Standards (NBS). In the late 1960s, the NBS was trying to find a way to update its national time-and-frequency broadcasts—previously disseminated through a short-wave radio station WWV in Fort Collins, CO—for the digital computer age. As Danielson and Howe explained, "The age of the computer and high-speed communications systems has brought about the need for more accurate time information, preferably in digital form. WWV's shortwave radio signals are characterized by noise, fading, and other atmospheric limitations," and thus "the signal that finally reaches us is no longer precise enough."[91] Precision was crucial for both economic and military reasons, as these signals were not only used by "electric power companies, radio and television stations, telephone companies, and navigators of ships and planes" but also by cold war monitoring systems.[92] The NBS discovered that the plain old analog television signal was actually a very robust place in which to hide their new digital time-and-frequency data. The hiding spot they chose for their TvTime system was known as the vertical blanking interval (VBI), "that 21-line portion of the 525-line NTSC television signal which does not contain picture information."[93] The NBS noted, "The national television distribution systems are of a quality and enjoy a maintenance standard not matched by other systems." In other words, the initial and ongoing capital and labor expenditures to maintain a geographic network of TV transmitters, paid for in the end by advertising revenues to media firms, provided the most robust infrastructure for sending digital signals of high accuracy and precision—"to millionths of a second"—for public and military purposes.[94] In December 1970, the NBS drafted a petition to be sent to the FCC requesting "a permanent authorization to place a coded time signal on Line 16 of the television frame," to provide "a high quality time and frequency source to the total population."[95]

Selling the TvTime system involved more than just submitting a technical proposal, however. The NBS had to convince both the FCC and the television networks that using up precious space in the twenty-one lines of the VBI would be worth it. The NBS first made a presentation to the Office of Telecommunications Policy where advocates argued, "TvTime can bring an atomic clock to anybody."[96] But a more convincing argument emerged in October 1971 when the NBS demonstrated TvTime for ABC network executives in New York City. As Danielson and Howe reported, "In addition to the time being displayed on the screen, written messages were being sent to other ABC affiliates and to the NBS in Boulder."[97] As a result, ABC executive Julius Barnathan and his staff "determined that in addition to presenting the time of day, they could insert captions on the television screen."[98] Here was an application that could sell TvTime to the FCC—text captioning for those who could not hear television audio, easily hidden from those who could. But the idea would have to be sold to the D/HOH community first.

Fortunately there was a perfect venue through which to sell TvTime captioning to deaf educators: the previously scheduled first National Conference on Television for the Hearing Impaired, which was to be held at the Southern RMCD in December 1971, only two months after the ABC presentation.[99] As Norwood later described it, "The purpose of this meeting was to seek ways and means of allowing the deaf and hearing impaired to obtain their rights to the invaluable educational, social, and cultural benefits of broadcast television."[100] Conference participants had already been scheduled to view a prerelease screening of the first WGBH-captioned *French Chef* episode; now they would be treated to a demonstration of TvTime captions as well. The NBS teamed up with the ABC for a dramatic demonstration. During the regularly scheduled broadcast of the youth-oriented crime drama *Mod Squad,* prewritten captions (encoded on punched-paper computer tape by an NBS secretary, Sandra Howe) were entered into the television signal at the New York City ABC studios and then decoded at the conference site in Knoxville using a prototype TvTime decoder.[101] The demonstration was so successful that Norwood asked the NBS and ABC to repeat it for the students at Gallaudet College in February 1972 (an audience already hungry for more captions ever since Torr's *Wild Wild West* screening two years earlier).[102] Barnathan later wrote of that moment to Gallaudet's Donald Torr: "I remember the day we captioned *Mod Squad* and presented it through the facilities of WMAL-TV to the students at Gallaudet College. It was thrilling to see their response. It inspired me to 'try to make it happen.' "[103] And Norwood, coining a new term for the hidden captioning technique, declared that "this demonstration proved the feasibility of a 'closed caption' system which permits captions to be seen only by a viewer who has a specially equipped television set."[104] From then on, the technology of television captioning for the deaf came in two competing forms: Norwood's closed captions, hidden in the VBI and only revealed through the use of a set-top decoder, and the newly renamed "open captions" that WGBH had been perfecting for *The French Chef*, visible to both hearing and nonhearing viewers without any extra equipment. Each pointed to a radically different technological system, and only one could be chosen in the end.

## Standardizing and Legitimizing Closed Captioning

Stakeholders in the closed-captioning debate sprang into action. At the close of the December 1971 Knoxville conference, representatives from the D/HOH community led by Mac Norwood pushed for the NBS closed-captioning system, which was seen as "immediately feasible and available."[105] By January 1972 the National Association of Broadcasters (NAB) had appointed ABC's Barnathan "chairman of a special subcommittee to develop standards for a television caption system for the deaf," hoping he would represent industry views (and protect industry profits).[106] After the February 1972 demonstration at Gallaudet, John Ball of PBS drafted a memo in which he "recommended that the organization, with its public interest orientation, become involved in developing this new form of television for the deaf."[107] Congressman Bob Dole, longtime advocate for the disabled, wrote to the NBS requesting information on TvTime's ability to caption for the deaf. And the NBS redrafted its TvTime petition to the FCC to highlight the captioning angle, a justification absent from its original 1970 draft.[108] One year after the Knoxville conference, in December 1972, the NBS filed its formal petition with the FCC to permanently allocate Line 21 of the VBI for "NBS TvTime signals, digital captioning information, digital channel identification, and such other forms of digital information . . . as the Commission finds to be in the public interest."[109]

As it turned out, the FCC's decision depended less on NBS than on PBS. The summer of 1972 had not been kind to public broadcasting. Soon after Ball had recommended PBS take a leading role in closed captioning, in June 1972, President Nixon had vetoed an expected, and increased, appropriation for CPB, stating that "an organization, originally intended to serve the local stations, is becoming instead the center of power and the focal point of control for the entire public broadcasting system."[110] According to Hoynes, Nixon argued that public television had become a "fourth network," exceeding its original mandate: "From Nixon's perspective, particularly as the Vietnam War continued to drag on, public television was a home for liberal journalists who produced biased news and public affairs programs with the help of federal funds." Nixon's veto had the intended effect: CPB management was replaced at the highest levels, and only then did Nixon release the funding.[111] While all this was happening, PBS, eager for some positive press to demonstrate its public-service mission, petitioned the FCC for temporary permission to test NBS Line 21 captioning against a competing technology offered by Chicago-based Hazeltine Research, Inc. (HRI), which used a "submerged interleaved subcarrier within the active portion of the video signal" rather than the VBI to carry hidden captions.[112] The FCC granted PBS this testing authority in August 1972, just as the first open-captioned *French Chef* programs were being aired.[113] The competing systems would be tested on the air over a period of fourteen weeks at twelve public television stations, screened for in-studio audiences of D/HOH viewers, with results evaluated by Gallaudet College researchers.[114] The cities themselves were selected, according to Torr, "on the basis of their geographic distribution, the availability of deaf subjects, and the interest of representative PBS television stations" (see figure 2.2).[115] Although there had been no money for audience research on the reception of the *French Chef*, testing and research on "Phase One" of the closed-caption trials would be funded to the tune of $301,000 by HEW.[116]

As with *The French Chef*, preparing the closed-captioning test took not only technological innovation but also labor innovation. Rather than contracting with the existing Caption Center staff to write the titles for the test—or even drawing on the roster of MSCF captioners who had subtitled films for the deaf throughout the 1960s—PBS hired Doris Caldwell from the Southern RMCD to caption thirteen different programs, ranging in length from thirty to ninety minutes each, at a sixth-grade reading level and a speed of 140 words per minute.[117] Caldwell had a master's degree in professional education from the University of Tennessee and experience with "the psychological/social/educational problems of hearing impairment."[118] By March 1974 both the systems and the programs were ready, and about two hours of captioned

Fig. 2.2. Closed-captioning decoder test sites, 1974–1976. Chosen through a combination of "geographic distribution, the availability of deaf subjects, and the interest of representative PBS television stations," the closed-captioning decoder tests in
the mid-1970s represented an uneven geography of captioned broadcasting which
correlated neither to population numbers nor per-capita rates of deaf and hard-of-
hearing individuals. Source: Map created from data in Michael P. du Monceau, "A
descriptive study of television utilization in communication and instruction for the
deaf and hearing impaired, 1947–1976" (Ph.D. thesis, University of Maryland,
1978), 61, and Thomas Freebairn and Martin Sternberg, "Television for hearing-
impaired audiences," *DA* (Dec. 1974), 23–24

material were broadcast every two weeks throughout the summer.[119]
No individual households could view the captioned broadcast, as each
household would have required its own caption decoder; instead, the
chosen audiences viewed the closed-captioning together in a studio at
each of the twelve local PBS affiliates.[120] The pilot test was able to reach
some 1,400 viewers in this way. All were all aged seventeen or older,
noninstitutionalized, and all had "at least minimal hearing loss" as defined by the National Center for Health Statistics. However, the vast
majority of participants had "profound" hearing loss (1,073 out of
1,412, or 76%). The fact that the viewings took place in the evening
may have limited the experiment's accessibility to elderly viewers, who
would have been more likely to think of themselves as "hard of hearing"

rather than "deaf."[121] Nevertheless, the pilot test was judged a success, with the NBS system chosen to move forward into "Phase Two" research and development under a new $322,000 grant from HEW.[122] By November 1975, PBS felt confident enough to petition the FCC for permanent Line 21 authorization for closed captions.[123] PBS had been so successful in pushing the captioning agenda forward that the FCC, still sitting on the January 1973 petition by the NBS to reserve Line 21 for TvTime plus captioning, decided to request comments on the PBS proposal first.[124] Then the real debate over closed captioning began.

## Debating a Technological System and Its Labor Costs

To many in the mid-1970s, "television for the deaf" referred to a wry skit on the new late-night NBC comedy show *Saturday Night Live* in which actor Garrett Morris presented an "interpretation for the hearing impaired" of the news as it was read by fellow cast member Chevy Chase: sitting beside him, Morris repeated exactly what Chase was saying, only louder. At the same time, especially at FCC hearings from May through September 1976, industry representatives and D/HOH advocates wrestled over the details of the costs, benefits, and workings of a proposed nationwide closed-captioning system. PBS wanted the FCC to permanently authorize the system as soon as possible "to stimulate industry development of the hardware and software." But the consumer-electronics industry lobby group, the Electronic Industries Association (EIA), recommended that the FCC only allow closed captioning on a three-year temporary basis "until there is more field testing of the proposed method and consideration of other systems."[125] Similarly, the broadcasting industry lobby group, the National Association of Broadcasters, argued that "there are simply too many unresolved questions" to support the Line 21 captioning rule. Most of the EIA and NAB objections were focused on technological considerations. For example, the NAB (and CBS in particular) feared that "exclusive and limited use" of Line 21 for captioning as proposed by PBS "would greatly deter, if not completely undermine, further application of the vertical interval technology for other uses." The FCC agreed that the VBI was "both valuable and scarce" and worried about assigning exclusive use to PBS's captioning system. Similarly, the EIA (and Eastman Kodak in particular) objected that "the PBS plan . . . can't be used with most of the networks' prime-time programming" because "64 percent of the prime-time network programming is originated on film."[126]

Equipment costs to the networks for encoding and transmitting captions were another point of contention. When the NBS had first filed its TvTime petition with the FCC in 1973, it had optimistically estimated that "a complete network installation to generate, transmit, and decode captions will cost approximately $3000."[127] Now PBS had increased this estimate tenfold, citing the capital cost to each network (and perhaps to each station which wanted to insert captions into local, original programming) at between $30,000 and $50,000.[128] While this might seem expensive, supporters pointed out that "a broadcast quality videotape machine can run as much as $140,000."[129] The networks, however, did not accept the PBS estimate, raising it another tenfold: NBC predicted a broadcast investment of $500,000 per station just to carry captioning, and CBS put the figure at $250,000.[130] PBS president Lawrence Grossman was outraged, calling the CBS estimate "absolutely off-the-wall."[131]

However, the most strident opposition to reserving Line 21 for closed captioning—and thus paving the way for a national text-on-TV system—came as a result of the labor implications of captioning. Outside observers had long underestimated the time it would take to caption a typical television show. E. Jack Goforth's original 1968 proposal for hidden captions assumed a two-to-one ratio of labor time to program time: "Approximately one hour would be required to prepare captions for a half-hour show."[132] But in November 1975, PBS captioner Doris Caldwell, one of the few persons with firsthand experience in the time and costs of this labor, put that ratio at *forty*-to-one: "a qualified caption-editing team at a median salary of $15,000 + 22% for fringe benefits" spent a combined forty hours of labor at a wage of $8.80/hour in order to caption a one-hour program, for a total labor cost of $352 per program/hour. Engineering labor and production supplies added another $490, so after factoring in capital equipment amortization costs, Caldwell pegged the total cost to caption one hour of video at just under $1,000.[133] Caldwell's time estimate of captioning labor alone was enough for the networks to criticize the petition, as they predicted there would always be "insufficient time to caption because program suppliers are sometimes unable to deliver programs sufficiently in advance of air time" (and never mind trying to live-caption the news). But the networks also contested Caldwell's labor-cost estimates. CBS claimed to have tested captioning on an episode of its hit family drama *The Waltons*—captioning performed to "CBS standards of artistic quality, including doing justice to the original author, screen-play and producer"—and came up with a cost estimate of $3,800 per broadcast hour.[134] NBC's estimate was double that again.[135] PBS president Lawrence Grossman later taunted CBS to back up its claims: "Bring us that *Waltons* episode; let us demonstrate our captioning systems to you; let disinterested judges—technicians, educators, the deaf themselves—view the results. I'm willing to bet in advance that we can do the job in half the time and at less than half the cost."[136] But even if CBS was right, according to an editorial appearing in the *Washington Post* that September by Senators Patrick Leahy (D-VT) and Charles Percey (R-IL), "a network spends an average of $270,000 to produce an hour of prime-time television programming."[137] Captioning would at most run between 1 and 2 percent of the entire production cost.

The real question for many was not "what will it cost?" but "who will it benefit?" Here the problem of defining the audience for captioning emerged once again. PBS was still using the HEW estimate of "at least 13.4 million" persons without access to TV in the United States, even though, as described above, that number clearly included persons who were hard of hearing, not deaf, and those limited in only one ear, not both.[138] Yet industry opponents claimed only 335,000 "profoundly deaf" citizens would actually benefit from captioning—a difference of two orders of magnitude—and that "the remaining 13 million plus could be assisted by sound amplification" rather than captioning.[139] In this way, the television industry of the 1970s was advocating for the same sort of audio-only solutions to media accessibility that the film industry had embraced in the 1930s—the equivalent of Garrett Morris shouting in the viewer's ear. Deciding which side was right meant making a judgment about who would need or desire captioned TV: only the relatively small deaf community or the much larger hard-of-hearing community as well? PBS had researched this very question during its closed-captioning tests but did not use the results in its campaign to the FCC. In the spring of 1974, Gallaudet's Donald Torr had surveyed 1,412 in-studio viewers of Caldwell's closed-captioned programs. Out of this population, 94 percent indicated they would like to be able to view such captioned programs at home. This aggregate number was further broken down into percentages from subgroups based on age, education level, and degree of hearing impairment (slight, moderate, severe, and profound). Torr acknowledged that the data might be biased because "the responses were obtained from people who cared

enough to come to a demonstration." Nevertheless, he took the 94 percent positive-response rate and used current HEW estimates of the
number of hearing-impaired individuals in the United States to make
(admittedly rough) estimates of the total hearing-impaired and deaf
populations (at different age, education, and hearing-impairment levels) in the United States who might purchase a caption decoder. In
doing so, he put the total number of possible decoder consumers at
about 4 million—well above the industry figure of 335,000 but well
below the official PBS figure of 13 million.[140]

A few people who provided comments to the FCC argued that, in
fact, the size of the D/HOH community should not even matter—
broadcast justice was a right, and rights are not based on population
size. "Size of the hearing-impaired community is not truly a factor—
all citizens are entitled to adequate service," wrote one petitioner.
Another argued that because broadcasters "enjoy use of a public
resource"—the airwaves—"and profit thereby," then "they have a responsibility to contribute toward serving the hearing-impaired."[141]
This echoed the recurring claim of D/HOH activists that denying captioning amounted to audism, in that "the deaf and hard-of-hearing
person has been deprived of the cultural awareness and knowledge
gained through the medium of television and taken for granted in a
hearing world."[142] Even as early as the 1971 Knoxville conference, deaf
educators like Jerome Schein had been arguing, "Is it not strange that
access to a public resource like the micro-airways can be governed by a
Nielsen rating? . . . [T]he size of the deaf audience alone should not be
the measure of its rights to participation in our society."[143] But such
appeals made little headway, either in the press or in the government.
Perhaps the most effective strategy was to begin to claim that closed
captioning was not only for the deaf and hard of hearing. For the first
time, PBS argued that "the total market for decoders will be greater
than just the hearing-impaired, suggesting that captioning in a foreign
language and the usefulness of captions for persons learning the English language may expand the market."[144] Linking captioning to a
wider set of interests would turn out to be a crucial strategy for captioning advocates over the next two decades.

In the end, helped by a last-minute lobbying effort by PBS president Grossman, deputy commissioner of the Bureau of Education for
the Handicapped, Edwin W. Martin, and NAD executive director Frederick C. Schreiber, the FCC balanced the demands of the captioning

advocates with the reluctance of industry by passing a contradictory regulation, which would have ramifications for years to come.[145] Calling it "a most difficult and troublesome decision," in December 1976, the commission decided to "permit the use of Line 21 for captioning, but not reserv[e] it for the exclusive use of such captioning." Those at the FCC realized that the rule—intentionally lacking any language obligating stations to produce or carry captioning—would create an uneven geography of captioned programming at best: "The PBS proposal would go far in making some programming available, incorporating closed captioning, wherever licensees voluntarily provide such service and people are willing to invest in a decoder to receive such programming."[146] Gallaudet College tried to put a positive spin on the ruling, saying its "flexibility" would "permit later development of Line 21 for transmission of weather information, standard time, sports headlines, stock market quotations" and other services "which should be of benefit to all viewers, the deaf included."[147]

## Developing an Infrastructure for Closed Captioning

Even before the FCC had given its official approval to use Line 21 for captioning, PBS had acquired $656,500 in "Phase Three" funding from HEW.[148] This money was less for research and more for development. PBS was charged with producing two key captioning components: on the supply side, "a computerized, stand-alone caption editing/insertion device that will be suitable for use by any producing agency" with the first unit "scheduled for completion in early 1978"; on the demand side, "a commercial decoder/receiver for home use" looking toward "the first production line units" in summer 1978.[149] Unlike the case of open captioning, where studios could simply use off-the-shelf character-generation hardware (like the roughly $50,000 CBS Vidifont machine), both pieces of the closed-captioning technological system had to be built from scratch.[150] Recalling the network predictions of a $500,000 capital-equipment bill for each station, PBS was determined to have their caption production system come in under budget at $50,000 apiece.

Linking capital costs to labor costs, those at PBS had originally envisioned a minicomputer-based system in their 1975 FCC petition, such that "one non-technically trained person could efficiently and easily perform the captioning function." The prototype used a DEC PDP-11 sixteen-bit processor and a teletype keyboard without a screen to automate the labor process: "The captioning process consists of entry and editing, timing the captions, and recording the captioning information on the helical work copy which may subsequently be used to place the encoded captions onto the quadruplex master tape. All of these tasks require only a single person."[151] The PBS engineers understood enough about the captioning labor process to envision a workflow similar to that described for film subtitling in chapter 1: transcription, spotting, writing, proofing, and production.[152] However, the designers of the console did not envision (or acknowledge) the various divisions of labor that captioners in the deaf-education community had already developed to streamline their work. At the National Technical Institute for the Deaf, workers discovered that "an experienced transcriber is one of the most valuable persons on our captioning staff" who "saves the caption writer hours of time in the editing process"; at the Kansas School for the Deaf, captioners realized, "Noting scene changes is most effectively accomplished when one person snaps his fingers at each scene change and a second person marks the audio script with a slash mark at the appropriate place."[153]

The system PBS eventually came up with was based on microcomputer hardware (a Zilog chip and cathode ray tube), not minicomputer hardware (the PDP-11 and teletypewriter), and because floppy disks could store the captions in ASCII, the division of labor between caption editing and caption transmission could better be separated in time and space. A report commissioned by Bureau of Education for the Handicapped and performed by the Westinghouse Evaluation Institute described the process in detail: "The text to be converted into captions is typed into the system via the ASCII keyboard and stored on one of three floppy-disk memories associated with the microcomputer. The text then can be recalled and displayed on the monitor in units of several lines or paragraphs at a time, as desired." In other words, the PBS engineers basically built a word-processing system for captions. Next, the captions were spotted using a copy of the program on three-quarter-inch tape played on a helical-scan videotape recorder, and the operator used special "Appear" and "Clear" buttons to cue each caption manually in time with the broadcast. This recorded time codes onto the floppy disk with the captions. The floppy disk could then be sent to the television studio, where "an accurate time code reader and a microcomputer with two or more disk drives, in addition to a Line 21 encoder" were used to transfer the captions automatically to the video stream at the appropriate moment in the program. Using micro-computers in both caption editing and caption transmission reduced the costs to each site and opened up the possibility that caption creation could be performed entirely apart from the watchful eye of the networks. (It also answered the objection that captions could not be included in filmed content as easily as on videotaped content, although a human would have to be available during program transmission to manually invoke the captions as the film rolled.) In the total price breakdown of a single caption-editing console, the microcomputer cost only $8,300, the helical-scan videotape recorder $9,000. Labor for assembly and testing of each workstation added about $10,000, but the total still came in at under $50,000 apiece, just as PBS had predicted it would.[154]

The caption-production system was expected to be the greatest hurdle, but it was in fact the production of an affordable caption decoder for the consumer that would prove to be a more difficult challenge. Original estimates were almost comically optimistic. After the 1971 Knoxville conference, "NBS engineers estimate that the simplest caption display model, which may be available as early as 1973, would cost less than $20."[155] By the time of the 1975 FCC petition, PBS estimated that "decoders, in quantity, may cost about $100."[156] However, PBS recognized that this was only with mass production (100,000 units), and only after a considerable development investment of "several hundred thousand dollars," a task that fell to the Texas Instruments corporation.[157] Not all of the decoder cost was due to new technology, however; for ease of use, external set-top decoders had to have their own TV tuners built in, a necessary redundancy because it allowed for more flexibility in attaching the decoder to a television set or personal videocassette recorder.[158] After soliciting bids from potential manufacturers in March 1977, a deal was reached for Sanyo to manufacture the decoders at its Arkansas plant, with Sears retailing them.[159] By this time, PBS admitted that the decoder would probably cost between $200 and $300 if purchased separately, with the hundred-dollar figure redefined to mean the added cost of a television with a decoder built in.[160] But PBS assumed that D/HOH households would snap up the decoder at almost any price. Even the executive director of the NAD said, "Hearing impaired persons [already] spend up to $700 for movie projectors to see captioned films."[161] He failed to note, however, that captioned films were screened before deaf clubs, which pooled expenses, not in individual households. Government demographic data from 1971, which revealed that "there were proportionately more hearing-impaired persons with family incomes under $5,000 and fewer hearing-impaired persons with family incomes over $15,000" than the general population as a whole, were also ignored.[162]

## Whatever Happened to Open Captioning?

Even while substantial time and money was being devoted to the hardware development for closed captioning, PBS was continuing to open-caption certain on-air programming—relatively inexpensively and with no cost to the viewer—through what it called the Interim Captioning Service. Federal funding for this service—originally through HEW, and later through CPB—had been awarded under "Phase Two" of the closed-captioning program and was later renewed under "Phase Three."[163] In April 1975, PBS began feeding about five hours per week of open-captioned programs nationwide to member stations.[164] Although later analysts characterized this approach as a technological dead end, it is significant because its purposes and practices shed light on the way closed captioning was later instituted—and why it met with so much resistance from both television producers and television consumers.

The open captioning at PBS was still performed almost entirely by Doris Caldwell, although by this time she had two assistants. In fiscal year 1976–1977, the whole operation only cost $156,000, about a third of which represented materials and shipping costs for videotapes. Caldwell herself earned a salary of $44,188 as "principal captioner"; one assistant had a salary of $24,054; the other, $20,044. One assistant, Vira O. Milbank, had a bachelor's degree from Gallaudet and experience in the captioning of films—presumably at MSCF. But previous captioning experience was not a requirement of the job. Rather, in addition to a "minimum BS degree" or "five years' experience as an executive secretary," candidates needed "complete accuracy in spelling, punctuation, semantical and grammatical construction"; "experience in editing, script writing, and/or caption-production"; "flexibility in scheduling"; and "speed and accuracy in typing."[165] The job was rewarding but extremely tedious. The Westinghouse report on PBS captioning argued, "As currently structured, the captioning process incorporates a large amount of essentially clerical effort, in addition to the more highly developed editorial skills. Either rotation among tasks or breaking up the discrete elements of the process and their assignment to appropriately suited employees is recommended to overcome the fatigue and boredom evident in one project."[166]

Part of the purpose of Interim Captioning was to continue testing the transmission of closed-caption signals through the network. In the first phase of interim captioning service, PBS used its "seven decoders in the field" at seven selected PBS stations to decode prototype closed-captioned broadcasts and transmit them with "open" captions to all households. But "as station requests for decoders continued to mount, far exceeding the limited supply of prototype decoders, PBS decided to serve all stations by feeding a repeat of each weekly episode with 'open' captions (the closed captions were decoded at the national origination center in Washington DC and subsequently distributed over the entire interconnection system)." The first round of programming to be shown in prime time included the best of PBS's trademark education-and-uplift programming: seven episodes of *Feeling Good* (health); thirteen episodes of *Nova* (science); seven episodes of *Tribal Eye* (travel); seven episodes of *Masterpiece Theatre* (drama); and thirteen episodes of the *Adams Chronicles* (history). By January 1976, about 70 percent of all PBS affiliates were screening open-captioned programming in this way.[167]

Besides stress-testing the delivery of a closed-captioning signal, a PBS internal report from January 1976 listed several objectives of the Interim Captioning Service: "To maintain and expand interest among the target audience in future purchase of commercially available home decoders"; "To broaden contacts with producers and broadcasters of primetime TV programs"; "To refine and delineate basic caption-production techniques through continuing experience with problems."[168] But soon PBS and WGBH were debating whether *too much* open captioning would eventually dampen audience enthusiasm for closed captioning. In April 1976, the argument came to a head. Michael Rice, vice president and general manager of WGBH, wrote in a heated exchange with John Montgomery of PBS that "we [WGBH] have no dispute with PBS on the long-term objective of providing a closed-captioned service," but "to argue that you should withhold available programs from the hearing-impaired public *at this time* in order to stimulate individual purchases of home-decoders *at some unspecified future time* doesn't make sense." WGBH's Rice countered, "It is advantageous to display open-captioned programs *now* so that hearing-impaired viewers will know the value of what they will be able to get by closed-captioned service *later*."[169] Montgomery and others at PBS were clearly worried their support for open captioning would somehow undermine their petition to the FCC for permanent use of Line 21 for closed captioning (still under review during the summer of 1976). But Rice was only asking PBS to screen one more half-hour of captioned programming per week. Rice wrote to Montgomery at the end of April, "We are asking PBS to do no more or less than what PBS was set up to do: to give to its member stations access to programs in which they have shown interest and which they would like to have the right to use. That we are even having to extend this correspondence is surprising and disappointing to us in light of PBS' own public statements to Congress."[170]

Part of this debate over the first goal of the Interim Captioning Service (sustaining momentum for closed captioning) was likely related to its second goal (refining captioning labor practices and determining captioning labor standards). A vigorous argument had emerged by the mid-1970s over whether television programming captioned for deaf audiences should be edited for a lower reading level or presented verbatim for equality with hearing audiences. Deaf educators on all sides had long acknowledged that D/HOH students scored statistically lower on reading tests across a wide variety of grade levels. As one author put it in the early 1990s, "The average deaf sixteen-year-old reads at the level of a hearing eight-year-old. When deaf students eventually leave school, three in four are unable to read a newspaper. Only two deaf children in a hundred (compared with forty in a hundred among the general population) go on to college."[171] But what to do about this statistical difference was a point of debate. In the mid-1970s, the four main sites of deaf captioning—two universities, Gallaudet College and the National Technical Institute for the Deaf, and two public broadcasting entities, WGBH Boston and PBS—each took different positions on this issue.

Gallaudet, where Torr's group of two-and-a-half full-time staffers had been open-captioning videotaped material for in-house educational use since 1970, held down one end of the debate. Because "it would never be practical to caption at different reading levels," Torr "asked that every word be captioned, including those in the commercials."[172] Explained Torr, "To be sure, students might not comprehend certain structures or words, but neither do hearing infants and children, who gradually master the language through continual exposure to it." This justification in Torr's mind married equity goals with educational goals. But Torr also articulated an aesthetic goal, which only verbatim captioning met: "I was also concerned about the intent of the original producer, director, and writer. Anyone who has ever had any connection with the development of a film knows that every frame and word receives intense scrutiny. I would prefer to rest upon their selection, and I am certain that they would prefer that I do so."[173] Torr was in fact so concerned about artistic integrity that he personally lobbied the Senate to change the laws on copyright "to permit educational institutions serving deaf people to videotape broadcast and cablecast television programs without payment of a fee for the purpose of adding captions."[174] In October 1976, Torr got his wish: the new copyright law signed by President Carter mandated that "non-profit educational institutions for the hearing-impaired be permitted to copy television programs for captioning and use within the institution."[175] Preserving the integrity of the original text affected the nature of all captioning at Gallaudet: "The captioners strive for accuracy and must research technical terms, names of people and places, and foreign terminology."[176]

At the other end of the spectrum was the WGBH Captioning Center, which edited an audio track heavily when creating captions in order to reach a speed of 120 words per minute. Sharon Earley, who became the center's director in February 1977, defended this editing philosophy in 1979: "Deafness, particularly prelingual deafness, seriously impedes the acquisition of English skills . . . As a result, the deaf child is often several years behind his hearing peers in developing reading skills. Unfortunately, this gap does not close, and the average deaf adult is reading at a fourth-grade level." Editing was thus necessary because "unlike a normal printed text, captions exist only temporarily. They appear for a few seconds and then vanish. There is no opportunity to refer back to a caption for clarification, as there is when reading a book." The Caption Center's editing philosophy fed directly into the skills that their captioners required: "The captioner must know a good deal about deafness, be aware of the limitations of the medium, and be a resourceful writer and an analytic thinker. The process itself requires more time than would entering verbatim captions."[177] As a result, it was reported, "selected candidates [at WGBH] undergo training for eight months in all phases of captioning and broadcast work."[178]

The reason the WGBH editing policy was so well known—and so controversial—was that WGBH had taken the lead in providing one of the most critical captioning products available in the 1970s: the nightly news. Their effort had begun in late 1972, just after their *French Chef* open-captioning pilot had aired. Seeking to push the boundaries of captioning to near realtime performance, the Caption Center developed a plan to open-caption President Nixon's second inaugural address for rebroadcast the same day it occurred in January 1973.[179] Phil Collyer and his team of captioners ran into only one problem: "The network pool refused to release the audio portion." Fearing that if the rebroadcast lacked any sound, viewers would think their TV sets were broken, the group decided to broadcast a Spanish-language translation over the audio.[180] (Thus, although two different media-disadvantaged audiences were served simultaneously—deaf viewers and Spanish-speaking viewers—Spanish-speaking deaf viewers were still not served.) WGBH's successful experience with captioning this live speech after only a six-hour delay led them to believe that they could perhaps caption the nightly news in the same way. As Earley put it, "With a same day turn-around of a 21-minute program proven feasible, a 23-minute news program suddenly appeared within reach."[181]

WGBH turned to ABC as their patron for several reasons. First, ABC's Julius Barnathan had already been publicly involved in the closed-captioning efforts; it was obviously good press for the network. Second, ABC had a long history of titling emergency broadcasts for the deaf. Third, Collyer knew that "ABC had already allowed the public television station in Rochester, New York, to rebroadcast the evening news with a sign language interpreter"—previously the only way to interpret news live for the deaf.[182] Soon a deal was reached: ABC would allow WGBH to tape the six o'clock national news each evening, quickly caption it, and then rebroadcast it at eleven o'clock over PBS affiliate stations on the East Coast. But there was a catch: each local ABC affiliate had to give approval for local PBS retransmission (after all, ABC affiliates might not want the national news competing with their own late local news offerings). And ABC would not pay for the captioning; a $106,000 grant from HEW took care of that.[183]

The Caption Center already had the equipment and the staff to do the captioning. However, the tight turnaround time of five hours demanded a dramatically different division of labor—and the complex syntax and vocabulary of the evening news (as opposed to cooking in the *French Chef*) demanded a policy on editing. The key innovation was to use several captioners working in parallel; however, this division of labor opened up additional questions.[184] First was the problem of standards. As Earley described, "In order to maintain a consistent captioning style from story to story, editing guidelines had to become much more formalized, and the decisions that captioners needed to make in the situations when guidelines conflicted or did not exist, standardized." Next was the question of workflow: "A transcript of ABC's audio had to be typed. That had to be edited into captions. The captions needed to be checked and placed, then entered into the Vidifont system, checked once more and corrected. Because there was no time to control caption display with the computer time-code system, captions had to be changed manually, live on air. This required a rehearsal. In addition, we had to prepare insert material to fill ABC's commercial breaks and set up for a live broadcast." Then there was the ever-present problem of fatigue: "One person could transcribe at peak efficiency for about an hour before fatigue set in and slowed the typing. So an editor relieved the transcriber, who moved on to another task." There were seven news captioners trained in all, each who could fill any role in the division of labor so that they could run the news with only five captioners on hand at a time.[185]

Finally there was the question of constructing the captions. Citing a Gallaudet study indicating that "the average graduate from an educational program for deaf and hard of hearing students read at about a third-grade level," WGBH extensively edited the news script (see figure 2.3). Captioning researcher Carl Jensema later recalled, "The word count was cut by about a third and the reading level was cut from roughly the sixth-grade level to the third-grade level. All passive-voice sentence construction was removed, nearly all idioms were removed, contractions were eliminated, clauses were converted into short declarative sentences, and even jokes and puns were changed if it was felt the deaf and hard of hearing audience would not understand them."[186]

Viewers were not deterred by the editing. The *Captioned ABC Evening News* was a hit. After its debut on a handful of Eastern Educational Network stations in December 1973, local affiliates signed on rapidly.[187] Almost a year later, when PBS went national with the program (ironically, three days before President Nixon resigned), more than fifty

| Original audio track | Edited captions |
|---|---|
| "Now credit card companies testifying before the Federal Privacy Protection Commission have disclosed they also give out the details of what their clients buy and pay for, or don't pay for, to government agencies and private attorneys on presentation of subpoenas. American Express says it does not even tell its clients when this happens." | NOW CREDIT CARD COMPANIES HAVE TESTIFIED THAT THEY ALSO GIVE OUT THE DETAILS OF WHAT THEIR CLIENTS BUY AND PAY FOR TO GOVERNMENT AGENCIES AND PRIVATE ATTORNEYS WHEN SUBPOENAS ARE PRESENTED. AMERICAN EXPRESS DOES NOT TELL ITS CLIENTS WHEN THIS HAPPENS. |

Fig. 2.3. Example of caption editing at WGBH, ca. 1977. The *Captioned ABC Evening News* was edited for vocabulary, reading speed, and sentence structure to a third-grade level. Source: Redrawn from du Monceau (1978), 90

stations had signed on.[188] The show continued to earn HEW support and was carried by 118 stations in thirty-four states by the start of 1977 (see figure 2.4).[189] Surprisingly, this success was not only due to D/HOH viewers. In some areas, the local ABC affiliate demanded that the captioned program air not at eleven o'clock, but at eleven-thirty or twelve, to avoid competition with the affiliate's local news program.[190] The *New York Times* noted, "Many of the ABC stations are reluctant to have their local late-evening news compete with Howard K. Smith, Harry Reasoner and the rest of the network's Evening News team."[191] In New York City, NYU's Deafness Research and Training Center "acted as intermediary between the local PBS station which wanted to broadcast the Captioned News and the local ABC-TV affiliate station which did not wish to grant permission for the program's release in New York City without important conditions."[192] Local affiliates feared that even hearing viewers would prefer watching the open-captioned network news rather than the noncaptioned local news, an assumption that ran counter to the conventional wisdom that hearing viewers would never willingly choose to view open captions. Thus, when the WGBH Caption Center attempted to promote captioning standards, it spoke from a position of both intensive daily experience and unquestioned national success with both hearing and nonhearing audiences.

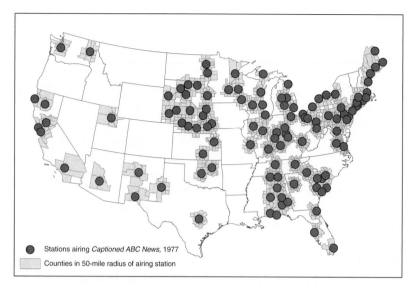

Stations airing *Captioned ABC News*, 1977
Counties in 50-mile radius of airing station

Fig. 2.4. Sites of open-captioned news delivery, 1977. Representing a negotiation between local PBS stations, local ABC affiliates, and local deaf and hard-of-hearing viewers, the geography of captioned news broadcast through the 1970s was still an uneven patchwork but exceeded the range of previous PBS closed-captioning tests. Source: Map created from data found in Gallaudet College, Public Service Programs, *Now see this: A survey of television stations programming for deaf viewers* (Washington, DC: Gallaudet College, 1977)

To others, the WGBH caption-editing strategy was least controversial when applied to children's programming (where the caption presentation rate was further reduced to sixty words per minute).[193] In fact, when the Caption Center was charged with captioning thirty-nine episodes of the PBS kids' show *ZOOM* in the summer of 1975, they let their creativity run wild: "When the Zoomers [kids] jump up and down, the words jump up and down. They grow BIGGER, smaller, and if the Zoomers tear across the stage during a musical number, the words race across the screen as if they had personalities of their own."[194] Perhaps because of the extra-spicy captions, the cost to title each episode could reach $2,000 per broadcast half-hour.[195] For the captioning community, *ZOOM* offered a compromise position on the captioning debate: verbatim for adults, edited for children. This was close to the policy adopted by the National Technical Institute for the Deaf, where 97 percent of the school's nine hundred students had lost some or all of their hearing either at birth or prior to age four.[196] Like Gallaudet, the NTID captioned video in-house to screen for its students. But its captioning producer, Linda Carson, used a hybrid editing strategy: "Programs captioned for entertainment purposes will not be edited (verbatim captioning). Programs captioned for instructional purposes will be edited when necessary to conform to the reading and language level requirements of the intended audience," usually 140 words per minute.[197] Thus two possible compromises on the editing debate existed: one based on audience (adults versus children) and one based on program content (entertainment versus education). This last option was the strategy Doris Caldwell took at PBS. In 1973 she had argued that in teaching deaf children to read, "graded [edited for reading level] captions permit the use of language controls and the application of developmental reading principles."[198] But she also maintained, "Editing, to a professional caption producer, means keeping the script as verbatim as possible while condensing to a grade readability level and reading speed appropriate for the target audience (adults or children)."[199] Thus Caldwell, much like the NTID captioners, took the Gallaudet position for adult entertainment programming and the WGBH position for children's educational programming.

As the Interim Captioning Service attempted to integrate material captioned by Caldwell at PBS and material captioned by the Caption Center at WGBH, a serious rift developed between the two organizations—one that would remain for years. It began with the April 1976 argument between WGBH's Rice and PBS's Montgomery, in which Rice had asked Montgomery for PBS to carry an additional half-hour of WGBH-captioned programming. When Montgomery declined, Rice responded, "Your decision to limit the distribution of programs captioned by WGBH protects and encourages a PBS decision about the nature of good captioning [meaning, verbatim captions] that has by no means been universally accepted."[200] Soon after, WGBH's Phil Collyer began a correspondence with PBS's Doris Caldwell on the nature of "good captioning." Collyer argued, "I don't believe precision is that important . . . You state that oversimplification of the content abuses the integrity of the producers of the program. I do not believe this is necessarily so. The program producers I have dealt with have welcomed our efforts to extend the audience for their programs to the hearing-impaired." Collyer claimed, "Our objective is to serve as great a number of the audience as possible by seeking a more common denominator: a lower reading level and reading speed."[201] In May 1976 Caldwell replied, asking Collyer, "Does this mean you are opting to serve only that relatively small segment commonly referred to as the 'low-verbal deaf'?" She then countered, "The low-verbal group (largely, the prelingually deaf with inadequate schooling) is continuing to lessen due to improved educational strategies." Caldwell felt that "rewriting of content for purposes of simplification may be both unwise (posing the danger of distorted information) and unwarranted (serving well only a minority of our target audience.)" She used PBS research to argue that her verbatim captions "were overwhelmingly acceptable to sample audiences who completed over 1,400 opinionnaires administered after each viewing of programs selected for captioning during Phase I of the PBS closed-captioning project."[202]

In the end, the two captioners, and the captioning agencies they represented, had to agree to disagree. WGBH's Collyer ended his letter saying, "I hope you will agree that you and I, both as hearing people and caption producers, are not the ones who should determine that approach. The decision should be made by the deaf themselves. I hope that somehow this will be done."[203] PBS's Caldwell expressed the similar sentiment that "open availability of PBS-designed encoding units and expanding developments such as real-time captioning will make many of our present concerns obsolete."[204] In April 1979, WGBH departed from its editing philosophy to provide near-verbatim captions for the PBS miniseries *The Scarlet Letter,* reportedly because in Nathaniel Hawthorne's prose "the language was so much a part of the message" and because "the dialogue moved at a slow enough rate."[205] But the debate had real consequences for the way captioning labor was perceived. The 1978 study by Westinghouse of the PBS captioning process tried to resolve the editing debate once and for all, through "group sessions" in five cities where a total of 375 "hearing-impaired" persons "viewed either edited or verbatim captioned segments from television programs [both WGBH-captioned and PBS-captioned] and rated them in terms of satisfaction, likeability, understanding, and clarity," and through "individual sessions" where eighty people participated in "a true experimental design, with random assignment of participants to conditions (viewing verbatim versus edited segments) and random sequences for viewing the stimuli." Their recommendation—"Rather than varying both vocabulary and syntax, perhaps editing would be easier for the captioner and more beneficial for the deaf viewer if the reduction of the text occurred in just one of these two dimensions"—reflected the underlying cost consideration in the debate.[206] If verbatim captioning could be justified, perhaps both editing time and editing costs (and the associated familiarity with the language levels of the deaf) could be trimmed from the captioning labor process.

## An Institutional Framework for Closed Captioning

Though the debate over captioning standards was left unresolved, the Interim Captioning Service pressed on, the development of the caption-editing console and set-top decoder continued on-track, and it seemed that some sort of national closed-captioning system was just around the corner. An internal PBS memo from January 1977 predicted the mass production of caption decoders by "the next fiscal year."[207] PBS president Grossman hoped to begin "Phase Four" of closed captioning—final development—in July 1977 with a last $450,000 from HEW, and PBS assured its Board of Governors and Managers that "a viable working system should be in place in the course of 1979."[208] In January 1977 the executive secretary of the NAD, Frederick C. Schreiber, wrote to Gallaudet's Donald Torr to ask, "What can be done to expedite bringing decoders and captioning to the deaf, who are champing at the bit—raring to go—if only someone will tell them where or how."[209] Even President Carter chimed in with "letters to the networks on the subject of the problems of the hearing impaired in benefiting from television," though he did "not urge any course of action on the networks."[210] Yet gaining the final support of the television networks, who had argued so vigorously against FCC allocation of Line 21 for captioning in the first place, was a harder task than the D/HOH education community ever imagined. In the end, it took the creation of a whole new agency for the siting and management of captioning labor, underwritten by a public-private partnership of funding and affiliated neither with PBS nor with the Big Three television networks, to convince the broadcasters to participate even voluntarily.

At least the D/HOH community was still united behind captioning. Early in February 1977, the National Association of the Deaf convened a meeting at the Bureau of Education for the Handicapped including representatives from Gallaudet, New York University, A. G. Bell, and PBS (although apparently no one from WGBH was there). Even the president of the National Association of Broadcasters attended—this was a meeting to strategize with the broadcast industry, not against it. According to Torr, "It was decided that the NAD would coordinate efforts to get things moving," especially "a meeting with the networks, then to talk with independent producers and possibly corporate sponsors." These deaf educators were trying to walk a fine line with their constituents, though. On the one hand, they agreed that "it would be best for the deaf community to avoid pressure tactics at this time." But on the other hand, deaf citizens were urged to write letters of support to the networks, "due to the fact that there is a great deal of skepticism as to the numbers of deaf people who would buy such decoders."[211]

The D/HOH captioning advocates underestimated the objections of the networks. The March 14, 1977, issue of *Broadcasting* magazine reported that ABC president Frederick S. Pierce "cited copyright and labor union problems in transferring written dialogue to captioning, which uses fewer words than are written in a script"; CBS president John D. Backe argued against " 'freezing' the state of the art at this juncture" before a better television-based information system known as teletext might be ready; and NBC president Herbert S. Schlosser had the audacity to claim that his network already provided for the hearing impaired by using "full-face shots of newscasters . . . for the benefit of lip readers."[212] (Even oralist D/HOH associations agreed that lip reading was no substitute for captioning in television accessibility—at best, only 25 percent of what was said might be gleaned this way.)[213] In a letter written a few months later, Torr lamented that the efforts of the D/HOH community were not working: "I do not sense any movement to educate the networks and enlist their support or to field their questions and attempt to resolve the problem. I sense that BEH/PBS are moving to get decoders made, but no action is being taken to design an overall operating system." Torr suspected that the networks were not addressing the real issue: "If the networks were to be straight with us, perhaps we could get things moving." He had previously offered the labor of his own Gallaudet captioners, just as with MSCF in the 1960s: "Gallaudet College has had considerable experience in captioning and the staff there would be available for this purpose if needed." But this time Torr also suggested an idea that was a startling departure from previous assumptions that the networks would handle all captioning labor and costs: "Can we set up nonprofit captioning centers in, say, L.A. and N.Y.C. which prepare the captions and deliver them to the network in a format which can be easily transferred to their video tape record for subsequent transmission? How can that be funded? Can it pay its own way through advertising revenue?"[214]

This idea for a nonprofit captioning center, apart from both PBS and the networks, was the beginning of a solution. By autumn of 1977, HEW secretary Joseph Califano had organized a meeting with the network heads to convince them to adopt the PBS closed-captioning system. Publicly the government was putting increased pressure on the industry. Califano was quoted as saying, "The cost of this program for the networks, when measured against their unprecedented profits [$295 million in 1976], will not be too great; I believe they can afford it."[215] And FCC commissioner Robert E. Lee, speaking at the NAB meeting in New Orleans, cited the figure of 13 million hard-of-hearing Americans who could benefit from captioning and made a veiled threat to take more "affirmative action" with broadcasters if they failed to "make modern communications systems available to the deaf."[216] But behind the scenes, a compromise along the lines that Torr had suggested that summer was in the works. PBS's John Ball later recalled, "The national television networks agreed to participate voluntarily in the service on the condition that a responsive, free-standing nonprofit organization be established to provide the captioning."[217] And Norwood recollected that ABC agreed to participate on the condition of "captioning at a rate of no more than $2,000 per hour."[218]

Finally in March 1979, a deal was announced by HEW Secretary Califano. Proclaiming "we celebrate the immense good that can come about when government, private industry and the voluntary agencies join hands and cooperate in the public interest," he explained that PBS, ABC, and NBC would participate in a Line 21 closed-captioning system (network participation was voluntary, and CBS was conspicuously absent).[219] The plan seemed simple: the government would create a new, private, nonprofit corporation called the National Captioning Institute (NCI), which would produce the actual captions, manage the development and marketing of decoders, and research future possibilities in captioning.[220] By the end of that year the plan was moving ahead. NCI had found facilities on both coasts, one in Falls Church, VA, and one in Los Angeles, CA; a board of directors had been named, headed by Don Weber, a Texas businessman who was also on the board of PBS; two advisory boards, each consisting of "seven people active in deafness issues," were to be created; and, finally, John Ball of

PBS had been named president of NCI.[221] Califano's press release had rewritten captioning history a bit, however. No mention was made of the efforts of D/HOH educators over the previous decades in imagining, prototyping, and lobbying for accessible television. Instead, Califano gushed, "We can trace these developments directly to a visit President Carter made to a Federal Department soon after he was inaugurated [in January 1977]. On that visit, he met a woman who is deaf and who told him about the concept of closed captioning. President Carter was intensely interested in the idea. Very soon he wrote to the Presidents of the three networks and the Public Broadcasting Service, asking them to work with him and with HEW to make this important concept a reality."[222]

## Making NCI Work

Legally separating NCI from both HEW and the networks was the first step toward this reality; making this public-private partnership work was another thing altogether. NCI would charge the networks $2,000 per broadcast hour for captioning services; if each network paid to caption an expected five hours per week of programming, the total cost to each network would be roughly $500,000 per year (although NBC and ABC both predicted that "additional taping and transmission costs" would raise the total annual expense to $750,000).[223] Networks would bear the labor costs, but each local affiliate would have to "purchase one or more broadcast decoders (expected to cost approximately $1,000–$1,500 each)" to ensure the captions were being passed through at the local level. However, labor costs were not the only form of income for the institute. Each time a home decoder was sold, now at an expected price of $250, NCI would reap an eight-dollar royalty fee. The chairman of NCI's board assured the networks that "the manufacturer of the adapter units is taking a low mark-up on its product" and "the retailer [Sears] has indicated essentially that it will set a price so as to obtain no profit."[224] With this dual revenue stream, NCI expected to be self-sufficient after four years. Until that time, it was to benefit from a third source of funding in the form of continuing grants from HEW: $3.5 million in startup funds during the first year and an additional $3.4 million spread over the following three years. This was on top of an additional $3.5 million loan "from four insurance companies," which had financed the initial mass production of decoder chips.[225] Thus the ability of NCI to wean itself from loans and grants depended on the willingness of the networks to continue to caption programming, and on the willingness of consumers to purchase caption decoders.

By 1979 NCI was busy preparing for its scheduled March 1980 debut, organizing a division of captioning labor the likes of which had never been seen in the United States—neither in subtitled film nor in educational television. NCI spent $3 million outfitting its two offices, installing sixteen "captioning booths" in Falls Church (with fifty-five employees handling mostly PBS programming) and another thirteen in Hollywood (where twenty-five more employees handled mostly commercial programming). Ball estimated a labor requirement similar to that described by Doris Caldwell at PBS, where "35 hours of work by NCI's highly trained staff of caption editors go into captioning a one-hour television program."[226] But it was the turnaround time for captioning an individual program, not total labor time per broadcast hour, that caused the most fear. Captioning would be the last step in a television-production process that routinely worked up to the last minute before airtime; NBC in particular worried that the institute would not be able to caption shows fast enough.[227] Unwilling to ship their "quad tape" master copy to NCI, the networks sent a dubbed copy of each program on low-cost videocassette to the captioners, which then returned a floppy disk of captions and time codes to the network.[228] NBC granted NCI only forty-eight hours to perform the captioning, "without a script" if one was not available from NBC.[229] Thus at least three temporalities—the television production schedule with its make-or-break air date, the caption labor demands with a nearly forty-to-one labor-time to content-time ratio, and the inevitable delays of physical transport of videotapes and floppy disks—had to be negotiated if closed captioning was to succeed.

With such tight production deadlines, NCI might have made use of all the captioning talent and facilities that it could, including the Caption Center at WGBH Boston. But in a move that shocked the WGBH staff, the institute refused to license the Line 21 Caption Console technology to any outside vendors—even though WGBH experts had regularly consulted with PBS on the design of the system.[230] While WGBH could continue to produce open captions as before, it was now shut out of closed captioning. Whether this move had any link to the debate between PBS's Caldwell and WGBH's Collyer about caption editing is unclear. NCI still could not live-caption the nightly news, and in 1980 WGBH's director wrote to assure readers of the *American*

*Annals of the Deaf* that *The Captioned ABC News* would continue on nearly all of the 180 PBS stations that carried it.[231] But even with this ongoing assignment WGBH needed to have Line 21 capability soon, with or without NCI's help. Recalled WGBH's Loeterman, "Almost immediately, The Caption Center cannibalized its open captioning system and developed limited Line-21 capability. But it would be four-and-a-half years before it employed captioning technology that would be accepted by most broadcasters and cablecasters."[232]

With WGBH excluded from Line 21 captioning, the spotlight was firmly on NCI as the first closed-captioned programming hit the airwaves, and the first caption decoders rolled into Sears. On March 16, 1980, the three participating networks began transmitting words: on PBS, the high-culture *Masterpiece Theatre* and *Once upon a Classic;* on NBC, *The Wonderful World of Disney* for kids; and on ABC, an action-packed *ABC Sunday Night Movie* featuring the military adventure *Force 10 from Navarone.*[233] Ironically, the service was launched "before the first batch of special TV set decoders had even been delivered to the public."[234] By June, however, about twenty thousand decoders had been sold, and it seemed that closed-captioning was well on its way.[235]

The pace of decoder sales would soon begin to haunt NCI, however. Sometime before September 1980—six months after the debut of closed captioning—Doris Caldwell (who had moved to NCI along with Ball) asked Gallaudet's Edward Merrill to help her estimate how many decoders would be sold each year. (Amazingly, such an estimate had not been computed yet.) Merrill turned to Gallaudet's captioning expert, Donald Torr—the same person who in 1974 had predicted for PBS that the entire captioning market would only encompass about 4 million D/HOH persons, not the 13 million that PBS had chosen to promote instead. Torr's calculation of likely decoder sales followed a similar logic. Starting with the same 4 million figure, Torr wrote, "I reason that the individuals who will really be most likely to buy decoders will be profoundly deaf persons." This would limit the number to "266,000 persons age 17 or older." Torr then estimated that roughly 65 percent of these 266,000 persons would be married, which would yield 172,900 married and 93,100 unmarried persons. Thus, "the estimated 172,900 married profoundly deaf persons would represent 86,450 married couples who would need only one decoder." This led to a total of 179,550 possible unique decoder households. If only 93 percent of these households owned televisions, it would reduce the

number to 166,982 households. Finally, Torr added the 30,004 deaf
students counted in a 1980 survey by the *AAD* (since deaf parents
rarely had deaf children, these were not counted in the first figure) to
come up with a ballpark figure of 200,000 households willing to purchase a caption decoder. Torr reasoned that those purchases might be
stretched out over four or five years, yielding an expectation of forty to
fifty thousand sales per year.[236]

Unfortunately, NCI had been proclaiming that it would manufacture and sell a hundred thousand decoders per year, for "a total of
about 400,000 sales over the next three or four years."[237] Thus when
closed captioning hit its one-year mark in March 1981, NCI was in
trouble: only about 35,000 homes were estimated to have captioning
decoders (a figure below even Torr's estimate). Similarly, the total
amount of captioned programming was lower than expected at only
twenty-five hours per week—with PBS, ABC, and NBC participating,
this meant an average of a little over one hour of captioned fare on
each network per day. NCI president Ball admitted that captioning suffered a "chicken and egg problem"—people would not buy caption
decoders without more captioned programming, but programmers refused to caption more content without a larger audience of decoder-
owning viewers.[238] NCI research supported this argument. When the
agency conducted interviews with 113 "core" deaf persons—"relatively
young, socially active, prelingually deafened individuals with below-
average incomes"—they found that even though each "watched an
average of 3.3 hours of television per day," only 20 percent of them
had purchased a caption decoder. Holdouts cited "too expensive" and
"not enough close-captioned programs" as their main reasons for not
purchasing one.[239] NCI received the same response when they surveyed people who had previously contacted the organization to request information about closed captioning.[240] If the "core" of the deaf
community and the people who had already asked about decoders
could not even be convinced to buy one, how did NCI ever expect to
reach the entire 13 million-strong D/HOH population?

NCI researchers Jensema and Fitzgerald tried to answer this question
with a more detailed set of data. Every single caption decoder sold except for the first run of 3,700 had included a consumer questionnaire
in the box. By the end of 1980, "of the 27,561 postcards distributed,
12,528 were returned to NCI, a 45% response rate."[241] Initial analysis
revealed a pattern of "relatively high income among households with

closed-caption decoders" as well as a higher percentage of persons with high school diplomas and college degrees compared to all TV households.[242] But most deaf households were not high-income households, and even though the IRS agreed to treat decoders as a medical deduction for tax purposes, the $300 price tag was a deterrent to many.[243] In addition, the hard-of-hearing population, which tended to be older, was not represented in proportion to its numbers in the D/HOH community: "In spite of the fact that over 40% of the hearing-impaired population is 65 years of age or older, only 16% of all hearing-impaired closed-caption viewers reported are within this age group."[244] So far, only a select subset of the total D/HOH population in America had either the means or the interest to own a decoder—even though it had been well established that, for example, deaf children actually watched significantly more television per week (23–29 hours) than hearing children (13–19 hours).[245] And all of these numbers were only made worse by new, larger estimates of the size of the D/HOH population, which as a result of a 1977 survey by the National Center for Health Statistics put the number of "hearing-impaired" persons in the United States at 16.2 million, 3 million more than PBS's previous working estimate.[246] A decoder market that should have been growing was instead looking perilously small.

NCI was in trouble. In 1981 it reported "captioning revenue" of $3.1 million but incurred operating expenses of $4 million. Some $900,000 in Department of Education grants made up the difference in 1981, but this pattern could not be sustained.[247] With decoder sales so low, Sears lost its exclusive right to distribute the set-top boxes in 1982.[248] Captioning was also experiencing quality problems. Even though the Line 21 signal, with its low data speed, was designed to be "recovered error-free by home decoders even in fringe reception areas," with over a thousand local stations, there were bound to be problems passing the signal through to households.[249] "The most common reception problem is dropping of a frame of video, which takes out the two characters contained in that frame, so missing letters usually come in pairs."[250] Such technical glitches were long-standing annoyances to both NCI and its audience. Yet NCI could do little to enforce signal quality at the local-station level.[251] Other errors came not through technological networks, but through the temporal and spatial division of labor. As one captioning expert later described, "When operating under a tight schedule, a program may go to the captioner on videotape before all of the final editing is complete. The dialog may be changed and re-dubbed after the captioning is done. This can lead to captions for dialog that doesn't match, or even doesn't exist in the final tape."[252] To audiences, of course, whether mismatches between audio and text were due to humans or machines made little difference; both kinds of errors reflected poorly on the new closed-captioning system. Nevertheless, at its one-year anniversary NCI claimed to be "on target toward becoming self-sufficient by 1983."[253] As many could already tell, they were wrong.

In 1932 a commentator in the *Auditory Outlook* reported, "The telephone companies are working on something now that ought to be of great interest to all who have an auditory defect—television. Two-way television, in which images are transmitted by wire just as the telephone transmits sounds, makes it at last possible for the hard of hearing to converse at a distance."[254] From the vantage point of the 1930s, what we might today refer to as the videophone was seen as the ultimate promise which television held out for deaf communities—an answer to their disconnection from the auditory-only telephone.[255] Deaf access to the telephone eventually took another direction entirely—leveraging outdated teletype technology in the 1970s to create the TDD/TTY system—but television would remain enticing to generations of deaf educators as the apparent technological fix to the inability of the deaf to use the radio for entertainment, news, or education. Yet it would take half a century from that original 1932 introduction before television became a useful technology for deaf communication.

For most of that time, the deaf and hard-of-hearing audience remained virtually invisible to television-program producers, distributors, networks, and advertisers who bought and sold "eyeballs" (economically desirable viewing audiences). Other analysts who have looked at the history of closed captioning have noted the basic difficulty that "the television industry, based on network competition for the largest share of a mass audience, had little motivation to consider the interests of the small percentage of viewers with communication-related disabilities."[256] But this explanation ignores the fact that other "small percentage" audiences—such as teenaged children and African Americans—had already been courted by television by the time closed captioning emerged in the 1980s. What was really at work with deaf audiences was the triple dilemma of not only being too small in aggregate but also being insufficiently desirable to marketers and overly dispersed for local stations to bring innovations to all geographic markets equally. Thus the counting, mapping, and classifying of deaf populations became increasingly important as accessible TV projects moved from conception to reality.

The result of such counting and recounting, through a decade of experimentation, agitation, and negotiation between educators, engineers, and lobbyists, was the National Captioning Institute—a new model for public-private partnership in addressing social goals through market mechanisms. But many of those involved simply saw NCI as an institution for managing a technology—the Line 21 system of caption distribution and decoding—rather than managing both a labor process and a consumption process. Decades of film captioning and video experimentation had taught the deaf-education community that fundamental issues of edited content versus verbatim content, time pressure versus quality control, and art versus science would not simply evaporate once captions were encoded on floppy disc. And decades of debate over the proper place of English text in a D/HOH community divided between oralists and manualists, medical diagnoses versus social understandings of deafness, had shown that issues of caption consumption would not be as simple for this media-minority community as government and industry might believe. Many D/HOH households were desperate for media equality and eager to participate in the cultural life of the nation through television, but they continued to balance the costs of bringing closed captioning home with the benefits of viewing only the occasional highbrow documentary or lowbrow sitcom.

Just as the actions of D/HOH consumers were not as predictable as government and industry thought, the move from open to closed captioning was not simply a technologically driven "advance" immune from historical explanation. Though there was a long history of both cinematic and educational experience with open captioning (inherited from film subtitling and deaf schooling), and even though the Caption Center had such a success with its universally accessible open-captioned *French Chef* pilot, the new closed-captioning systems fit not only the legibility limits of 1970s-era TV transmission but also the government and corporate assumptions about (hearing) television-audience behavior. Perhaps the potential solution of open captioning—artistically more adaptable, economically more palatable, and arguably more democratic—was written off too soon by deaf activists and network executives alike. Perhaps hearing audiences would have embraced text-on-television in at least some contexts—sports, news, education—providing a simpler and broader starting point for the ultimate goal of universal broadcast accessibility. Today's littering of American television screens with textual channel logos, program advertisements, and news "tickers" makes one wonder what might have been. In the history of closed captioning that actually unfolded, the success of NCI's new system would depend both on bringing captions to the nightly news and on bringing hearing audiences to those captions. But before either of these could happen, a third speech-to-text practice—court reporting—needed to come of age as well.

# Chapter 3

# Stenographic Reporting for the Court System

Court reporting is a mystery profession. Attorneys and, even, judges oftentimes are ignorant of the actual work involved in producing a transcript.

*Sandra McFate, court reporter, 1973*

The year 1959 was a busy one for Alaska. Having just attained statehood under the approval of the Eisenhower administration, the former territory was now faced with the daunting task of quickly setting up a modern bureaucratic apparatus that would meet the requirements of federal law. In particular, Alaska needed to construct, equip, and staff an entire state court system. But besides building courtrooms, appointing judges, and hiring lawyers, Alaska court administrators would have to attract and employ a very particular, highly skilled, and relatively obscure group of laborers: court reporters. These were the men (mostly, at that time) who, using either manual (pen-based) or machine (keyboard-based) stenographic skills, sat silently at the side of the bench and diligently recorded every word that was said in a courtroom case. But their work did not end there; often, after the taking down of testimony during working hours at a salaried rate, came the writing up of that testimony during off-hours at a per-page rate, selling the resulting "verbatim transcript" to plaintiffs, defendants, the media, or the court itself.[1] The dual-income structure of court reporting was a long-standing practice that, by midcentury, many court managers sought to end—and the situation in Alaska offered them a unique opportunity to do it.

Early in 1959, Buell Nesbett, chief justice of the Alaska Supreme Court, traveled to Washington, DC, to discuss the situation with administrative director of the federal courts Warren Olney III. Just a year earlier, the Judicial Conference of the United States—the policy makers of the federal judiciary, comprising judges from the Supreme Court, the US district courts, the US Courts of Appeals, and the US bankruptcy courts—had conducted an "independent investigation of the court reporting system in the federal courts," published by Olney's office, which came out "strongly in favor of the use of sound recording" to capture the legal record.[2] Thus Olney suggested a bold technological (and spatial) fix: instead of importing court reporters to the remote northern state, Alaska could purchase tape recorders.[3]

Soon after this meeting, the Connecticut-based Soundscriber corporation, makers of a new electromagnetic recording system that could capture sixteen hours of speech on a slow-moving, three-inch-wide belt, was awarded a thirty-thousand-dollar contract to wire new courtrooms in Juneau, Fairbanks, and Anchorage. Together with the new technology came a new physical courtroom design. The Anchorage court was designed as a "nearly acoustically perfect recording studio," a circle full of carpeting, sound-deadening ceiling tiles, rubber-mounted doors and wood-slat baffled walls.[4] But this architecturally integrated recording system was hardly labor-free: the Anchorage court required the labor of three full-time "young girls" as transcribers, each turning out an average of fifteen pages of transcript per day and earning between $485 and $550 per month. Labor turnover was endemic: "A transcriber on the average works for 4 to 6 months and then leaves."[5] But the labor of supposedly "unskilled" female transcribers was cheaper and easier to control than the labor of self-avowed professional male court reporters who could command a wage of five to fourteen thousand per year—not including transcript fees.[6] Not surprisingly, Olney praised the Alaska experiment in a forty-three-page report "sent to all US circuit and district judges," arguing that "when the service provided by conventional shorthand reporters is inadequate, overly expensive, or otherwise unsatisfactory, electronic sound recording is a practical alternative."[7] Not surprisingly, the National Shorthand Reporters Association (NSRA) claimed just the opposite: "Any system of recording proceedings in substitution or in lieu of the shorthand reporter is unsatisfactory."[8] Electronic recording, or ER as it was called, was certainly not perfect; the Soundscriber soon acquired the nickname "sound scratcher."[9] But the court reporters had reason to worry. "Make no mistake about it," wrote one reporter in 1967, "the future is not going to see a happy combination of recording machines in courtrooms alongside of official court reporters; rather, it is going to be either one or the other."[10]

The taking down of live testimony in coded form, to be turned into a written record some time later, might not seem to have much to do with either film subtitling or television captioning. However, all three practices dealt with the tricky boundary between speech and text, and the combination of technological tools and labor skills necessary to translate one to the other. Especially since the early-modern period, increasingly complex political-economic organization has demanded some way of recording speech. In particular, both the growing size of society and the broader geographical scale of control demand the taking down and writing up—or "time- and space-shifting"—of a variety of live events such as political speeches, judicial proceedings, organizational conferences, and classroom lectures. The court reporters who performed such tasks in the 1950s and 1960s would become crucially important to closed captioning in the 1980s and 1990s. But before this could happen, these speech-to-text workers would face a series of technological crises, enduring first the mechanization and then the computerization of their profession, all the while trying to stave off the threat of electronic recording. Although they would successfully preserve their professional identity, a startling gender transformation would take place as well, as court reporting followed office stenography to become a feminized occupation. To understand the history of television captioning, then, it is necessary then to know something about the history of courtroom reporting—another hidden world of information labor crucial to the temporal and spatial transformation of voice into print.

## From Female Office Stenographers
## to Male Court Reporters

Court reporters have long traced their basic skill of "stenography"—"the rapid writing of words, using phonetic symbols or signs . . . instead of alphabetically spelled words"—back to ancient Rome.[11] Although it may be true that former slave Marcus Tullius Tiro, around 63 BC, became personal secretary to the orator Cicero by inventing "a difficult and complex system of note-taking, which took approximately ten years for others to master," the early-modern period is a more direct antecedent of today's practices.[12] In England especially, with growing colonial administration in the seventeenth century, the state increasingly required stenographic services. A system authored in 1672 by William Mason was adapted around 1750 by Thomas Gurney, "first official reporter of parliamentary debates in England." Gurney's system was "hard to learn, extraordinarily fast and a jealously guarded secret" when it was officially recognized by the government of England in 1772.[13] It was still in use in 1837 when author Charles Dickens worked reporting sessions of Parliament; the eponymous narrator of *David Copperfield* related that shorthand was "about equal in difficulty to the mastery of six languages."[14] Each new stenographic system claimed to offer a better way of recording (and re-creating) rapid, complex speech. But all were constrained by the basic idea of combining a fixed set of "phonograms" or "stenoforms"—characters or symbols used to represent a word or word part—into a written transcription that was both physically recordable and mentally decodable.[15]

By the nineteenth century, such systems were crossing the Atlantic to America. Some one hundred systems were published in the United States between 1880 and 1890; but after the early years of the twentieth century, the Pitman and Gregg systems were the main survivors.[16] In 1837 British educator Isaac Pitman introduced a shorthand system requiring lined paper where "positioning a sign above or below the line indicates omitted vowels." Pitman's "was the premier system of shorthand for decades," growing even more in popularity when his brother Benn brought the system to America in the 1840s.[17] However, it was another British-born system that would win over the states in the end. John Robert Gregg learned the Pitman shorthand system himself before establishing his own system in 1888 in Glasgow. Gregg would bring his "cursive, as opposed to geometric" shorthand method with him when he emigrated to the United States in 1893, and his system "became the first shorthand to be taught in [US] high schools."[18] After the turn of the century, shorthand reporters using the Gregg system began to win state and national speed contests, popularizing the method.[19] (A 1933 article in the *American Annals of the Deaf* even advocated the teaching of Gregg shorthand to deaf students before teaching them written English.)[20] Nearly a century later, Gregg was still the standard manual system in the United States, "taught in over 20,000 schools and used daily by two million stenographers and secretaries."[21]

That Gregg and Pitman shorthand could "scale up" in speed helped bring a skill, status, and gender polarization to the field of stenography.

In the 1880s, "stenographers were usually male, well paid . . . , and free to develop their skills of shorthand dictation, transcription, and office management. They stood at the top rank of office workers, above clerks, copyists, bookkeepers, and messenger boys."[22] Historian Lisa Fine cast such work as entrepreneurial: "Self-employed [male] stenographers often rented offices in downtown buildings that contained either governmental agencies or private businesses requiring stenographic and transcription services."[23] However, in a rapid reversal, by the turn of the century, 80 percent of all stenographers and typists (jobs which increasingly blurred together) were women, as were "the majority of students studying shorthand and typing in the private business schools."[24] The occupation had been "feminized" in two senses: numerically and culturally. As historian Angel Kwolek-Folland explained, "Gender segregation meant different promotional tracks, pay scales, and definitions of skill and status, as well as different behaviors for those in management or executive capacities (men) and those in clerical positions (women)."[25] This broad and rapid feminization held contradictory meanings for the women involved. On the one hand, as historian Margery Davies pointed out, "the proletarianization of clerical employees" meant the new women were, unlike the men they replaced, "members of the working class."[26] But on the other hand, as Fine described, "Women's entrance into clerical positions posed a direct challenge to the commonly held belief that not only was the office a male space, and office jobs men's work, but also that all sorts of urban settings—elevators, street cars, restaurants, boarding-houses—were inappropriate for working women."[27]

The new gender division of labor was related to both a social and a technological division of labor. Socially, as Fine noted, "between 1890 and World War I, stenography-typing, the fastest growing clerical job, was increasingly the province of younger women of foreign stock," a different demographic from the rest of the older, mostly native-born female clerical labor force.[28] Technologically, as the typewriter became a more common fixture in the American office, the need for shorthand in the office rose as well.[29] When acoustic recording machines for dictation emerged after 1900, companies marketed them first as replacements for narrowly skilled female stenographers and then as enhancements for more broadly skilled female secretaries.[30] Female stenographers were generally able to preserve their positions; as Davies pointed out, they "tended to add individual quirks or shortcuts to the [shorthand] system being used" such that "the stenographer might be the only one who could read his or her notes."[31] However, the shorthand these women learned and practiced was only adequate for taking down slowly spoken business letters or office memos from single speakers, not for recording more rapid or less formal speech. This did not stop some experienced female stenographers from following the example of their male antecedents in setting up independent "public stenography" offices. Fine described how these "older women, many living on their own," successfully "rented their own office space, determined the conditions and pace of their work, and remained stenographers for a significant period during their working lives."[32] But as Davies pointed out, most female office stenographers were thought of as somewhere between typists and secretaries in the office pecking order: "In 1916 one author described the work of a typist as 'purely mechanical.' But 'the stenographer's work,' [she] argued, 'comes a little higher because the stenographer executes the thoughts of someone else. A secretary must think independently, and at the same time execute the thoughts of another.'"[33]

Courtroom stenography was characterized as different from office stenography in almost every way. First of all, the skill level of a courtroom stenographer had to exceed the "secretarial speed" of 125–135 words per minute, as court speech came from many different persons, sometimes over two hundred words per minute.[34] Second, court reporters had a legal responsibility before the bench to certify that the trial record they produced was accurate (and could be subject to testifying in court themselves).[35] Finally, court reporters occupied a contradictory spatial and temporal position in the courtroom: during trial, they were expected to be "invisible" to the proceedings, as neutral and unbiased (almost mechanical) observers; but during recess, they were expected to quickly convert their backlog of recorded notes into printed transcript, as no case could be appealed without this official record. While female office stenographers fought the popular image of the "sexy stenographer"—committed to film in the 1915 Charlie Chaplin short *The Bank*—male court reporters lamented that "most people neither understand what the reporter is, what he does, how he does it, nor how 'he got that way.'"[36] Thus the functional expectations of greater speed and greater responsibility in court reporting, coupled with the cultural and numeric feminization of office stenography, helped reproduce courtroom stenography as a male profession against office stenography as a female one. Through the twentieth century, as electromagnetic dictation machines and expanding secretarial duties eventually eroded the separate job category of the office stenographer, the "immense growth of the bureaucratic legal and administrative apparatus" after midcentury led to an increased demand for courtroom stenography.[37] The court reporter's status was maintained even as much of the labor in court reporting actually moved out of the courtroom and into the offices of private and corporate lawyers, where taking depositions in informal settings replaced taking testimony in formal courtrooms.[38] Even today, court reporters "resent comparison to stenographers and other clerical employees of the court," claiming to possess "the knowledge which is required to understand the complex testimony which is presented in most courts today."[39]

The court reporter's claim to professionalism depended on the physical commodity resulting from that "complex testimony": the verbatim transcript. Faith in the reporter's typed record of events as an accurate and objective representation of reality is one of the key requirements of nonprejudicial justice: "Nothing levels or equalizes litigants in courts with different educational, social, and economic backgrounds more than a faithful verbatim record prepared by a neutral court reporter."[40] Yet much like speech-to-text translation in film and television, court reporters must adapt their written representation from its spoken origins.[41] The difficulties of recording live speech will inevitably introduce moments of decision, editing, and selective attention, a practice well recognized at least a century ago. As one court reporter described in 1903, "Frequently three or four attorneys and the witness are all talking at once, while the court, too, is trying to get in a few words. It must be apparent that on such occasions no human hand can take down every word uttered, and the stenographer must decide with lightning rapidity what is important and what is not."[42] Another reporter revealed, "The correction of misstatements or grammatically incorrect usage by attorneys or judges is mandatory."[43] Critics have pointed out that when court reporters unconsciously correct the English usage of powerful and professional speakers but preserve the non-standard usage of powerless or working-class speakers, "discrimination and prejudice against the speaker of a nonstandard dialect" may result.[44] In all these ways, argued scholar Anne Walker, "Not only does [the 'verbatim' record] not exist, but it would be unacceptable to the very community who requires it if it did."[45] Court reporting ends up resembling that other speech-to-text practice, language translation.[46] Even if the transcript is understood to be fallible, however, court reporters have long claimed that the very act of taking the record serves as a sort of "discipline" during the trial itself: "The fact that a verbatim record of proceedings is being made has a salutary effect on the conduct of the persons in the courtroom."[47] Although fallible and partial, the semantic and symbolic aspects of the "verbatim" transcript underpin the court reporter's claims to status and legitimacy within the legal process.

## Divisions of Labor in Court Reporting

Court reporting is not without its own internal status divisions, however—long-standing differences in working conditions and occupational meanings that remain today, and which are discussed openly in court-reporting instruction manuals and journals. The biggest polarization has been between permanent and contingent employment relations, with full-time salaried "combination reporter/court-clerk/ secretaries" on the one hand and part-time "independent contractors paid on a per diem basis and performing no other court functions" on the other.[48] Court reporter John Reily's memoir of his thirty-one years in the occupation illustrates this dichotomy. Salaried reporters— "officials"—are usually attached to a single courthouse or, in the case of a rural district, a single judge traveling between courthouses. In contrast, independent reporters—freelancers—are usually hired out to courts or lawyers through intermediate, privately owned reporting agencies, especially in urban areas. And although officials tend to take courtroom testimony almost exclusively, freelancers might be hired either to substitute for a day in court, or to take a deposition outside of court.[49] Freelancers lack the promise of regular work that officials possess, but at the same time, freelancers are not subject to sudden increases in demand when previously recorded testimony needs to be transcribed for an appeal, as officials are. Even on the job, warns the author of a current court-reporting textbook, "freelance reporters experience a great deal of downtime commuting to and from an assignment, on a lunch break, or simply waiting for the job to begin."[50] Thus the division of labor between officials and freelancers is not only institutional but spatial and temporal as well.

Within the private reporting agency, relations with workers range all the way from self-employed individual "exchange" reporters who hire themselves out to many different agencies, to "on staff" employees who work for a single agency. Although agencies occasionally hire staff reporters outright as employees—with salaries, benefits, and tax deductions—contingent staffing is more common.[51] Reporting educator Dana Chipkin described the difference: "Being *on staff* entails giving what is known as *first call* to an agency. First call means that each day at call-in time you will call this agency first for an assignment. If the agency has no assignment for you, then you are free to call another agency. If the agency you are on staff with does have an assignment for you, then you are expected to accept work from that agency rather than from another agency that may have already contacted you for a job."[52] A range of practices, from actual contracts to unwritten rules of conduct, have emerged to keep reporters who are working on staff with a particular agency from accepting (or acknowledging that they accept) work from a different agency. Such practices, which tend to benefit the agency more than the reporter, may today run afoul of IRS rules on "independent contractors."[53] Nevertheless, noted Chipkin, "many agencies prefer new reporters to be on staff because it takes a while to train the new reporter and teach him or her agency protocol and preferences."[54]

There are benefits to being on staff, especially for new reporters who have not built up either a reputation or a client list of their own. Those on staff usually get first chance at jobs, especially the inevitable "spot call" from an attorney at the last minute for a job beginning within the hour. Although accepting work in this way from a single agency can become a "comfortable habit" in the words of one reporter, the rules of who gets which calls first—usually, but not always, based on seniority—can be a source of friction. On-staff reporters walk a tense line: they have no written contract ensuring employment but at the same time must schedule vacations and other time away from the agency beforehand to retain their place in line for job calls. Reporters tolerate such conditions because they know that throughout the year, their job tends to be "feast or famine."[55] As one reporter put it in the *Journal of Court Reporting*, "What notion scares you more? Not having enough work to support yourself and your family, or having so much work you feel crushed by the weight of your job responsibilities?"[56] With such variability, ad-hoc subcontracting arrangements where freelance reporters, never willing to give up a job, parcel out overflow work to each other, are common.[57] Insiders have long recommended that "the very first step is to identify other court reporters in your area. Call them on the telephone and arrange to have lunch with them. Become socially acquainted with them, if possible. You can rely on these folks for work that they can't get to."[58]

Once a job is assigned, income is still not assured. Like officials, who are first paid a salary to sit in court "taking down" testimony and then paid a page rate by lawyers, the media, or the state for "writing up" the testimony, freelancers are paid for both in-office stenography and off-hours transcription. But both forms of labor do not always materialize. A freelancer may be sent to a "bust" assignment, where a client or witness fails to appear; the reporter receives a small "appearance fee" or "per diem fee" for wasted time. Even if testimony is taken, it may only amount to a few pages of transcript, in which cases a "minimum" will be charged rather than a per-page rate. Sometimes a deposition is held, and the reporter gets a good "take" (records a substantial set of steno notes) but no transcript is ordered afterward. In this case, the reporter may estimate the number of transcript pages that the take would have yielded and ask the freelance agency to bill the client for half this amount. In any case, it is the reporter who bears the risk of such contingencies, not the subcontracting freelance agency.[59]

At the head of this subcontracting pyramid of fifty or so reporters are the principals of the freelance firm—seasoned owners likely to have started out as court reporters themselves. The goal of an agency is similar to the goals of its reporters: find low-risk, long-term work in a highly competitive and contingent environment. Agencies seek long-term contracts with large corporate clients who demand regular legal work. Agencies will even set up subsidiaries under different names to serve competing corporate clients (sometimes with the same reporters working for both clients). But for every job the source of profit is the same: the "reporter/agency split" resulting from the amount of finished transcript pages the reporter produces. A seventy-thirty split would mean that the reporter receives 70 percent of the transcript fees, and the agency 30 percent. Although 30 percent might seem like a small percentage, the agency pays no reporter training costs, taxes, health care, equipment fees, or disability insurance for what Chipkin called "your most precious work tools, your hands."[60]

Why would reporters choose freelance work over official work? The freelance agencies cite flexible workloads, a diversity of job sites, and a more casual work atmosphere as key attractions.[61] But there are structural reasons as well. Official jobs have historically been few and far between—only from 1 to 3 percent of federal court-reporting positions opened up per year in the 1960s and 1970s, and in most cases, such jobs demand seasoned, skilled, and credentialed reporters who have passed state examinations or national certification tests. Once someone becomes an official, he or she tends to stay there; in 1960, half of all US district court reporters had over a decade of service.[62] Thus a new freelancer needs both on-the-job experience and "speed-building" practice in order to pass certification and apply to be an official. The idealized career path runs from freelancer to official.[63] But freelancers might grow their careers in other ways, by specializing in a particular area of litigation (medical malpractice, patent infringement) or starting their own firms. In many jurisdictions, official reporters are permitted to take on the occasional lucrative freelance job themselves—even subcontracting their own courtroom work out to a lower-paid freelancer![64] In these ways, although the two types of reporting are conceptually distinct, the fates of publicly employed officials and privately employed freelancers are intimately related.

The wage structure of court reporting has tended to reflect this relationship between officials and freelancers. In the early 1960s, around the time that Alaska was choosing tape recorders over supposedly scarce court reporters, Warren Olney's Administrative Office of the US Courts sponsored a survey of 254 official reporters in the district court system and found that, on average, they earned from $6,500 to $7,100 per year in salary, plus another $4,000 per year in transcript fees.[65] In his 1964 dissertation on court reporting in New York City, former court reporter Morris Fried categorized such officials as being in the lower part of the "elite," a minority comprising "the Congressional reporters, the United Nations reporters, those reporting in the highest courts in the country, the most successful free lance employer-principals, and the most highly skilled free lance working reporters" who may have total incomes ranging from ten to twenty thousand dollars per year.[66] The salary of Edward Van Allen, chief court reporter for the New York Supreme Court, matched this high end in 1969, "not an inconsiderable sum in the framework of Civil Service" as he put it.[67] Fried speculated that most other reporters sat a notch down from this elite, earning seven to ten thousand dollars per year. Finally there were the "beginners," freshly minted reporters with "little or no practical experience," who "work for the less reputable freelance agencies" that "require no experience in their reporters, generally charge rates lower than the 'going' rate in the field, and are looked upon by those in the field as substandard." Beginner income ranged from five to seven thousand per year.[68] Although the income figures in each category have risen with the dollar over the decades—by 1975, the government estimated starting salaries for court reporters at ten to fifteen thousand per year—the general three-tier hierarchy has held constant.[69]

A court reporter's wages could vary greatly with geography, however; for example, Van Allen claimed that official salaries ranged "from a high of $17,800 in New York State to lows of $3,000 in Arkansas, Florida, and Kentucky" in the late 1960s.[70] Officials and freelancers have never been spread evenly across the landscape. Part of this is due to the geographical hierarchy of the court system itself, iterated through each of the three thousand or so counties in the nation, whether those counties are heavily or lightly populated—"from Los Angeles County or Cook County, to a rural county in western Tennessee."[71] By midcentury, a distinct urban/rural split existed between officials and freelancers. More freelancers worked "in large cities where business conferences, meetings, and conventions are held."[72] However, "in suburban and rural areas, most of the [deposition] work" was performed by moonlighting officials.[73] Official reporters in urban areas faced an interesting contradiction: "In the more metropolitan areas, where salaries are higher, court hours tend to be much longer and the opportunities to do private reporting are correspondingly less."[74] In the 1960s, New York City, a center of court reporting practice with levels of litigation unparalleled in the nation, was also one of the only sites where court reporters actually joined labor unions in large numbers.[75] Both freelancers and officials were subject to a constantly changing patchwork of state regulations and requirements. In 1965, only ten states demanded "Certified Shorthand Reporter" status for official jobs; by 1975 that number had increased only slightly, to thirteen. This geography roughly matched the location of the three dozen or so NSRA-approved court-reporting training programs at the time.[76] Court reporting as a career, then as now, depended not only on who you worked for and what you did, but where you did it.

The uneven geography of legal work was one of the main factors that had motivated court reporters across the United States to organize into statewide and nationwide professional associations. The National Shorthand Reporters Association (NSRA) dates to 1899, when 156 charter members elected a New York court reporter their first president.[77] By the 1920s the association was growing through new state affiliates and began concentrating on three interrelated issues that would continue throughout the twentieth century: wages, training, and certification.[78] One reporter complained in 1921 that "while the law emphasizes the requirement that the official must be an expert court reporter, the compensation for his valuable services is placed at a figure that would excite the risibilities of a journeyman plasterer or a brick or stone mason."[79] New York, Colorado, and Iowa became the first three states to pass "mandatory CSR laws" NSRA pushed in the 1920s, while attendees at the association's conference debated the question of whether "a four-year course leading to a college degree in reporting was sufficient" for such certification. Through the 1920s and 1930s the NSRA lobbied for a federal law to certify shorthand reporters, only to have its 1936 bill vetoed by President Roosevelt. Nevertheless, the following year the NSRA held its first Certificate of Proficiency (CP) tests, awarding the designation to twenty-seven members.[80] Through such efforts to standardize education and certification, association membership grew slowly to about 1,500 at the end of World War II. But in the postwar years, with reinvigorated economic activity and litigation, membership doubled to nearly three thousand by the time of the 1960 electronic-recoding dilemma. Reporting was poised to double its numbers again, to six thousand NSRA members, by 1970 (see figure 3.1). Just as in the original construction of court reporting as distinct from office stenography in the late nineteenth century, this dramatic increase after midcentury was related both to new technology and new gender dynamics.

## The Stenotype, the Dictaphone, and the Notereader

At the time of the NSRA's founding in 1899, an estimated 95 percent of its members used Pitman "manual" shorthand on the job, if the winners of national speed contests are any indication.[81] The other 5 percent had abandoned writing-based "phonography" for keyboard-based "phonotypy," using a variety of machines from different inventors. A turning point came in 1914 when a team of reporters specially trained on the new Stenotype-brand keyboard so soundly defeated the "pen and pencil shorthand writers" at the annual NSRA meeting that the speed contests were suspended, and when they were finally reinstated, machine writers were banned.[82] One reporter who began using

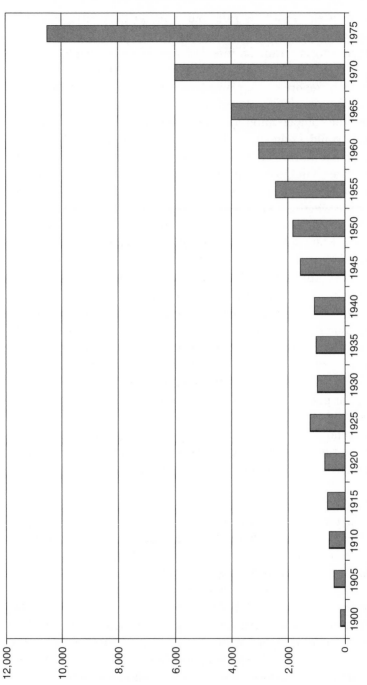

Fig. 3.1. NSRA membership, 1900–1975. The postwar years saw a steady rise in professional court reporters, as more women entered the field and machine stenography became the norm, with membership more than doubling between 1965 and 1975. Source: Graph redrawn from NSRA, *Celebrating our heritage* (Arlington, VA: NSRA, 1976), 71

the new keyboard in 1923 recalled, "I did feel more or less ostracized by some [manual] shorthand reporters, who somehow or other could not seem to realize that Stenotypy was merely another system for recording speech."[83] The generic stenotype machine would return to the contests in the 1950s, becoming the method of choice for most new reporters by the 1960s.[84] Not only would this new technology open the door to a new generation of reporting students—many of them women—but it would affect the division of labor between "taking down" and "writing up" by allowing reporters to spend more time in the courtroom and less time on the typewriter, through the creation of the new job category of "notereader."

The history of the stenotype goes back to the same period as the invention of the typewriter. One of the first versions, patented in 1879 by Illinois court reporter Miles Bartholomew, combined a shorthand system based on two- and three-character abbreviations for words with a ten-key machine that punched differing patterns of holes on paper tape. A few years later, George Kerr Anderson of New York patented a "word-at-a-stroke" shorthand machine using "English characters—not arbitrary symbols," where "one key, any combination, or all the keys, could be depressed at one stroke." These precedents—a new shorthand theory, the ability to depress several keys at once, and the use of a single "stroke" to represent either a short word or a single syllable of a larger word—came together in Ward Stone Ireland's twenty-key Stenotype of 1910, the direct ancestor of today's machines. After its victory in the 1914 speed contest, however, Ireland's company folded during World War I and was not resurrected until the 1920s. By 1922, the NSRA had agreed to admit stenotype reporters as members, and a decade later, during the 1935 "trial of the century" of Bruno Hauptmann—accused kidnapper and killer of the Charles Lindbergh baby—four Stenotype-using reporters were able to keep up with the daily media demand for the salacious details of the trial. By 1939 a functionally equivalent competitor to the Stenotype, the Stenograph sold by Stenographic Machines Inc. of Chicago, had cemented the design's form and acceptability.[85]

A stenotype reporter uses both hands at once and presses several keys across the keyboard at a time to create one phoneme per stroke: "The fingers of the left hand record the initial consonants, the thumbs capture the vowel sound, and the fingers of the right hand record the following consonants."[86] A thin paper tape scrolls out of the machine as each phoneme is printed. Individually adjusted keys allow reporters to operate their stenotypes rapidly while keeping their eyes on the person speaking, rather than on their own pencil and paper.[87] But as with manual shorthand, physical skill is only part of mastering the stenotype. A new set of steno representations must be learned for each sound in the English language. The complete stenotype code for an English word, represented by one or more keyboard strokes, is called a "stenoform." A typical stenoform for "computer" might take three strokes and print on three lines (one for each syllable) spaced across the tape like so:[88]

K O P L

P  U  T

E  R

Even punctuation is represented one stroke at a time: "a period is FPLT . . . A comma is RBGS and a question mark is STPH."[89] But there are limits to stenoform standardization; two reporters might employ different stenoforms for the same word. Because all reporters, over the course of their careers, tend to develop their own modifications to the steno theories that they initially learn, no two reporters can be counted on to use the same stenoforms for the same spoken utterance.

By the 1960s, most new entrants to court reporting were learning stenotype-compatible shorthand theories, rather than pen-compatible ones.[90] Experts estimated that education turnover time could be shortened from five years of manual training to only two years of machine training.[91] Some worried that faster training, more widely distributed across the nation, might have a negative effect on the profession, as beginners would "enter the occupation with sufficient speed, but with little of the additional training that manual writers were able to receive during the long years that they practiced to achieve those speeds, e.g., facility with grammar, ability to edit, familiarity with legal procedures and terminology."[92] One reporter argued (nostalgically) that longer training helped novices learn "that reading and acquiring a broad body of general knowledge was indispensable to a reporter."[93] In reality, even the streamlined skill and speed training on the stenotype was arduous, with dropout rates sometimes reaching 90 percent.[94] But many reporters felt that to keep court reporting a legitimate profession, training needed to remain lengthy, a "higher academic education rather than extensions of commercial courses in high schools or business colleges."[95]

Part of this legitimacy was tied to gender. Until midcentury—and the increased use of the stenotype—court reporting was a male-dominated field. Female reporters had existed since the 1900s, but it was not until 1951 that the first woman president of the NSRA, Ollie E. Watson, was elected.[96] As late as 1961 the court-reporting journal *Transcript* could counsel the assumed wives of its members, "We hope all of you are planning to accompany your reporter-husbands to the NSRA convention in Philadelphia."[97] As one reporter remarked at the time, "There is apparently more prestige for an attorney to have a male reporter taking a deposition of a witness than to have a female reporter, who may appear to his opponent to be a secretary."[98]

Yet by the mid-1960s, an estimated 30 percent to 40 percent of all court reporters were women (though they were largely absent from the elite levels of the profession).[99] And for the first time, a woman—Alberta R. Buster of Chicago—had won the NSRA speed contest, "proving that the female of the species can compete successfully with the male."[100] This change had begun during World War II, when female reporters replaced males gone off to war.[101] But it was also related to the stenotype and the new geography and temporality of training. Stenotype-based reporting schools were enrolling more women and fewer men. One reporter writing in 1967 objected to "too many girls attending our court reporting schools—bound to lower the standards for what is really a man's profession."[102] But these women were to be the future of the field, as suggested by a popular book on court reporting published in 1965—the text was written by a male, pen-writing New York Supreme Court reporter, but the cover of the book featured a picture of a female stenotype reporter.[103] A future NSRA president would later look back and write, "When I came into the profession [around 1960] it was about 75 percent male, and most of them weren't youngsters. The court system I went into was mostly pen writers. It's just in my time that the changeover has occurred from pen writers to machine writers. Since then the world has changed. Now we're a profession dominated by females."[104]

In the 1960s, the rising number of female stenotype students affected their profession in another way as well. These students provided the labor market for a new niche within the court reporting division of labor known as the "notereader." A notereader is not yet a court reporter, but knows enough general steno theory to be able to read the character-based, paper-tape output of a stenotype machine, translate it to English, and then type it up as an official transcript.[105] This subcontracted job resulted directly from the diffusion of the stenotype, as "use of the stenotype affords a standardization of notes not possible with any manual system, permitting others to read the notes and transcribe directly from them."[106] Even with this standardization, though, a notereader had to take time learn the steno theory idiosyncrasies of her employing reporter—up to "approximately one year to train a note reader, where the notes are unorthodox and execution is inaccurate."[107] Skilled notereaders were hard to find—and to produce—especially outside of major cities.[108] For this reason, they often came to the job through kinship relations, "often a wife or relative or friend" of the reporter who trained them.[109] Court-reporting schools were prime sources of notereaders, especially with respect to those 80 to 90 percent of reporting students who dropped out of the program before reaching acceptable speed on the stenotype.[110] Wherever they came from, notereaders were almost always assumed to be female and usually cast as performing notereading and transcribing labor in domestic space and time, "at home during the day or evening or weekend, at her own convenience."[111]

A court reporter who used a paid notereader gained a competitive advantage: the notereader freed the reporter from the onerous task of producing typed transcript ("pounding it out," as it was called) and allowed more time for the reporter to take testimony or depositions.[112] After all, the stenographic part of the process was the reporter's unique (and highly paid) skill. One court reporter from the 1950s was famously quoted as saying, "Five percent of reporting is the taking down and ninety-five percent is the writing up."[113] The actual figures were not quite this stark, but they were significant: a 1960 court-management report calculated that since a single transcript page of twenty-five lines equaled approximately 250 words, "about 40 pages of transcript would result from an hour's courtroom record" and "a reporter would be able to personally transcribe an average of about 10 pages of this per hour," resulting in a courtroom-labor-to-transcript-labor ratio of one to four.[114] Any technological or labor innovation that helped reporters pound it out was generally welcome.

Because pen-writers could not use notereaders—not only their steno theory but also their handwriting style varied too much to be consistently readable—they required other means to speed their transcript production.[115] Before the turn of the century, a reporter might dictate directly to one or more typists in order to speed transcript production.[116] But after 1900, reporters began using the same acoustical dictation machines that threatened to replace female office stenographers— "the greatest forward step in the development of shorthand reporting as a profession" according to one reporter in 1921.[117] By the time the electromagnetic Stenorette reel-to-reel machine became available in the 1950s, the practice was well known: reporters dictated their steno notes in English to the machine, and then passed the recording to a trained typist or "transcriptionist."[118] Again, this method allowed a reporter to spend more time "taking down" and less time "writing up," although the time needed to dictate one's notes out loud made this method less efficient than using a stenotype and notereader. One reporter in the 1960s estimated that with dictation a reporter "gets his transcript out in half or even a third of the time."[119] Later studies found that one hour of court testimony took two-and-a-half hours to properly dictate, but this still saved another two-and-a-half hours of typing time.[120] Reporters knew that finding an effective transcriber was difficult, as the job demanded more than just typing. Wrote one reporter in 1944, "These typists must be far better trained than expert Dictaphone typists familiar only with general office work. They must know how to 'style' transcripts. They must be masters in spelling and punctuation. Exceptional speed and accuracy are essential. They must know legal forms."[121] Just as with notereaders, reporters regularly lamented that there were not enough competent transcribers to go around: "I would like to see a greater influx of literate, well-backgrounded typists into the transcribing field."[122] And transcriptionists, like notereaders, were always understood to be women working part-time at home— assumed to be, as one reporter put it, "married to long-suffering husbands who must tolerate their wives typing sometimes until two or three in the morning."[123]

Stenotype court reporters might dictate their notes in English to recording equipment as well, if they could not find or train a notereader. But in many cases, a reporter's notes might not even be transcribed. Around 1960, although almost all depositions were transcribed, only about a quarter of US district court cases were.[124] Criminal cases were almost always transcribed because appeals required official transcripts. But preliminary hearings that did not go to trial needed no transcription, and the demand for transcripts in civil cases varied.[125] Thus any court reporter—machine or manual—had to be ready for sudden transcription demands that might interrupt the normal routine of taking testimony. Although ordinary delivery of a transcript was accepted to be from four to five weeks around 1960, many transcripts were requested as "expedited" or "daily" work—and these carried significantly greater rewards for court reporters if they could deliver on time.[126] Whatever the time frame, figures from the mid-1970s through the mid-1980s suggested that over 75 percent of all transcripts produced were paid for by the court, not by private litigants or media outlets.[127] To shirk such transcripts meant not only a loss of income but also possible legal sanction, so reporters, depending on the technology they used, increasingly turned to outside subcontractors—notereaders, transcriptionists, or even other reporters—to manage peaks of transcript demand.

## Electronic Recording and the "Transcription Crisis"

With the importance of the transcript to both the legal process and the court reporter's income, it is no wonder that when electronic recording suddenly became feasible in 1960—because of the Alaska experiment—the court-reporting profession was troubled. The 1960 report issued by the Administrative Office of the US Courts recommending electronic recording in district courts was quite a switch from just a decade before, when legal scholars assumed "mechanical instrumentalities will not displace the court reporter in American practice but rather will be utilized by him in the execution of his ultimate and independent responsibility of obtaining and preserving the record."[128] Even though the report recommended "gradual" phase-in of ER rather than "wholesale change," it also stated that "a compensation system should be developed which more closely relates a reporter's income from official court work with the amount of time he is required to devote to it."[129] In other words, court reporters in the district courts should no longer be allowed to charge for transcripts, a practice "recognizably open to abuse."[130]

The reporters feared that if such recommendations were written into federal law, the states would quickly follow suit, mandating ER instead of human reporters. So they fought back: "Let it be known to those throughout the land, who labor to discommode our very means of economic and professional expressions, that we shall meet the challenge with equal and even surpassing vigor, ruthlessly without quarter."[131] The NSRA defended its interests in its 1965 *Handbook of*

*Electrical Recording,* which claimed to "eliminate any label of 'court reporter propaganda.'"[132] The association countered the news from Alaska with reports from Washington, DC, and New Jersey, where ER was deemed an unacceptable alternative to human reporting.[133] At first, association members attacked the recording machine, arguing that its fidelity could not possibly capture the chaotic condition of courtroom speech.[134] Wrote one reporter, "Everyone in the court room—judges, counsel, witnesses, jury, and spectators—must be extremely guarded in their actions, so that no noises are created that might, and at times, do interfere with recordings."[135] Paradoxically, this same detractor also argued that ER systems were unacceptable because they were *too* sensitive to spoken audio: "There is the constant danger of privileged communication between lawyers and clients at the counsel table being recorded and later appearing in the transcript."[136] Whether too coarse or too sensitive, the mere presence of the recording equipment in the courtroom was also derided. One judge argued that "the presence of an 'iron monster' in the form of a microphone immediately in front of the witness' face serves usually to compound his discomfort and inhibit his delivery."[137] Yet this same kind of recording presence was one of the aspects of discipline that human reporters touted as keeping witnesses honest.

Further arguments pointed not to the technical abilities of the machine but to the motives of the machine's advocates. One detractor cited "the tremendous sales pressure being exerted upon our judges, lawyers, legislators and court administrators by those selling electrical recording equipment."[138] Another argued that any cost savings from eliminating reporters "would be placed back into the coffers of the federal government for redistribution to some other state or county."[139] And ER recordings were said to be prone to tampering, as it was discovered in Anchorage that "tapes are not locked up, in fact, but are stored in cardboard boxes in one of the offices in the transcribing section."[140] Finally, some arguments took issue with the essence of the machine, characterizing it as "dehumanized" and thus inappropriate to capture such a culturally important record as the legal transcript. One reporter wrote, "A new and modern electronic innovation cannot compete with the triune combination of a perfected shorthand system, a human brain, and a human personality."[141] Another held that "to bridge the gulf from sounded utterance to word-for-word comprehension takes the on-the-spot coordination of the eye, the ear, and the brain."[142] Machines were by definition incomplete stewards of the record, "worthless without the application of human intelligence."[143]

These reporters were ignoring one fact court managers knew well. Human intelligence *was* involved in interpreting the record from tape recorders—just not the intelligence of the reporters. Removing the reporters meant hiring the people who previously worked for those reporters. The 1960 report advocated instituting a training program "which would assure adequate transcription performance and help achieve standard transcription practices in all courts."[144] Court reporters found themselves in the awkward position of arguing against the competence of the very labor force they themselves relied on for producing the transcript: notereaders and transcriptionists. One Anchorage judge argued against ER by ridiculing his transcribers, saying, "The slaughtering of the English language that takes place is phenomenal. They produce some pretty shocking transcripts."[145] Transcriber errors were not just an education failure but a moral failure: "Generally speaking those who are engaged in the electronic recording business do not have this same degree of dedication and attentiveness and do not have the same degree of respect for the type of work they are doing."[146] Reporters emphasized their own special knowledge by arguing "the ordinary typist never encounters words such as those used in the normal lawsuit."[147] Such arguments against a labor force upon which the court reporter's income had long depended masked an opposite possibility: that in some cases it was transcription labor that was superior to court-reporter labor. As one former reporter argued, "A skilled typist or notereader can construct an *acceptable* transcript from the poor notes of an incompetent reporter. The reporter's level of competence thus might be known only to his typist for some period of time."[148] Either way, this was a gendered debate, as most "typists" were female and most court reporters still male. An Anchorage lawyer quoted by the NSRA opined, "The girls [who] they have to transcribe [the tape-recorded record] get sick of hearing it. If I had to make the decision, I would definitely decide on reporters, because I have found that the ability of a certified man as compared to that of a machine is superior in all respects."[149] As in the rest of the debate, the man's ability was really being compared to that of the girl, not the machine.

Not all reporters dismissed the machine (and the girl) so easily. Reporters had long found both acoustical and magnetic recording adequate for their own dictation to transcriptionists. Some brought this technology into the courtroom on their own. In the district courts in 1960, "about one-third of the reporters . . . use some type of sound recording as a 'backup' to their hand or stenotype notes."[150] There was some recognition that the increasing use of tape recorders could be an incentive to reporters to keep as close to "strict verbatim reporting" as possible.[151] In fact, all through the 1960s, "side by side" tests of reporters versus tape recorders regularly assumed that the tape-recorded version of events was the "touchstone used . . . to determine accuracy when the transcripts made from the notes of the official stenographic reporters and those from the electronic tape were in disagreement."[152] As one reporter pragmatically put it in 1964, "If electronic recording aids us in attaining such a result, common sense tells us to use it."[153] Another reporter admitted in his memoirs, "I was very glad to have a tape recorder working *with* me through long court sessions and reams and reams of deadline transcription. It saved my life and reputation. In my court, the tape recorder never threatened my job."[154] Reporters who admitted to the usefulness of ER were subject to scorn by their peers, however, as advocating "the self-administered poison which foredooms shorthand reporting as a profession."[155]

In September 1965, President Johnson signed HR 3997 into law, allowing for US district court proceedings to be recorded by both ER and a live reporter.[156] The NSRA supported this bill because it permitted audio recording only with the presence of a salaried court reporter making a simultaneous record.[157] By 1966, only three states besides Alaska—Indiana, Illinois, and Virginia—allowed ER in the court "without a shorthand reporter present."[158] As one reporter crowed in 1965, "If you've heard stories about shorthand reporters being consigned to the limbo of unnecessary, unemployable human beings because of automation, set your fears at rest."[159] By 1971 the NSRA had adopted a rather contradictory policy on audio equipment in the courtroom; the association was "unalterably opposed to the sole use of electrical recording as a replacement for the shorthand reporter," but those reporters could use "electrical recording as a matter of reference" if they wished.[160] Court reporters returned to their original argument about ER: that it revealed a shortage of trained reporters, not a problem with cost or performance. "Jurisdictions that have installed recording machines have done so because the supply of court reporters was insufficient to satisfy the needs of expanding court systems," explained an NSRA report, referring directly to Alaska.[161] Reporters reminded managers that "only one per cent of those who attempt to take up shorthand reporting actually persevere through the long hours of study and practice."[162] By such arguments, reporters acknowledged among themselves that "if we do not train enough reporters in the future, sound recording of necessity will have to make great inroads."[163]

But court managers who were unable simply to replace reporters with ER systems were not finished yet. In the 1970s, they turned instead to the question of changing the wage structure of court reporting. To do so, they pointed not to local shortages of court reporters or to the high costs of reporting labor (estimated at "$300 and $400 million annually") but to the turnover time involved in producing the transcript.[164] Especially after the 1971 founding of the National Center for State Courts (NCSC), "improved court management [was] the primary focus of many reports and state projects."[165] Court officials lamented that "long delays have been experienced, due in part to backlogs in transcript production."[166] In 1973, the National Commission on Criminal Justice Standards and Goals officially recommended that transcripts should be produced within thirty days of the close of a trial.[167] Researchers for the NCSC responded, "In too many courts, reporters are unavailable to record court proceedings, transcripts are consistently late, transcript quality varies greatly, and transcript costs are continually increasing."[168] As a writer for the Federal Judicial Center observed, time was the key complaint: "The contribution to delays in the appellate process caused by the time required for preparation of transcripts."[169] Two related problems with court reporting were identified, both due to the individual variation in steno style among court reporters (the personal choice as to which stenoforms represent which words): "Recorded notes are comprehensible only by trained reporters and remain unintelligible to all other court personnel"; and "the court reporter must be heavily involved in the preparation of transcripts, requiring that he spend significant periods of time out of court."[170] As one writer for the *State Court Journal* later put it, "Knowledge is power. In too many courts reporters have power because they have a monopoly on information about their craft."[171] Through the 1970s, it was the nature of the court-reporting skill and labor process that was seen as the root of the "transcript crisis."

## Computerizing Stenography in the
## Cold War "Translation Crisis"

The eventual solution to the transcript crisis in the court system would come as a direct result of another crisis, this time in the military-industrial-academic complex: the cold war "translation crisis."[172] Early in 1947, mathematician Warren Weaver had written to Massachusetts Institute of Technology professor (and founder of "cybernetics") Norbert Wiener, asking him "if the problem of translation could conceivably be treated as a problem in cryptography." Wiener said he was "afraid the boundaries of words in different languages are too vague and the emotional and international connotations are too extensive to make any quasimechanical translation scheme very hopeful."[173] But later that summer, UK researcher A. D. Booth suggested "a digital computer having adequate memory facilities could perform the operations necessary to translate a text written in a foreign language (FL) into the . . . target language (TL)," although he warned, "I make no claim that a literary quality in the result of the translation is to be hoped for."[174] In other words, a word-for-word dictionary substitution might be easily computable, but translating meaning was not. This was enough for Weaver, though, as in July 1949 he wrote a memo to two hundred colleagues on "the possibility of contributing at least something to the solution of the world-wide translation problem through the use of electronic computers of great capacity, flexibility, and speed."[175] By 1952, the first conference on the question had been held at MIT, and a few years later the field had a journal, *Mechanical Translation*.[176] Thus the research area of machine translation (MT) was born.

This was no pure academic pursuit, however. University of Washington researcher Erwin Reifler argued that "the ever-increasing volume of important publications in many languages, the insufficient number of competent translators, and the time consumed in translation all justify a search for a mechanical solution to the problem of high-speed mass translation."[177] With the cold war barely begun in the early 1950s, almost all the government-sponsored MT projects focused on the problem of translating Russian to English, often simply using "word-by-word" translations, pushing the limits of data-storage capacity and data-lookup speed.[178] Soon after the MIT conference, IBM and Georgetown University had teamed up on one such effort, and in

January 1954 were able to demonstrate their proposed solution on a standard-issue 701 computer at IBM headquarters in New York: "With 250 Russian words and their English equivalents stored in the rotating magnetic drum and with six rules of operational syntax, the 701 computer quickly printed out correct English sentences when simple Russian sentences were inserted."[179] Funding agencies like the National Science Foundation and the Central Intelligence Agency were eager to see the system scale up, but as one IBM researcher explained, "The mechanized translation scheme is practical only if the electronic system includes a rapid-access storage with capacity greatly in excess of customary mathematical computers."[180] An answer to this bottleneck came with "the truly ingenious photoscopic disc designed by Dr. Gilbert King"—a precursor to later optical-storage methods like the CD-ROM—which, by June 1959, had been used by Reifler's group at the University of Washington to create "an MT-operational lexicon containing almost 170,000 Russian-English entries" for their sponsors, the US Air Force.[181] All the components now seemed to be in place for a working system.

Gilbert King moved to IBM in 1958 and the Mark I USAF Russian-English translator moved with him.[182] By June 1959 the Mark I was translating articles from Russian newspapers at a rate of about ten thousand words per day.[183] In May 1960, when Nikita Khrushchev spoke before the Supreme Soviet on American U-2 flights over the USSR, it was a Mark I translation that was delivered to the Congressional Committee on Science and Technology two weeks later.[184] Soon King's group unveiled the Mark II language translator, built around "a 10-inch diameter read-only optical storage disk containing 55,000 Russian words and their English translations," which translated Russian "at the rate of thirty words a second."[185] At the 1964–1965 New York World's Fair, the system was demonstrated as working across a vast distance in near-realtime. Typists using combined Roman-Cyrillic typewriter keyboards entered Russian text into the system, which was "sent, character-by-character, over a telephone line to the Language Processing Laboratory at the IBM Kingston, New York location," where it was transformed into a "usable, but not perfect, translation" and then sent back to the fair site to be printed after only a short delay.[186] The system entered daily operation that same year at Wright-Patterson Air Force Base in Ohio, capping a decade of USAF-sponsored MT research at a cost of nearly ten million dollars.[187]

The World's Fair demonstration left out a crucial part of the story, however. A profound irony of these machine translation projects was that, in order to be productive, they relied heavily on human labor in both "pre-editing" data for input to the computer and "post-editing" the rough translation that the computer produced.[188] As early as 1960, many MT researchers were already speaking of "machine aids to translation" rather than "machine translation."[189] In fact, in the operation of the Mark II in the mid-1960s, forty-four people were required for a throughput of a hundred thousand words per day: fourteen Russian-text input typists, three trained machine operators, and twenty-eight post-editors who needed to be proficient in Russian![190] The text-input problem in particular had vexed Reifler in 1957 and continued to vex King years later.[191] While the hoped-for solution of an "automatic print reader" for Cyrillic text remained under development, King relied on his dozen or more typists, "with no knowledge of the input language," who transcribed Russian characters into "a phonetic-mnemonic alphabet" that could be encoded on punched paper tape.[192]

A new potential solution soon came from an unlikely source: stenography. In 1959, Gerard Salton at Harvard University surmised, "Since stenographers are trained to record spoken information at high speed, it may be possible to solve a substantial part of the problem of transforming spoken information into written form by using a stenographic transcript as machine input." Salton's key insight was that "the problems raised by the transcription of machine shorthand are [largely] identical to problems encountered in the automatic translation of languages."[193] IBM's E. J. Galli learned of Salton's work and speculated that an automated stenotype transcription system could help solve the pre-editing labor problem in machine translation of language: "The two operations of foreign language stenocode transcription and subsequent English translation could actually be combined into one and accomplished simultaneously."[194] Galli teamed up with Patrick O'Neill, an official court reporter hired onto the project around 1960.[195] Soon they came up with a Stenoform-English translation system based on the IBM Mark II, including 95,000 English words and their stenoform translations.[196] By 1965, "Error rates were between two and ten percent of the words transcribed," and the "throughput was 60 words per second, theoretically sufficient to transcribe the output of 30 Stenotype reporters."[197]

It is important to understand how this original IBM computer-stenography solution worked because it set the pattern for all of the projects to follow in the 1970s. Like all MT systems, stenoform-to-English translation involved keyed text input (in this case, from a specially wired stenotype keyboard), data-dictionary lookup from source language to target language, and printed text output. But because different stenographers might use different stenoforms for the same English words and because different English words might share a single stenoform, this data-dictionary lookup was rarely perfect. Each lookup resulted in one of four possible outcomes: (1) a correct English translation, where the properly entered stenoform was matched with the unique, intended English word; (2) a "conflict" where the stenoform resulted in two or more possible English words, usually homophones like "buy" versus "by" (in which case, all options would be printed in red, awaiting human proofreading); (3) an "untranslate," where the stenoform had no corresponding English word present in the dictionary (in which case, a phonetic transliteration of the word might be printed in red instead); and finally, (4) a "mistranslate," where either the stenoform was actually miskeyed by the stenotypist upon input or the translation algorithm was unable to understand a word boundary, as in "I scream" versus "ice cream" (in which case, an incorrect English word would be returned, but no error would be indicated).[198] Through all of this, the basic problem of different stenotypists using different stenoforms and other idiosyncratic shortcuts remained. Galli recognized that "stenotypy is not a completely defined language" and that "words may be spelled differently by different operators," but he believed that this problem could be solved with sufficiently large, and sufficiently rapid, data storage, where "one can store all possible abbreviations as well as alternate acceptable spellings."[199]

Limited as it was, machine-translated stenography still offered a potential solution to the general MT input problem. In King's plan, "documents were read slowly in Russian to stenotypists, who wrote phonetically. The stenotypists ultimately obtained a speed of 120 to 130 words per minute, having never heard Russian before."[200] King even hoped that "there might be some way in which he could encode and decode the two languages syllable by syllable and get some correlation perhaps in this manner, between, say, Russian and English."[201] Unfortunately for the field of MT, none of these labor-saving solutions on the pre-editing side was able to address the post-editing labor question.

By 1966, the Automatic Language Processing Advisory Committee (AL-PAC)—an investigative committee of the Department of Defense, the NSF, and the CIA—issued a report entitled "Languages and machines: Computers in translation and linguistics," which killed nearly all funding for MT over the next decade.[202] One of the committee's biggest complaints was the labor force necessary for the USAF Mark II translator to operate. ALPAC found that using the Mark II cost thirty-six dollars per thousand Russian words, more than both "contract translation" ($33/1,000 words) and government translation ($16/1,000 words). It also concluded that Mark II translations were "slow, expensive, of poor graphic arts quality, and not very good translations." The report even questioned the basic premise of a "translation crisis"; the committee was "puzzled by a rationale for spending substantial sums of money on the mechanization of a small and already economically depressed industry [translation] with a full-time and part-time labor force of less than 5,000."[203] Although the troubled Mark II translator remained in daily use at Wright-Patterson Air Force Base until 1970, IBM's research on MT ended in 1966 with ALPAC, and King left IBM for defense contractor Itek Corporation.[204]

Research on computerized stenography, however, continued. In 1964 IBM had "contracted to lease their system to the Central Intelligence Agency for a one year trial" under the coordination of Herbert Avram, "who had done special work in both machine translation and Stenotype computer transcription at the National Security Agency since 1959."[205] Court reporter Patrick O'Neill joined Avram on the project, and the system was ported to the flagship IBM System/360 computer using a "new search algorithm."[206] Now it could translate stenoforms at about thirty-five words per second, from a dictionary of 150,000 words.[207] Around the time of ALPAC, according to Avram, "IBM lost interest in the project," but the CIA "saw more promise."[208] IBM had decided that with their current 5 percent error rate—"on a typical page of transcript of 25 lines you would have an error every other line"—they would "wind up taking more time to make the corrections than it would take to [dictate and] type with a good typist in the first place."[209] But an important shift had occurred by 1968: "The CIA's own interest was for recording conferences and meetings," not language translation, and the agency even partnered with the Justice Department to test the system "in a criminal court in Washington."[210] Both Gilbert King at Itek and Ed Phillipi of Philtron Corporation had already spoken publicly about plans to bring such a computer-aided stenography solution, based on the IBM project, to the court-reporting industry.[211] But neither product materialized, and court reporters were already skeptical of such projects: "I'm a doubting Thomas," wrote one; "this method is a long way off and more likely will never become the standard," wrote another.[212] Now Avram and O'Neill took their turn—5 percent error rate or not, they formed their own startup company named Stenocomp in December 1968 to market a computer-aided stenographic transcription service to the court-reporting industry.[213] Their success—and the success of the tool that would come to be known as "computer-aided transcription" (CAT)—depended not only on convincing court managers that this new technology could help alleviate the transcript crisis, but also on convincing court reporters that their best weapon against replacement by tape recorders was rapprochement with computers.

## Selling CAT to Courts and to Court Reporters

By about 1970, although both Itek and Philtron had abandoned computer stenography, a slew of companies with high-tech names like Interactive Information in Troy, NY, Stenoscope in East Orange, NJ, and Stentron in San Jose, CA, had already been formed to compete with Stenocomp in bringing a CAT solution to market.[214] But figuring out whether that market meant the court managers or the court reporters was the tricky part. Overcoming their initial skepticism, the court reporters of the NSRA formed a Committee on Computer Transcription in January 1970 to follow events in the field, grudgingly realizing over the next few years that computerization might offer "the only real solution ever proposed to the problem of transcript delay" and "the reporter's best hope to continue in the future the role he has played in the past." Yet NSRA members worried about courts mandating "universal" computer use through "administrative pressure."[215] The court managers, for their part, worked through their professional and administrative bodies—including the National Institute of Law Enforcement and Criminal Justice, the Law Enforcement Assistance Administration (LEAA), the US Department of Justice, and the Federal Judicial Center (FJC)—to include a pilot study of CAT in a larger review of manual, machine, and ER court-reporting methods, conducted by the National Bureau of Standards. Their 1971 report, "A study of court reporting systems," concluded that CAT was "subject to a number of deficiencies which must be corrected before its potential can be realized." The bottleneck was a temporal one—although producing a "first run" rough transcript using CAT only took one-tenth of the time as standard dictation and transcription, post-editing of this transcript took so long that overall CAT turnaround time was two to four times greater than normal methods.[216] Besides, as one observer from the FJC noted, "the federal courts do not have a computer large enough for the translation programs."[217]

Stenocomp's next move after the failed test with the court managers was to approach the court reporters—O'Neill was sent "on the road for several years demonstrating the Stenocomp system" to reporters across the United States.[218] In June 1972, at IBM's New York City office, Stenocomp demonstrated its system for the NSRA committee and proposed "a joint business venture under which NSRA would offer a computer transcription service to its members, with IBM providing the computer hardware, and Stenocomp, Inc., furnishing the software." In this arrangement, transcript costs to reporters would be thirty to fifty cents per page, "provided a minimum volume of 10,000 pages per day could be generated and maintained."[219] As it turned out, the eighteen-month contract offered by IBM would have cost $500,000, and the NSRA could not afford such a pricey experiment.[220] But the terms of the offer fit in with prevailing ideas about computing at the time, popularized in Martin Greenberger's 1964 article in the *Atlantic*, which predicted the rise of the "information-processing utility" on par with electric or water utilities—an infrastructure of hardware, programming labor, and communication lines accessible to all for a low per-use cost.[221] If Stenocomp could not make its "transcription utility" work on a national scale with a nonprofit association as a partner, maybe it could succeed on a regional scale with a for-profit freelance agency as a partner.

Stenocomp soon found the perfect partner in freelance-agency owner (and future NSRA president) Frank Nelson of Los Angeles. Nelson excused himself from the NSRA Committee on Computer Transcription in order to take up a "business relationship" with Stenocomp in 1973, in which he was a sort of test client.[222] In order to both provide for data communications between his West Coast office and Stenocomp's Bethesda, MD, IBM mainframes, as well as to address the post-editing problem with a reasonable turnaround time, Nelson acquired a refrigerator-sized Data General minicomputer system of his own on a "lease-purchase agreement." Nelson's new computer-aided workflow had five steps: (1) input the stenoforms on a special steno-type keyboard adapted with a magnetic-cassette-recording interface; (2) use a modem and a dedicated phone line to transfer the cassette-based stenoforms from the Los Angeles minicomputer to the Bethesda IBM System/360; (3) wait for Stenocomp to perform the computer translation of stenoforms to English (for a per-page fee) and to modem the raw translated text (full of conflicts, untranslates, and mistranslates) back to California; (4) post-edit the raw "first run" translation right in the office on the CRT-equipped minicomputer; and (5) print the finished transcript out on an upper-case-only computer line printer. Nelson bragged, "It takes approximately seven minutes to send one hundred pages of raw steno to the host computer. It takes another seven minutes to retrieve it in English. It will take another six or seven minutes to print it out, rough."[223]

This impressive turnover time required not only a new technological infrastructure but also a new labor category—the scope operator or "scopist." This was the person who, using the CRT monitor, or "scope," attached to the minicomputer, performed the post-editing on the raw computer-translated transcription. Like a notereader, a scopist had to know a reporter's unique steno theory; like a transcriptionist, a scopist had to be familiar with legal terms and the correct format for final legal transcripts. Nelson revealed, "My former note reader operates my CRT and she doesn't have to make a rough printout if she doesn't want to. She can call a job out on the CRT directly, without printing it, by scrolling it up the screen, reading it as it goes by, and making the corrections on the screen immediately. This is possible because she can read stenotype notes."[224] Just as "machine translation" was in reality "machine aids to human translation," Stenocomp offered not "computer transcription" but "computer-aided transcription," as NSRA president Irving Kosky noted in 1975.[225] This division of labor made sense for the reporter, according to Nelson: "If you only have to proofread and not dictate, you will be on calendar more often [available for reporting jobs] because proofing can be done at times when you can't dictate." But because of the large capital costs of CAT equipment—up to sixty thousand dollars—a team of scopists would ideally have to work "around the clock," with the computers "utilized on a twenty-four hour basis because it's more efficient if it's never shut off." Nelson claimed that overall, this new division of labor would be cost effective for moderately sized freelance offices. But at the same time, he never actually revealed his per-page payment to Stenocomp.[226]

The question of the cost-effectiveness of CAT interested the court managers even more than the court reporters. By 1973 the NCSC had decided that CAT was now "technologically feasible" and sought another pilot test.[227] Stenocomp was chosen as the vendor, and the Philadelphia Court of Common Pleas (PCCP) was chosen as the test site.[228] With ninety-one judges and ninety "official" court reporters handling ten thousand criminal cases and 4,500 civil cases each year, the PCCP was a bold choice. This court spent over $4 million each year on court reporting wages and transcript fees: $2.5 million for salaries ($22,700 each) and benefits (23 percent), $1 million for some 650,000 pages of transcripts ($1.60 per page), and $500,000 for overhead. On average, those ninety court reporters produced over seven thousand pages of transcript each per year. But what really made the site attractive was the fact that the PCCP already owned two IBM 370/145 computers for running the translation software (extra minicomputers would be leased for editing) and was physically close to Stenocomp for technical support.[229] Reporters would each be equipped with a $1,500 cassette-enabled stenotype (such as Nelson used) and the translation fee to Stenocomp would be eighty-five cents per page.[230] A success here would be a huge boost to CAT in general and Stenocomp in particular.

As it turned out, one of the most crucial requirements for CAT success was not the supposedly improved Stenocomp software but the careful selection (and construction) of "computer-compatible" court reporters. Here "compatible" meant "a reporter's style must be consistent (each stenoform representing an English word without ambiguity) and notes must be clean (without keying errors)."[231] This compatibility contradicted existing practice at the time, because inventing one's own idiosyncratic shortcuts with the stenotype actually made one a faster reporter. Out of about forty PCCP reporters who volunteered for the pilot, Stenocomp hand-selected fifteen to participate in the test.[232] The firm had each potential CAT reporter submit about a hundred pages of transcript along with the original steno notes on which that transcript was based. Then, Stenocomp proceeded to "re-key the reporter's notes on a Steno-Recorder and send the notes through the translation process." After "sixteen hours of analysis," only about half of all court reporters were expected to pass this compatibility test.[233] Recalled one who was rejected, "I had always been proud of my clean and down-to-earth type of stenograph writing. This took the wind from my sails as I realized that I was not up with the times."[234] For each reporter who did pass, Stenocomp would develop a unique computer dictionary, representing that reporter's special steno style, to supplement the universal stenoform dictionary that came with the mainframe system. At the same time, the reporter would be trained to modify his or her steno style to be more compatible with the demands of the computer software. This initial training cost five hundred dollars per reporter, and ongoing dictionary maintenance cost another fifty dollars per reporter, every year.[235]

From the court-management point of view, the PCCP pilot test illustrated that, just as with ER systems, the most important benefit of introducing new technology into the courtroom was rationalizing control over court reporters. The authors of the final report wrote, "Court reporters traditionally have been permitted to control the transcription process because usually only the reporter could translate his stenotype notes. With CAT, whoever controls and operates the CAT system can control the transcription process." In fact, before the CAT test, "the Philadelphia court had not adequately monitored, regulated, or assessed its current transcription process." Thus part of this pilot project was instituting performance procedures to figure out how many transcripts each reporter normally produced, with what turnaround time—otherwise there would be no baseline against which to measure any gains for the CAT system. The combination of increased monitoring, control, and technological intervention led to a more rapid turnaround time for CAT transcripts (at an average of eighteen days each) versus traditional transcripts (thirty-eight days each). The system would also help make reporters more easily replaceable, even in the event of "a reporter's protracted illness, death, or departure from the jurisdiction." Here costs mattered less than new rules, incentives, and penalties to eliminate the transcript turnaround time crisis. The report concluded, "CAT is economically competitive with traditional transcript methods under appropriate conditions and management controls such as proper selection of CAT service approach, sufficient transcript volume, reporter motivation and skills, comprehensive administrative procedures, and production norms." And if CAT helped keep ER out of the courtroom, then court reporters should be "willing to help defray start-up costs [for example, purchasing their own electronic-output steno keyboards], and to eventually pay the entire cost for the CAT service, from their transcript fees."[236] Inevitably, under such a model, some of that cost would go to scopists performing transcript post-editing on the computer, as the PCCP test demonstrated that "the single greatest cost and time factor is editing."[237]

With the court managers taking a rather contradictory position on CAT—acknowledging its benefit but generally refusing to fund its installation—it was left to the court-reporting community to debate the risks and benefits of the new technology, just as they had done with ER a decade earlier. As with ER, some of the criticism of CAT was directed at the "dehumanizing" notion of computerization itself. One forty-eight-year-old court reporter with twenty years of experience wrote, "The prospect of being wired to a cassette, in turn attached to a telephone, thence connected to a man-made brain buried somewhere in the wilds of Outer Mongolia (or, even worse, in Washington, DC), is terrifying to me."[238] A 1977 cartoon in the *National Shorthand Reporter* showed four persons seated around a deposition table, each with a microphone wired into a large machine that looked like a mainframe computer. The punch line? Hidden behind the machine, wired to it by headphones, was a female court reporter on a stenotype.[239] CAT pioneer and advocate Nelson cautioned his peers to avoid such extremes: "Let's be rational in our criticism; let's aim it at the demerits of the system, not at the people devoting their lives, and perhaps their small fortunes, to its development and especially not by way of some computer billing system that has goofed up our gasoline credit card account."[240] Later writers would stress that, rather than harboring "dehumanization" fears, reporters should hold "modernization" hopes: "There is a great deal of personal and professional prestige in being part of the 'brave new world' rather than just an observer. *You* are not being left behind in the race to become computer-literate, you're part of the race."[241]

Reporters who had been following the PCCP tests debated not the general question of computerization but the specific compatibility demands of CAT. Statements from CAT vendors that "most reporters in the field today can be programmed or tuned to write for the computer" only served to heighten anxiety, and in the end there was "widespread agreement that only some reporters will be compatible with the computer."[242] Long-standing stenotype tricks like "shadowing" (pressing gently on the keys to create a lightly inked stroke) and "shading" (applying more pressure to create a darkened stroke) were incompatible with the demands of CAT.[243] And pen reporters clearly were shut out from the CAT revolution.[244] Some reporters who had been compatible enough to gain CAT experience talked of the increased stress of writing in a conflict-free mode: "There have been times . . . when I haven't been writing well, or when the proceedings for one reason or another have not been conducive to writing with the precision necessary to use CAT, that I 'pull the plug,' so to speak, . . . and revert to the pre-CAT method of making the record."[245] One reporter admitted that "one of the small joys I find in this profession— and a large help to me in keeping what little sanity I still possess—is writing myself notes" during the taking of testimony, and feared that with CAT, "I'll be sacrificing another little piece of my *self*."[246] CAT vendors tried hard to allay such fears. Stenocomp competitor Baron Data Systems of Oakland, CA, gave out badges at the July 1977 NSRA convention in Chicago bearing the words "THE COMPUTER LOVES ME" and "THE COMPUTER IS COMPATIBLE WITH ME."[247] Another competitor, Stentran, offered to evaluate reporter stenoforms and transcripts free of charge by mail. One reporter wrote, "I was delighted and relieved to learn that I had made the grade—the computer had found me potentially acceptable."[248] Estimates for dictionary-building and skill-retraining, even for those reporters deemed "potentially acceptable," varied from two days to three months.[249]

Vendors also had to convince the reporters that CAT would not radically change their employment relations with the courts. One reporter warned, "Watch out once the twenty-minute or overnight service—which under optimum conditions the use of computer-aided transcription can provide—becomes the standard expected by lawyers and judges."[250] Another feared that "the computer industry can easily make its deal with the hierarchy of the courts, leaving us the servants of the mechanism we expect to master."[251] And a third critic predicted, "If our National and State Associations embrace CAT, if reporters begin to make the changeover, if more of our schools teach computer-compatible theory, there will come a time when the loss of transcript fees is viable."[252] The NSRA had long suspected that "the NCSC's interest in computer transcription is . . . motivated more by its desire to eliminate the fees connected with the preparation of transcripts by official reporters than its professed interest . . . of eliminating reporter-caused appellate delays."[253] These fears were not without basis. At a panel on "Reporting in the year 2000" at the 1975 NSRA convention, J. Michael Greenwood, one of the PCCP pilot-test researchers for the NCSC, predicted management changes would appear soon: "Reporters will be affected by new standards relating to pooling, control of the transcript, and the independent-contractor or employee status of the reporter."[254] Vendors like BaronData touted themselves as labor mediators in this debate, selling the "promise of no loss of transcript fees at a low per page cost to the reporter . . . *a fine example of the courts working with the reporting profession to provide a workable plan to solve an old problem.*"[255] But not all reporters trusted the CAT service bureau as mediator: "In cases of backlog, overload, or breakdown, we would have to count on an agent who is also an agent for our competitors, and for businesses of other types, to decide whose work goes first."[256]

## From Typewriting Notereaders to Keyboarding Scopists

Even if their relations with court managers did not change, reporters knew that their relations with their subcontracted assistants would. In the early days of CAT, reporters had speculated as to whether a computer-produced transcript with a 2 percent error rate (forty errors for every ten pages) would be acceptable to the bar and bench.[257] But by the late 1970s it was clear to most reporters that automatic transcription without post-editing—the perfect "first-pass transcript"—was impossible.[258] Especially if CAT was to speed transcript production and allow a court reporter to juggle more than one job at a time, a scopist would be mandatory.[259] Even vendors like BaronData advertised this new division of labor along with their labor-saving products: "Daily transcript with a computer? Yes, definitely. One reporter, one scope operator/transcriber, and CAT!!"[260] The question of the value of the scopist in the CAT labor process mirrored larger debates over office-based "systems redesign" projects—usually focused on new word-processing products—common to the late 1970s.[261] In one courtroom CAT experiment in 1977, the scopist was a certified court reporter and had an annual salary plus benefits cost of $30,360. But in another pilot project two years later, the scopist was a part-time student reporter, paid by the other court reporters at a rate of three dollars per hour.[262] Whatever the wage rate, the ideal scopist served the same function as the transcriptionist and notereader: to allow the reporter more "taking down" versus "writing up" (see figure 3.2).

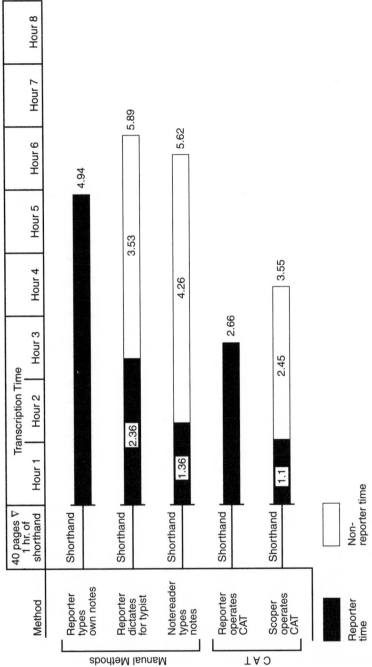

Fig. 3.2. Court reporter time versus subcontracted labor time, 1981. A court reporter's goal was to maximize in-court time "taking down" the transcript and minimize out-of-court time "writing up" the transcript. A social and technological division of labor, involving transcriptionists, notereaders, or scopists, enabled this. Source: NCSC, State Judicial Information Systems Project, *Taking the court record: A review of the issues* (Williamsburg, VA: NCSC, 1981), 35

Scoping inherited many of the artifacts of the dictation-transcription process, including the "dog sheet" on which the reporter noted unusual terminology or proper names needed to produce the final transcript.[263] There was some debate over how much steno theory scopists needed to know. BaronData claimed that with their software, "the scope operator (who now doesn't need to know stenotype theory) can review the transcript for punctuation, extra words, etc., and the scoping is done quickly and easily."[264] But as time went on, it became clear that effective scopists not only needed to know steno theory but, like notereaders, needed to understand an individual reporter's steno style.[265] The scopist could then take over the onerous task of updating a reporter's CAT dictionary with new words, improving the computer translation each time.[266] Scopists employed by large freelance agencies—used as a "load-leveling" strategy to both keep the agency CAT system running more hours and to relieve reporters of transcription duties—sometimes gained additional job functions akin to those of a computer-systems manager.[267]

With the meaning and value of scoping still so contested, in the late 1970s it was unclear who would actually fill these new jobs. As with transcribers, scopists were recruited from court-reporting schools: "You might look at those students who have finished their theory class, thereby having an understanding of how to read notes and at least a year and a half away from being a reporter, or those students in the higher speed level who have decided against becoming a reporter but want to use the skills they have learned so far."[268] Kinship relations were put to work as well, with plenty of stories of wives scoping for their reporter husbands (but none with the relationship reversed).[269] Some even recommended hiring retired reporters as scopists.[270] Yet the largest potential labor market resulted from the prospect that "the competent transcriber of today [would] become the skilled scope operator of tomorrow."[271] One scopist recalled, "Having been a notereader for over 25 years, I knew by 1987 that this method of stenotype transcription was becoming obsolete, so the most logical step was to open a computer transcription service."[272] By the late 1990s the conventional wisdom was that "most transcribers today have made the transition from typing tapes dictated by reporters, to performing the equally valuable service of 'scoping' rough drafts of transcripts for reporters who have adopted the system of computer aided transcription."[273] But those who did so had to purchase and master their own CAT-compatible equipment in order to stay competitive.

The rise of the scopist revealed that if anyone's work was being "automated" as a result of CAT, it was that of the notereaders—those trained transcriptionists who could read a reporter's idiosyncratic stenoforms without the reporter dictating notes into a tape recorder. As early as 1974 CAT was itself being described as "a mechanical note reader."[274] The analogy went both ways; one reporter began to refer to his notereader whimsically as a "PAT" system, for "people-aided transcription."[275] At first, such comparisons favored humans; one vendor admitted that although "the computer functions as an electronic notereader," it also had "the further difficulty that it can exercise no judgment—it can only obey instructions previously given to it."[276] However, eventually the CAT system was described as superior to the human: "A notereader that is always on the job, that never makes a typographical error, doesn't misspell, and types at extremely high rates of speed."[277] Either way, such comparisons begged the question: instead of replacing the notereader with CAT, why not just train and hire more notereaders? "If half of the money already expended on CAT were applied to the training of note readers," wrote one reporter in 1976, "backlogs would be a thing of the past."[278] Another offered, "We know of no concerted effort to recruit and train note readers, but this may be a large part of the answer."[279] One creative reporter even put his recommendations in verse:

> Proclaim that wasteful drafts by CAT
> add not the final touch.
> Work hard to build a better way—
> train notereaders and such;
> For when you gain stability,
> you've no need for a crutch.[280]

In the transition from transcriptionist and notereader to scopist, once again the changing gender mix in reporting lurked just beneath the surface of the debate. By the mid-1970s, the feminization of reporting that had begun in midcentury had reached a "tipping point": of the roughly seven thousand NSRA members listed in 1975, 55 percent were women.[281] This balanced ratio hid an age and experience disparity, however. In many reporting schools, female students outnumbered male students four to one, but men held most senior jobs and ownership positions.[282] The demographic shift had the field talking. Some reporters felt that the critical mass of women would both depress wages and keep men out of the field. One Arizona school owner lamented that "the men he interviews often consider reporting as a feminine job."[283] Another reporting professor complained, "You have only to look at the states that have 75 percent or more women reporters, as compared to men, and you will find that state is substandard in the salary it pays the court reporter."[284] Some assumed that women filled the lower-wage, contingent ranks of reporting by choice, arguing, "The 'part-time' reporter is nearly always female."[285] In any case, it was the older, more experienced, usually male heads of freelance agencies who were making implementation decisions about CAT. Younger reporters, either facing the new demands of "computer compatibility" or bailing out of reporting school to become scopists, would tend to be female.

As a result of this demographic polarization, the very public debate over whether to adopt CAT played out against a more subtle debate over the effect of women on the reporting profession. As early as 1974, one of the heads of the NSRA Committee on Computer Transcription was a woman: Doris O. Wong, a Boston freelance-agency owner (and early BaronData customer).[286] When Wong took over the "CAT corner" column in the *National Shorthand Reporter,* the page was revamped to feature a cartoon of an actual cat writing on a steno machine, with an obvious woman's figure and shoes. Such whimsy seemed to indicate that, rather than the typical gendering of computers as exclusive objects of male activity, CAT was an appropriate tool for reporters of both sexes. But a year later, when male reporter Richard Tuttle took charge of the committee, the lady cat was gone, replaced with the canonical technological representation of a minicomputer with dual reel-to-reel tapes staring out like blank eyes (see figure 3.3).

Such tension over how to gender CAT appeared again and again. Some hoped that holding up CAT as an answer to the ER crisis would attract men back to the field because, as one California reporting school owner surmised, ER "might be frightening men more than women, since a man's chief concern is his ability to sustain his family."[287] But often the arguments around adopting CAT were wrapped up in a discourse of uplift and professionalization linked to university education, as opposed to high school education—a strategy that also had a gender dimension. One reporter wrote, "We are perhaps 'unwittingly' practicing discrimination against men by recruiting from girls' high school shorthand classes 'rather than' college English classes," a

Fig. 3.3. Gendering of CAT logos, 1975. Court reporter Doris Wong's "CAT Corner" illustration and the logo for court reporter Richard Tuttle's renamed "CAT Talk" column point to the contested gendering of court reporting. Source: Gilbert Frank Halasz, "CAT at work," part 1, *NSR* (May 1975), 26, reprinted with permission of Richard M. Bumpus; Stenographic Machines Inc., "Stenograph's computer transcription system now ready to produce transcripts," *NSR* (Nov. 1975), 40–41, reprinted with permission of the National Court Reporters Association

sentiment echoing CAT pioneer Frank Nelson's worries that CAT would only be successful if "future recruitment by reporting schools [is] directed to mature college-type professionals."[288] Responding to such fears, the NSRA "launched a campaign to convince prospective male students of the attractiveness of reporting, its advantages, and the possibilities of high remuneration."[289] Some NSRA reporters decried this move as "affirmative action in the wrong direction."[290] Yet a decade later, private court-reporting schools still sometimes admitted that "one goal of reporting firms is to attract more men into reporting." One school director in Denver—herself a woman—said, "The caliber of the profession is a little higher when there are a lot of men."[291]

Some reporters clearly benefited from the new combination of CAT and scopist: "I am now effectively working on four cases at one time, counting the deposition on which I am free to be working."[292] Sophisticated CAT users could quickly provide clients with keyword indexes to transcripts, or even provide transcripts on computer-readable media.[293] As even the NCSC acknowledged, "Successful CAT reporters find that they can get the bulk of their work done during office hours, and no longer spend evenings, weekends, and vacations working on transcripts."[294] But reporters with less capital to spend on hardware, less time to spend on retraining, less freedom to remain at the CAT site, and less tolerance for the inevitable glitches of early CAT systems did not fare as well, as revealed by a 1981 NSRA survey of early adopters across four different vendors. "I have become increasingly concerned over the length of time it is taking to correct the problems with the software," complained one CAT user.[295] Another wrote, "The system is at the office and if you're out of town on a job, you can't do as much work as if you were dictating with a dictaphone right there out of town with you."[296] The opinions of the court reporter who wrote "I could say more, but it's late, I've been fighting the 'damned machine' as we call it, most of today," probably revealed a widely shared sentiment.[297] In the end, the NSRA's assessment was a positive but sober one. The association's CAT expert, Jill Berman Wilson, wrote in 1981, "Yes, it has changed the way we do business and our office procedures—you might say exchanged one set of problems or headaches for another set. But the advantages clearly outweigh the disadvantages."[298]

## An Uneven Geography of Court-Reporting Technology and Labor

By the end of 1980, fewer than a dozen public courts around the nation had purchased CAT installations, even though over three hundred freelance agencies and nearly two thousand freelance reporters were using the system.[299] Unlike the NSRA's enthusiastic assessment of the benefits of CAT to the court reporter, the NCSC's assessment of the benefits of CAT to the courtroom was more conservative, as "only one of the eleven courts presently using a CAT system has been able to achieve a cost-effective operation." The problem, unsurprisingly, was attributed to a management failure: "In a court that does a good job of managing its reporting resources, CAT can be smoothly integrated into court operations and can be expected to achieve the intended goals of time and cost savings. In a court that either does not manage its reporting resources or does it poorly, a successful CAT operation is not likely."[300] Similarly, a Federal Judicial Center study in 1980 concluded that 40 percent of the federal court reporters using CAT "found no appreciable improvements in transcript efficiency," and another 20 percent "reported an increase in transcript preparation time after adopting CAT" due to factors such as "the availability and location of the CAT system, the unreliability of the computer printers and computer failures, the inefficiency of the translation software (computer dictionary), and excessive text-editing time."[301] Thus the management strategy of ignoring CAT dovetailed with the reporter strategy of adopting CAT: "The best way to respond to the threatened loss of transcript-related income is to develop CAT in the private sector so it can be used to reduce backlog and hence reduce pressure for installation of court-owned-and-operated computers."[302]

Although courts were not eager to own and operate CAT computers, they remained eager to own and operate ER systems—ignoring the growing argument of the court reporters that widespread adoption of CAT would remove the need for such measures. After examining seven of the ninety-five federal district courts in 1981, the General Accounting Office (GAO) issued a scathing report, "Federal court reporting system: Outdated and loosely supervised." While the report admitted that "court proceedings are being recorded properly and transcripts are prepared accurately," it claimed that "many court reporters have often overcharged litigants, used Government facilities to conduct private business, and used substitute court reporters extensively." But instead of even mentioning CAT as a solution, the report recommended ER, "which, if adopted, would result in annual savings of about $10 million."[303] The GAO report brought decades of arguments over ER to a boiling point and resulted in the Federal Courts Improvement Act of 1982, which mandated research pitting the stenotype against ER—again with no mention of CAT.[304] The resulting Federal Judicial Center study, published the following year, tested the stenotype against ER in twelve district courtrooms and rated ER higher in each of the three critical areas of accuracy, speed, and cost. For example, whereas 100 percent of the ER-based transcripts were delivered within thirty-five days, only 77 percent of the steno-based transcripts were. And the annual equipment and labor cost of one audio recording system, estimated at $18,604 per year (including backup systems), was less than half the estimated cost of comparable stenographic labor, at $40,514.[305]

The court reporters reacted with outrage. In autumn of 1983, New York court-reporting associations and unions met to consider a response to the report, and both the NSRA and the Stenograph Corporation (a stenotype and CAT vendor) commissioned their own rebuttal reports.[306] The court reporters were particularly angered that "no reporter on CAT was asked to participate." Their main accusation was that the raw data of all transcripts from ER versus all transcripts from the stenotype were not compared completely, but cherry-picked for evidence damning the reporters.[307] The rulemaking justices at the Judicial Conference of the United States, however, decided to leave the decision of which option to use up to individual courts. Whereas before, electronic recording had been allowed only if a court reporter was also present to take verbatim testimony by hand or by machine, now "individual United States district court judges may direct the use of shorthand, mechanical means, electronic sound recording, or any other suitable method, as the means of producing a verbatim record of proceedings."[308] CAT had not yet saved the reporters from ER.

The best that the court reporters could hope for, at least at the federal level, was an uneven geography of employment and electronic recording that might or might not be copied in the lower courts on a state-by-state basis.[309] A summer 1983 survey by the Conference of State Court Administrators provides a snapshot of the official reporter geography just as CAT was moving from the minicomputer to the microcomputer platform (and becoming more affordable in the process). Despite the efforts of the court managers since 1960, only four states reported overall dominance of ER—Alaska, Indiana, Massachusetts, and Tennessee—although New Mexico and South Carolina reported a fifty-fifty split between ER and stenotype methods. Surprisingly, eight states could not provide an accurate estimate at all, but the other twenty-eight responding states (and DC) reported some majority of stenotype reporting. However, the type of court and the type of litigation made a significant difference in the choice of transcription method: "The predominant method of court reporting in general jurisdiction courts was machine shorthand while audio recording was the method most often used in limited jurisdiction courts." As for computer-aided reporting, although this high-tech method was most likely to be used in the largest court systems "serving populations of 500,000 or more," out of the ten states that reported some CAT use, only one—Maryland—had more than 10 percent of their reporters on CAT.[310] The future of stenographic court reporting was by no means certain.

Over a century ago, in 1868, St. Louis court reporter Charles E. Weller was lucky enough to receive an early model of a brand-new piece of information-processing equipment as a favor from his friend C. Latham Sholes: the typewriter. Sholes and Weller were both curious to see how this revolutionary labor-saving technology might enhance the production of the transcript. As it turned out, it was Weller's wife, Margaret, who actually "learned the keyboard" and "assisted her husband in making the test by working the machine from dictation of his shorthand notes." As the story goes, "For many years afterwards Mrs. Weller was her husband's sole assistant, having in the meantime learned his system of shorthand, and during this time worked out thousands of pages of testimony, working from dictation until she became tired, and then changing seats with her husband and reading to him from his notes while he worked the machine."[311]

In the years since this first documented partnership between a reporter and a notereader, court reporting has weathered the further technological transformations of the mechanical stenotype, the electromagnetic tape recorder, and the digital computer. By the mid-1990s, in New York City, "almost all freelance firms require[d] their reporters to be on computer. Dictating and typing [had] gone the way of the pen."[312] A few years later, the venerable stenotype itself was being sold in a "paperless" version—lacking the ability to imprint stenoforms on a strip of fan-folded reporting tape—because of the ubiquity of computer-aided transcription.[313] But throughout every transformation in tools and practice, a division of labor between "taking down" and "writing up" has survived—a division structured by age and gender, status and skill, technology and theory, and time and space. Court reporters of 1890 would likely understand the sentiment expressed in 1990 that "with the well-known burnout rate in the reporting profession, it may well be that using a scopist regularly could add years of productive life to some careers."[314] The court reporters of yesteryear might also appreciate the contradiction that each new advance in reporting technology threatened this contingent, subcontracted job category. One writer in *Forbes* observed in 1993, "What the PC is doing is increasing the back-office productivity of the 40,000 or so U.S. court reporters, in many cases replacing the reporter's assistant. It isn't replacing the professional who sits near the witness box."[315]

Even if they were not being replaced, though, those professionals near the witness box were themselves valued and standardized in new ways, as stenotypes were connected to computers and notereaders gave way to scopists. As early as 1921, the NSRA had recognized that "if [manual] shorthand could be standardized and one universal system employed by everybody, it would be possible for one stenographer to read with facility the notes of another, thus obviating the retrial of many cases where death had intervened."[316] As manual shorthand began to yield to the stenotype, the hope was rekindled that "the reporter of the future will be thoroughly grounded in a truly scientific system of shorthand writing, sound in fundamentals, ever advancing. He will not over-indulge in shortcuts. He will not use arbitrary symbols. He will write phonetically."[317] The 1970s demand that reporters become "computer compatible" seemed to hold out the hope of finally standardizing education, practice, and artifact—a standardization that court managers probably hoped for even more than court reporters did. But as it turned out, the irrepressible transcriptionist-turned-scopist offered a way for court reporters to mitigate the pain of retraining, retooling, and reskilling. A good scopist not only helped to correct a reporter's errors in the first-run transcript but also managed the reporter's customized CAT stenoform dictionary. Like notereaders and transcriptionists, scopists sat at the uncomfortable boundary between "good enough" reporting skills and "verbatim" transcript demands.

Of course, not all court reporters used scopists, or used CAT, or even "sat near the witness box." Over the twentieth century, the institutional division between official reporters and freelance reporters seems to have grown more polarized with each change in reporting technology and labor. By the early 1980s, many private freelance court-reporting firms had embraced minicomputer-based CAT as a collectively shared competitive advantage, while many state courts were still experimenting with new wage divisions of labor as a result of CAT. In all, by the mid-to-late 1980s about half of the nation's estimated 28,000 court reporters were using CAT.[318] And even though freelancers made up roughly two-thirds of all reporters at this time, they had for years comprised the vast majority of CAT users.[319] But over time, these "early adopters" did not necessarily reap greater benefits. By the early 1990s, officials were earning a median yearly income at least 10 percent higher than freelancers.[320] The gap had widened by the late 1990s, according to a 1997 NCRA survey. The average freelance reporter made $51,524 per year including transcript fees, but had a personal capital investment in CAT equipment of $24,570, and spent nearly $6,000 in supplies yearly. On the other hand, the average official court reporter made

$61,100 per year including transcript fees, but carried a personal capital investment in equipment of only $13,991, and spent only about $1,600 on supplies yearly.[321] Questions over the proper compensation formula for the dual aspects of the reporter's work persist—such as recent suggestions that reporters charge for their now-computerized services "by the download or even kilobyte, by the hour or by some sort of blend of hour and page rate"—and are structured by this institutional division.[322]

Some of the disparities between the wages of freelancers and officials could be explained by age—especially since the customary career path in court reporting was to start out in the freelance labor market to gain experience, transitioning into the more competitive official labor market when (and where) possible. But this age and career path difference was also structured by gender, such that by the time CAT had become the accepted technological and labor practice for court reporting in the 1990s, the feminization of the field was nearly complete. According to the NCRA's own figures, by the early 1990s fully 86 percent of court reporters were female, with women making up 96 percent of all reporters aged eighteen to thirty-five, but only 48 percent aged fifty-five and over—meaning men were still more likely to make greater salaries, and "more likely than women to be agency owners."[323] Even some female NCRA members continued to lament that "reporters are primarily women at this time," fearing "I'm not sure that we will see the sort of full-time dedication to the career of court reporting we've seen in the past."[324] But any idea that the women in reporting were somehow less committed to the career—perhaps because they were more likely to occupy the part-time, contingent layer of the reporting hierarchy—was false. Although two-thirds of the female reporters were married, nevertheless nearly 63 percent of all female reporters were "sole or main support of [their] household," where fully 41 percent of all female reporter households included children to support. For these women, court reporting was a clear step up on a gender-segregated career ladder: "About two-thirds of female CRs were previously employed in a clerical occupation," and "almost no women came into court reporting from anything other than a traditionally female and low-paying occupation." Still, the NCRA found that "in states with higher proportions of female court reporters, court reporter earnings are lower, even after controlling for regional wage differentials that are found in all occupations."[325] Thus the NCRA continued its policy of "affirmative action" efforts to woo men, including a 1998 high school recruiting video highlighting male reporters, which, it was hoped, "helps give the male students a visual image that there is a place for them if this is something they might be interested in."[326]

These self-scrutiny and self-definition efforts illustrate one aspect of the history of court reporting quite different from the story of film subtitling and television captioning in America. The reporters, through their state and national associations and publications, delighted in their ability to keenly watch and debate the potential "impacts" of new information technologies on their (self-proclaimed) profession. Part of this behavior was rooted in the relatively rapid rise of court-reporting work and workers throughout the century, tied to the broad genesis of what many have termed the modern bureaucratic, indus-trial, urban "information society." Reporters eschewed the practice of "technological somnambulism" as philosopher of technology Lang-don Winner famously termed it; they did not want to sleepwalk through periods of technological change but to muster new technolo-gies to their profession and purpose.[327] For example, reporters coined and promoted the term "computer-aided transcription" (emphasizing the "offline" moment of transcription, often subcontracted out, as in need of technological aid) instead of the terms "computer-aided ste-nography" or "computer-aided reporting" (which might have high-lighted the "online" moment of capturing live speech as requiring a technological fix). Perhaps no one knew better than the court re-porters that such words mattered. But what nobody could predict was that this ability of the reporters to quickly identify and promote op-portunities for protecting and expanding their technological liveli-hoods would soon connect with the plight of the closed captioners in defining and defending their troubled technological system.

# Part Two

# Convergence in the
# Speech-to-Text Industry

# Chapter 4

# Realtime Captioning for News, Education, and the Court

I'm one of very few people in the world to be doing this and I really love it. I'm using my professional reporting skills, I'm involved with a new technology, and I'm providing a real service for a large segment of our society.

*Martin Block, court reporter and realtime captioner, 1982*

It was billed as the "Trial of the Century," just as others had been decades before, but the O. J. Simpson murder trial in 1994 would be the last claimant to that title for the twentieth century. Not only would the litigation be long and complex, lasting fifteen months and generating nearly fifty thousand pages of transcript as grist for the inevitable appeal and civil suit to follow, but the media interest in this sports superstar turned accused killer was unprecedented. The twenty-four-hour news network CNN was barely a decade old, the World Wide Web was just beginning to take off, and conservative-led talk ruled the AM radio waves. Big money would be paid for—and made from—every tidbit of the spoken testimony. How would the two women court reporters, Chris Olson and Janet Moxham, ever keep up with the scale and speed of such a trial?[1]

The solution was plain for all to see on Court TV: a laptop computer perched on presiding judge Lance Ito's bench. As the two court reporters traded off typing a stenographic record of the proceedings, their computer-aided transcription (CAT) system automatically translated their stenoforms to English, displaying the resulting transcript text live on a series of computer screens scattered throughout the courtroom, for both judge and counsel to view instantly.[2] Despite the presence of digital audio and video recording equipment, Olson and Moxham's "dirty copy"—as the realtime transcript was called before being post-edited later—commanded a high price from both legal and media clients.[3] Said Moxham, "We would hand-deliver a copy to the judge and each attorney, provide ASCIIs to each side . . . and provide condensed transcripts to the Judge so he'd have fewer pages to take home. Bench conferences were blocked out so the jurors couldn't see them, so everyone wanted those the most. We printed those excerpts separately for about 30 media clients." The two reporters had to wire the building for all of these services themselves: "I had to get phone lines, move equipment to another floor and re-install the modem line," explained Moxham. "Eventually I had four phone lines in the courtroom . . . I was paying for everything involved."[4] Each night after trial the reporters would work overtime to "scope" each other's work ("with a little scoping help from my mom" according to Moxham), post-editing some three hundred pages of transcript per day. But their labors paid off. The trial, which resulted in some 48,667 pages of transcript, may have grossed Olson and Moxham over $500,000 (well more than Judge Ito's $107,390 yearly salary).[5]

The twin abilities of court reporters, turning around such huge volumes of daily transcript with little post-editing and displaying a readable realtime transcript in the courtroom, came as a result of the latest technological advance trumpeted by the National Court Reporters Association (NCRA) to fend off accusations that court reporting was an obsolete labor practice destined for replacement by electronic recording.[6] Impossible with the mainframe-based CAT systems Olson and Moxham had witnessed as they started their court-reporting careers in the 1970s, "realtime CAT" was suddenly feasible as microcomputers entered the office in the 1980s.[7] But it was not just technology that brought CAT to realtime speed; it was a pair of social needs, in broadcast captioning and in deaf education, which motivated the development and dissemination of this technology. And before this new technology could be brought to bear on the endemic "transcript crisis" in the courtroom, a generation of court reporters who had spent the 1980s becoming "computer compatible" with the first round of CAT systems had to retool once again to become "realtime compatible" in the 1990s. CAT technology and CAT labor both moved through the contexts of television captioning and classroom interpretation in the 1980s, paving the way for the "courtroom of the future" in the early 1990s.

## Captioning the Nightly News at NCI

After only one year of closed captioning, the National Captioning Institute (NCI) was in trouble; decoder sales to deaf and hard-of-hearing (D/HOH) households remained low, caused in part by the meager hours of captioned material the networks provided, and the networks refused to provide more content unless decoder sales increased. One "killer application" in particular was missing: NCI could not live-caption the nightly evening news, meaning D/HOH viewers still relied on the time-delayed, open-captioned, PBS-aired version of the *ABC Evening News* produced by the WGBH Boston Caption Center (in those markets where it was available). A few D/HOH experts believed that live news captioning was unnecessary: "The newscasters," argued director of audiology at the New York League for the Hard of Hearing, "generally face forward so that lip-reading can be done."[8] But the bulk of the D/HOH community had long believed otherwise. At Gallaudet, Donald Torr's in-house captioning team had for years devoted one of their closed-circuit cable channels to an automated, on-screen scrolling textual translation of the UPI wire-service news, which they received through a dial-up line and fed into a character generator.[9] The Clarke School for the Deaf in Northampton, MA, employed a pair of work-study students to live-caption evening network programming over an in-house CATV system by typing on a character-generator keyboard "as quickly as it is humanly possible."[10] Even before NCI went on the air, in a 1979 mail survey of "viewing habits, preferences, and attitudes of 1,759 hearing-impaired adults" across the United States, the most frequently requested programming was news.[11] Further NCI research in the early 1980s verified that both decoder owners and nonowners wanted closed-captioned news.[12] NCI researchers surveyed decoder owners and even found that "82% [of respondents] would prefer a closed captioned news program early in the evening rather than an open-captioned news program late at night" (assuming that the quality of captioning in each case would be identical).[13] Yet experts in deafness had long recognized that since "writing captions is an art," captioning unscripted programs live "would be costly, unless engineering research could yield up some new devices to reduce labor costs."[14]

Over a decade before the founding of NCI, the D/HOH community had looked into computer-aided transcription in the courtroom as a potential solution to the realtime captioning dilemma—but for classroom use, not broadcast use. As early as 1969, Ross Stuckless of the National Technical Institute for the Deaf (NTID) in Rochester, NY, began to explore ways for deaf students in hearing classrooms "to get verbal lectures in 'real-time' and in printed form."[15] In 1970, before even the National Shorthand Reporters Association (NSRA) had witnessed its first CAT demonstration, Stuckless met with IBM and Stenograph representatives "to explore the classroom use of such a system." However, as Stuckless later recalled, "neither computer hardware nor computer software was up to the task and the development costs for the system we needed were beyond reach."[16] A mainframe CAT solution, which required both batch-processing translation and significant human post-editing, was just not feasible for realtime display of speech-to-text. But the idea of bringing stenography to bear on the captioning problem persisted. In 1977, Gallaudet president Edward Merrill took a trip to the Royal National Institute for the Deaf in the London and described an early experiment to display raw steno notes for a deaf member of Parliament: "A court reporter types out a speech or debate as he normally does. This types on a tape which he does not transcribe but the phonetic cues appear on a visual display. The deaf person learns to use the phonetic cues so he understands what is being said almost at the time it is said."[17] While training all D/HOH persons as stenographic notereaders was not feasible, interest was rekindled in funding a realtime-display CAT solution. In 1978 Malcolm Norwood and the Department of Health, Education, and Welfare (HEW) funded a prototype realtime CAT system developed by Stentran Systems and Teledyne Geotech, which they hoped would "make it possible to caption live programs" with 95 percent accuracy.[18] Unfortunately, because the software still ran on a remote IBM 370 mainframe computer, accessed by the stenotypist over a 1,200-baud dial-up phone line, it was deemed infeasible for widespread realtime use.[19]

Other vendors were moving CAT out of the mainframe environment, however. One was Translation Systems, Inc. (TSI) of Rockville, MD—a business grown from the ashes of original CAT pioneer Stenocomp in the late 1970s, still employing court reporter Patrick O'Neill, who had worked with IBM on projects for the US Air Force and CIA.[20] TSI renamed their CAT solution TomCat and by 1980 had ported it to a $15,000 128K Jaquard J-500 microcomputer—all of the hardware was encased with the CRT in a desktop box resulting in "a complete, in-house, stand-alone computer system."[21] Here finally was a CAT solution small enough, fast enough, and affordable enough to make realtime "stenocaptioning" feasible.[22] NCI worked together with TSI to create a custom system called InstaText, which could translate the stenotype input from an operator into broadcast-ready Line 21 closed-captioning signals with a total "throughput," or lag time, of only four to seven seconds.[23] The system appeared to work—now all NCI needed was a labor force skilled enough to operate it.

The person they found to act as what the NSRA later called the "first real-time reporter" had long experience working with CAT in a proto-type setting: Martin H. "Marty" Block, who for sixteen years had worked as a reporter in the Philadelphia Court of Common Pleas, and who had been involved in the crucial mid-1970s Stenocomp CAT test there (see chapter 3).[24] Block recounted that it was his experience meeting "a new acquaintance's young, and deaf, daughter" that convinced him to move his family from Philadelphia to Washington, DC, and go to work for NCI at about age forty, after two decades as a court reporter.[25] As the first NCI-employed operator of InstaText, he was instrumental in helping to develop the software.[26] But he was also influential in developing a new style of reporting beyond the previous "computer compatible" stan-dard. To reach the coveted 1 percent error rate deemed acceptable for broadcast television—no more than two errors per minute when writing at a speed of 180 words per minute—a reporter now had to become "re-altime compatible." This meant not only reaching a skill level where one's fingers always did as they were told on the stenotype keyboard but also purging one's CAT dictionary of "conflicts" or "stenonyms"—stenoforms used for more than one English word—and learning to key instead in a "conflict-free" style so that the CAT translation system would never have to guess which word the stenotypist meant. Block de-scribed how consistency, not speed, was the biggest requirement here: "I am not a shorthand speed champion. A CAT system like the one we have at NCI can be used successfully by almost any machine reporter of average skills. Just pay attention to long and short vowels, write as cleanly and consistently as you can, and don't change your shorthand writing system unless absolutely necessary."[27] Conflict-free CAT diction-aries inevitably grew larger than standard CAT dictionaries, on the order of seventy to a hundred thousand words.[28]

Since the computer could only translate a word properly if that word had previously been loaded into its stenoform-to-English dictionary, Block's live stenotype reporting was only part of the labor required to make realtime captioning work. After all, lacking a specific entry for ABC news reporter Ken Kashiwahara, InstaText might translate his name as "keen cashew what had a are a," no matter how quickly or cleanly the stenotypist keyed it in.[29] Since the news varied each day with new stories, places, and persons, NCI hired a work force of "text editors" who worked all morning scanning "major daily newspapers and the weekly news magazines for names and places likely to be mentioned in the program later that day." Text editors monitored the news wires "for late-breaking news reports" as they prepared "the daily dope sheet."[30] That dope sheet (the same as a court reporter's "dog sheet") would then be used to update the CAT dictionaries and brief the stenocaptioner before the live broadcast began. As Block later described, "The single most important aspect of captioning a live event is anticipation. The more the captioner knows about what is coming, the more accurate and prompt the captions will be."[31] Later stenocaptioners estimated that the prep time for a single half-hour daily newscast could run anywhere from one-and-a-half to three hours per day.[32] In effect, the "text editors" performing this prep work were really scopists, as they needed to know both current events and proper realtime stenographic technique.

For the first few months, Block and his team of text editors practiced their division of labor in a "pseudo-live" captioning process based on the WGBH Caption Center method. Starting in November 1981, Block captioned an early feed of the *ABC World News Tonight* live at NCI at either six o'clock or six-thirty, but his captions would be held back for some quick post-editing and then transmitted to ABC for rebroadcast during the later feed of the news at 7:00 p.m. (EST) (where some 40% of the audience, spread among forty-six ABC affiliates, could see the captions).[33] The team spent nearly a year working out the kinks of both the computer system and Block's steno style in this way.[34] Like the existing WGBH *Captioned ABC Evening News,* then, this was a delayed broadcast. But the NCI version was delayed only an hour or less, not five hours; it was shown on ABC stations during a normal news slot, not on PBS stations during late night; and it was closed captioned, so only decoder owners could see the results. Perhaps most significantly, NCI claimed, "Unlike PBS's opened captioned news, the captions were verbatim"—echoing the caption-editing debate between Phil Collyer and Doris Caldwell from the 1970s (described in chapter 2).[35] Concerned about

the higher speed of these verbatim captions—180 words per minute—
NCI surveyed decoder owners and found that 76 percent of respondents
said they could read all the captioned material.[36] InstaText appeared
ready for a realtime broadcast.

Yet InstaText found its first live test not on the airwaves but in the
courtroom. On March 25, 1982, deaf lawyer Michael A. Chatoff was to
argue a case before the Supreme Court—*Hendrick Hudson Central School
District v. Amy Rowley*—dealing with the question of what obligation
public school districts should have toward D/HOH pupils in terms of
providing language interpreters. The case was billed as "the Court's
first opportunity to define what Congress meant in the Education For
All Handicapped Children Act of 1975, which says that all handicapped children are entitled to a 'free appropriate public education.' "[37] Because Chatoff was "not proficient in signing or lip-reading"
himself (he had become deaf not prelingually as a child, but postlingually while in law school), in his oral arguments he was assisted by a
court reporter using realtime CAT so that he could respond rapidly
and accurately to the justices' questions during the limited amount of
time allotted for such exchange (a time limit the Court refused to extend).[38] The president of TSI, who provided the system for Chatoff to
use free of charge, hand picked a longtime TomCat user (and an associate of Block's), court reporter Joseph Karlovits, as the person who
would be the most "realtime compatible" in this showcase moment.[39]
The whole computer system had to be housed outside the courtroom
itself because of noise and bulk, but having been stress-tested by Block
and his team at NCI for months, it passed its first public performance
well.[40]

Finally, on October 11, 1982, NCI was ready to live-caption the
evening news without a net.[41] Managing to incur only a four-second
delay, Block and fellow stenotypist Bobbie Showers captioned the 6:30
p.m. feed of the *ABC World News Tonight* live, although their team of
post-editors still quickly cleaned up the captions to be rebroadcast during seven o'clock feed.[42] The editing, just as stressful as the stenocaptioning, took three text editors working at "triple time"—"if the
section is three minutes long, you must edit it in nine minutes or
less."[43] Together, the two feeds would reach nearly the entire national
viewing audience—at least, all those with caption decoders.[44] NCI
quickly surveyed audience reaction and found that 69 percent were

happy with the rapid-fire, near-verbatim captions (though 89% noted the occasional garbled caption, something that rarely if ever happened with the open-captioned, time-delayed ABC news from the Caption Center).[45] Despite the hopes of the broadcast industry, live captioning was no cheaper than prerecorded captioning—NCI estimated it cost $2,200 to live-caption one hour of TV.[46] For now, the effort was paid for by a $1.3 million grant from the Department of Education.[47] But regardless of the cost or quality concerns, the moment was a great victory, not only for the reputation of closed captioning at NCI but also for the profession of court reporting. According to Block, "What's really beautiful about this whole process is that only a shorthand reporter can do it. A tape recorder can't do it . . . It takes someone skilled in shorthand reporting to turn what [ABC news anchor] Frank Reynolds says into a format that a hearing-impaired person can read as he is saying it."[48]

If there was a downside to live stenocaptioning, it was the devaluing of the important contribution of non-realtime captioners—now rechristened "offline" captioners as opposed to their "online" realtime counterparts. By the mid-1980s, realtimers were referred to as the "elite" captioners in the NCI labor force, with the idea that "Captioning prerecorded materials, commercials and videos . . . is less stressful."[49] But as described in chapter 1, in translating any visual and auditory medium into text through subtitles, profound temporal and spatial limitations worked against artistic and educational goals, such that balancing the two became "a skill, requiring highly developed language, concentration, and timing," according to one D/HOH advocate.[50] The offline captioners did not suddenly become less skilled with the emergence of their online peers. In fact, captions for prerecorded material demanded different kinds of attention—to precise caption quality, timing, and placement—and were more likely to attain a verbatim standard (if that was the goal) than any captions a court reporter produced, no matter how skilled. Certainly much prerecorded programming remained to be captioned; by 1989 NCI was producing sixty hours of live captioning per week, but seventy hours of prerecorded captioning.[51] The devaluing of offline captions was perhaps epitomized in the fate of the WGBH Caption Center. Suddenly forgotten as pioneers in bringing a national evening network newscast to all D/HOH Americans, regardless of household ability to pay for decoders, their open-captioned, time-delayed broadcast was canceled in December

1982 "for lack of funds."[52] Fortunately, by 1984 they had reverse-engineered the Line 21 hardware and software under control of NCI and could participate in closed-caption production. Yet in the late 1980s, their offline captioning services still cost more per hour than their realtime captioning services.[53]

Through the 1980s, funds became available for an increasing menu of national network newscasts, not just on weekday evenings but on mornings and weekends as well. By 1986 ABC was having NCI caption *This Week, Good Morning America, 20/20,* and *Nightline;* the following year NCI added *NBC Nightly News.*[54] Part of the money came from the networks, and some came from donations, but the bulk of the costs continued to be covered by the Department of Education, on a three-year, $5.3 million contract.[55] Funding more news programs meant paying for more labor, as so many news shows occurred during the same time slot on the same day of the week. For example, in 1987 NCI used a $500,000 grant from the W. K. Kellogg Foundation to "purchase new equipment to expand its production facilities making it possible to simultaneously [live] caption the three network evening news programs."[56] The NCI live-captioning staff grew from ten stenocaptioners in fall 1987 to fifteen in fall 1988.[57]

Yet as network news captioning grew, another segment of the schedule was left out: local news. Finding not only the funding but also the labor to create live captions for hundreds of daily newscasts across the country each day suddenly emerged as a crucial closed-captioning problem. The first local news captioning came not as a result of NCI, but through the WGBH Caption Center. In March 1986 Boston's PBS affiliate managed to live-caption the six o'clock news from the local ABC affiliate, and the ten o'clock news from the PBS station itself, using the same realtime CAT methods that NCI pioneered.[58] Two months later, KDKA Pittsburgh and KAKE Wichita followed suit, and by April 1988 about a dozen stations were captioning at least one local newscast, according to NCI's John Ball.[59] The handful of local stations that could fund their own stenocaptioning by 1990 did so in a variety of ways: some secured sponsorship from local business; some paid the cost themselves; and some received foundation grants.[60] These early sites of local news captioning often had some connection to Deaf culture and often came about through a charitable venture. For example, when the Rochester, NY, ABC affiliate hired court reporter Linda Miller to caption its local news in 1987 in order to serve the "sizable deaf community" centered around NTID, "the local cable television company chipped in and offered decoders, free to deaf customers and at cost to others in the community."[61] That same year viewers in Washington, DC, finally received local captioning on a single eleven o'clock newscast "thanks largely to a WJLA executive with a deaf son and a Gallaudet University graduate who works at the station." The $100,000 yearly cost would be borne by corporate sponsors.[62]

Even when local television stations refused to caption live news, local governments sometimes could be convinced to pay for live captioning of public meetings. In 1992, at the urging of a local vendor of realtime captioning equipment and labor, the city of Fremont, CA— home of the California School for the Deaf and some three thousand D/HOH residents—became the first municipality to closed-caption its city council and school board meetings, which were already being aired over a local public, educational, and government (PEG) cable channel. The city invested $15,000 in capital equipment and hired a stenocaptioner at $100–$125 per hour, paying for the labor through "a monthly $0.07 per subscriber surcharge on cable television bills."[63] Similarly, after a twenty-six-year-old deaf man entered the race for city council in Tampa, FL, the surrounding county government was motivated to spend $26,000 on equipment and $20,000 for six months of stenographic labor (from three reporters) to caption local government meetings for the estimated sixty thousand D/HOH residents in the area.[64] In both cases, not only was a substantial population of D/HOH city residents necessary to legitimize each project, but an individual aware of the captioning costs and benefits also had to come forward and instigate change.

The difficulty in local news and government captioning was not just identifying an audience and attracting funding, though; it was attracting skilled captioners from the court-reporting community. In the May 1989 issue of *National Shorthand Reporter,* an advertisement from the WGBH Caption Center proclaimed, "Tired of the courtroom? Try television!" (figure 4.1).[65] By about 1990, NCI realized it had better begin to grow a cohort of skilled stenocaptioners for local markets around the United States.[66] Darlene Leasure, an NCI stenocaptioning manager, was designated "to work closely with ambitious individuals who are court reporters by day and train them to become real-time captioners by night." She described her job in a court-reporting journal in 1991: "I travel to the site to train the court reporters to caption

# Tired of the courtroom? Try television!

WGBH is seeking court reporters interested in live television captioning of News and Public Affairs programs. As a **stenocaptioner**, you will be working with a team of highly skilled reporters combining the two exciting fields of Computer-Aided Real-Time Translation and live television broadcasting. You will be using your skill to help make television accessible to the deaf and hearing-impaired community.

- RPR or CSR preferred
- computer experience preferred
- training provided
- full benefit package
- competitive salary

Applicants should send resume to:

Real-Time Captioning
**The Caption Center** – WGBH
125 Western Avenue
Boston, MA 02134

or contact Suzanne Astolfi
(617) 492-2777
ext. 3886

Fig. 4.1. Caption Center advertisement, 1989. After the development of realtime captioning based on stenographic techniques in the 1980s, caption agencies increasingly recruited employees from the court-reporting profession. Source: *NSR* (May 1989), 77

their local newscasts. The intensive training sessions usually last three to five days. For the next few months, training occurs much like in a correspondence course . . . When the captioners are ready to go on the air, I make a return trip to their location. I put the finishing touches on training, conduct dry runs, and deal with the television station, where personnel always like to do a story on the development, complete with a live remote of the neophyte captioner captioning that very first newscast on the air." She even trained some prospective captioners right in their own homes.[67]

Growing the captioning labor force was critical because of another technological development in local news captioning, the "electronic newsroom." Based on the stand-alone scrolling-text TelePrompTer, which had been a mainstay of television performance since the 1950s, this technique simply fed the same scripted text that the news anchors already narrated right into the closed-captioning stream.[68] Most of the time, the scripted text matched what the anchors said. But prerecorded "packages" from on-location reporters were often left uncaptioned, and any live deviation from the script (such as for breaking news) was uncaptioned. The advantage of such systems was cost. One manager of a CBS affiliate station in New York estimated in the early 1990s that the software and hardware costs of the electronic newsroom solution were between $2,000 and $7,000, compared to an estimated $100,000 per year for stenocaptioners.[69] By 1991, when Marty Block was praising court reporters at the National Conference for Closed Captioning of Local News, roughly 85 percent of all local captioning was already performed by TelePrompTers.[70] Court reporters were again faced with technological replacement in a market they had only just recently helped to create.

But growing their own captioning labor force was a slow process. In the mid-1980s, NCI's captioning supervisor Tammie Shedd reported, "Of the court reporters who apply to NCI, only 1 in 10 gets the job." Once there, NCI captioners estimated that "one hour of live captioning for NCI is as stressful as a day's work as a court reporter."[71] Court reporters who had already endured a decade of becoming "computer compatible" to stave off the electronic recording threat now faced becoming "realtime compatible" all over again. Part of the problem was a basic reorientation in the purpose of stenography, both to discriminate sound-alike words as they occurred and to key them differently to avoid stenonym conflicts: "Training your brain to write in context, not write phonetic sounds," as one realtimer put it.[72] "If you're a working reporter, don't expect to be able to change your entire writing style overnight," warned another.[73] Even the most optimistic estimates of retraining time ranged from six weeks to six months.[74] For this reason, new reporting-school entrants—by this time, almost entirely women—were prime candidates for realtime training: "They often come to us unencumbered with totally ingrained writing styles and are sometimes more receptive to writing style changes."[75]

But the most difficult transition to make for working reporters was the conceptual leap from one product of labor to another, from delivering an official, verbatim transcript on paper to creating an ephemeral, good-enough translation on the screen.[76] As one captioner later explained, "A court reporter's end product is your transcript, and the majority of effort in creating that product happens after your actual writing time. For a captioner, your end product is what you produce while you're writing, so the majority of your effort happens before your writing time." This made it hard for courtroom or deposition reporters to pick up extra work as television captioners: "It's almost a different mental mode that you go into, and I found it very difficult to switch back and forth and gave up verbatim reporting after captioning for a couple of years."[77] Marty Block described the two experiences as radically different back in 1982: "I call this process computer-aided translation rather than transcription because there is no transcript."[78] How could the nation's "transcribing" court reporters be convinced to retrain themselves into "translating" reporters?

## Capturing Classroom Lectures for the Deaf

Redefining captioning as "computer-aided realtime translation" (CART) soon revealed another nationwide use of realtime court-reporting skills, one with far greater diversity than the national or local evening news: on-the-spot translation of live events, in person, for mixed D/HOH and hearing audiences. The first context for CART was in the "mainstreamed" hearing classroom for deaf students. As described above, Ross Stuckless at NTID had first explored this idea in the late 1960s but ruled out a stenocaptioning solution because of the limitations of slow, expensive, and bulky mainframe computers. In the early 1970s, NTID simply settled for a fast typist hooked up to "a machine which transfers the typed words to a television screen in the front of the classroom or to individual television monitors on the deaf students' desks."[79] Even such limited measures were often an improvement on signed interpreting in the classroom; quality interpreters were hard to find (a National Registry of Interpreters for the Deaf had only just been established in 1964), used any of a number of incompatible signing systems (from ASL, a language in its own right, to Manually

Coded English), and could be largely unintelligible to late-deafened adult students (who were less likely to learn sign language than those born deaf).[80] More importantly, a signing interpreter using ASL had to understand the content of a classroom lecture in order to translate it from spoken to visual language—a translation that was not only difficult to perform in realtime for courses in fields such as science and engineering but inevitably limited because of structural differences between the two forms of communication.[81] As described in chapter 2, these were the same objections to sign interpreters that advocates of deaf-accessible television raised in the early 1970s. Besides, even with high-quality sign interpreting, separate notetakers (usually fellow students working for a flat per-course fee) were required to take down the lecture material for D/HOH students.

But by 1981 NTID had acquired one of NCI's InstaText realtime CAT systems, testing it with court reporter Jeanne Matter and thirteen deaf students in standard undergraduate business and psychology courses.[82] In 1983, NTID hired court reporter Linda Miller—who at the time had been "thinking of becoming a teacher of the deaf and leaving court reporting altogether"—to bring this "real time graphic display" (RTGD) system to more classes.[83] After some training by NCI's Marty Block and TSI's Patrick O'Neill, Miller was ready to go, lugging a "a 20-pound CAT writer and another 20 pounds of computer equipment strapped to a heavy-duty handtruck" to four courses per quarter, displaying each translated lecture on a television in realtime (like in captioning) and editing a take-home transcript for the students after hours (like in court reporting).[84] An NTID study of some of her students revealed that "students assigned higher ratings of understanding to realtime print on a television screen provided by RTGD than to [sign language] interpreting. Further, they rated the hard-copy printout provided by RTGD as more helpful than notes provided by paid student notetakers."[85] Miller would go on to live-caption a local Rochester newscast a few years later.[86]

Matter's and Miller's experiences at NTID defined CART as an easy way for court reporters who were tired of the legal system to move their CAT skills to the realtime level. The job had obvious benefits: a local setting easily accessible for part-time work; a welcoming and deserving D/HOH client community; a lower-stress, non-verbatim environment (compared to courtroom work and network captioning, at least); and a possible stepping-stone to local broadcast captioning. The

NSRA was quick to promote this new consumer market for realtime court reporters: "We are the only people in the world who can provide this essential service for this segment of society," wrote NSRA president Jerome Miller in 1989. "Even if helping to develop real-time captioning weren't to our benefit financially and in many other ways, which it certainly is, I believe that it would be our responsibility as ethical professionals to pursue and promote this aspect of our work."[87] That same year the NSRA, along with realtime CAT vendors Stenograph and Xscribe, participated in the Deaf Way conference organized by Gallaudet University to promote captioning technology and labor. Private captioning contractor American Data Captioning performed "live captioning of all plenary sessions in the main meeting hall."[88] Soon the NCRA launched a task force on using realtime CAT in the classroom, "with a goal towards developing a job description, minimum qualifications and salary ranges for this type of service at the college level."[89]

CARTing, like captioning, involved clear differences in experience and skill from day-to-day court reporting. And both the stenographic recording equipment (the keyboard) and the stenographic translation and display equipment (computer and screen) had to be mobile. This meant that CART reporters had to spend extra time and effort to make sure their technology meshed with their location. Even today, as one expert recommended, CART reporters must "go early or even a day ahead of time" in order to "scope out a room to locate the electrical outlets and to see what kinds of chairs are available" and "find a location away from the heavy traffic pattern of the room so that people won't trip over your equipment."[90] In terms of skill, speed was less important than a conflict-free style, requiring the construction and mastery of a much larger data dictionary of stenoform-to-English translations—well above a hundred thousand entries, more than twice the size of a typical dictionary required for courtroom or deposition work.[91] As one reporter explained, "A lot of the words aren't new, difficult or hard to write. They just never come up in the legal setting. A simple word like *cow* conflicts with *could you*."[92]

The passage of the Americans with Disabilities Act (ADA) in 1990 lent a boost to CART just as the NCRA was pushing the new service.[93] Although the act had little to say about captioning—mandating only the captioning of government-funded public-service announcements and the provision of caption decoders on hotel televisions—the act moved the legal language of disability firmly into the civil rights arena.[94] Modeled on both Section 504 of the Rehabilitation Act of 1973 and the Civil Rights Act of 1964, the ADA defined disability as "any condition that impairs major life activities such as seeing, hearing, walking, or working" and introduced the idea of "reasonable accommodation" for disabled workers, students, and consumers.[95] The D/HOH population was included in this definition, and observers pointed to the fact that out of the "43 million disabled" served by the provisions of the act, "the largest subset of this group is the deaf and hard of hearing, representing approximately 24 million people."[96]

The labor market for local television-news captioning remained tight, but stories abounded of court reporters finding CART jobs in their local colleges and universities. At Columbus State Community College, one reporter provided her services to fifty students free of charge for one quarter in 1992, hoping that the school would see the value of her translation work and find permanent funding.[97] Another freelancer broke into CART that same year in Seattle by contacting leaders in the local deaf community: "I began giving demonstrations, first to mainly the deaf organizations and then colleges and universities. I didn't understand what the deaf community and deaf culture were about at first, and I wasn't getting very positive reactions." Then she received a request from a deaf Seattle University student to CART classes for him, "paid by the University as a sign interpreter."[98] Such pay was erratic; in the late 1980s some university CART reporters made only fifteen dollars an hour, while others made up to fifty dollars.[99] In 1991 Ross Stuckless argued that CART reporters should be paid similar to certified sign-language interpreters: $75 for the first two hours (including setup time) and $35 for each hour after that.[100] Today's CART professionals recommend treating sign interpreters not as competitors but as potential colleagues: "Contact every interpreter in your area and exchange cards and brochures with them."[101] But part of the sign interpreter's value was based on scarcity. In 1994 it was estimated that there were only two thousand certified sign-language interpreters working in the United States.[102] Even a decade later, one reporter working in CART commented that "the pay is not as great as reporting depositions."[103] Another noted, "I am not aware of any educational realtimer who receives health benefits, paid vacation and sick leave, or office space from a school."[104]

Even for relatively low wages, reporting students soon learned that informal CART work was an effective way to hone their realtime skills for the professional job market. One student who gained a spot CART-ing English writing classes at a local college—sometimes together with a sign-language interpreter—admitted that because she had only taken one realtime class and had only built up a twelve-thousand word CAT dictionary, "at first I was overwhelmed."[105] Today, colleagues advise fellow CARTers to "appear confident in your skills, even if you're quak-ing inside."[106] An educational CART reporter needs no official certifi-cation (unlike with official court reporting), so individual reporters must judge whether they are breaking "the Golden Rule in realtime": "Don't do it if you're out of your league."[107] One CART advocate lamented that her Web site on the new service attracted the wrong sort of attention: "I received far too many e-mails that said in effect, 'It's so great to know I can have a career with my steno skills even if I can't (finish CR school, get past 180 wpm, pass the RPR, pass my state's cer-tification test).' "[108] Yet even today, according to one university disabil-ity coordinator, when schools create CART positions expecting to hire seasoned stenographers, they can end up competing for labor with the medical-transcription industry, not the court-reporting or captioning industries.[109]

Sites of deaf education for youth were the easiest markets for CART providers to tap, but clients could be found in many areas—especially late-deafened and hard-of-hearing adult clients who never learned a signed language and were not socialized into Deaf culture. Groups like Self-Help for Hard of Hearing People (SHHH) and the Association of Late-Deafened Adults (ALDA) could be found in most major urban ar-eas.[110] The cofounder of the Chicago chapter of ALDA recalled that at their first meeting, communication was so difficult that they set up their own do-it-yourself captioning solution using a hundred-dollar Radio Shack home computer: "We hooked the computer to a televi-sion set and let the [hired sign language] interpreter loose at the key-board. The words appeared on the television screen as they were typed and stayed there until the screen's 10 lines were filled . . . Literally overnight, certain members of the group began to open up and share their feelings about being deaf."[111] A few months later, the Stenograph corporation and the Illinois Shorthand Reporters Association had stepped in to donate equipment and labor for true stenocaptioning.[112]

But while such relationships are often initiated through pro-bono work, the intent is that D/HOH groups will eventually pay for CART service. One reporter from Florida wrote, "I gave first-timers a rate they couldn't refuse, and they were able to either obtain sponsors for following years or put it into their budgets."[113] Trusted reporters might even be hired for work at regional and national gatherings of these societies.[114] At the other end of the scale, individuals involved in local self-help groups might hire a CART reporter with portable equipment to translate for them one-on-one at particularly challenging appointments: "medical appointments, church meetings, funerals, programs such as Alcoholics Anonymous and even such things as police interrogations," according to one reporter.[115]

Such examples illustrate that, as one reporter put it, "being the ears for someone is pretty personal."[116] This was the greatest attraction for many reporters who turned to CART: "One thing about real time that I appreciate is, boy, it is a lot of fun getting away from attorneys. They don't even say thank you, and now I'm getting hugs!"[117] What began as volunteer, part-time work for many freelance reporters, both veterans of the early CAT days and new entrants to the field, now crowded out their regular deposition work.[118] Recognizing the powerful public-relations opportunity that their CARTers represented, the NCRA even began to act as a lobbying organization for students who desired to obtain publicly funded CART service in their schools.[119] But an important question remained unanswered. The D/HOH community, despite its differences and diversity, had joined together to embrace broadcast captioning as a media justice campaign throughout the 1970s. Was CART in the classroom being universally embraced by its clients in the same way?

For now, the answer seems to depend on where in the education system CART is used. The systems pioneered at places like NTID served an acute need in college education for the deaf, which could be traced back to the 1963–1965 rubella epidemic, "through which approximately 8,000 infants were born with some degree of hearing loss."[120] These students came of age just as university-based CART options were first opening up. By the late 1990s, researchers estimated that over twenty-five thousand D/HOH students of varying degrees were enrolled in higher education in the United States, a figure which had "increased dramatically over the past two decades." But since the late 1980s, in both two- and four-year programs, D/HOH students overall had a disturbingly low graduation rate of only 25 percent. CART was most commonly introduced in higher education, both at predominantly D/HOH schools such as NTID and Gallaudet (which together enrolled only about three thousand of those D/HOH students), and at predominantly hearing schools like state universities and local community colleges. In both cases, CART fit well into a long-standing practice of trying to bolster D/HOH graduation rates by providing both professional sign-language interpreters and student notetakers to serve the twin functions of realtime translation and subsequent transcription; according to the National Center for Education Statistics, in 1993, 67 percent of schools with D/HOH students provided sign interpreters and 75 percent provided notetakers.[121] CART was seen as serving a wider audience (clients did not have to know sign language) with a smaller labor force (a single CART reporter could both produce a live display in the classroom and printout a transcript later) so that by 1993, eight major US colleges were piloting CART provision.[122] About a decade later, CART had won enough acceptance that the University of California found itself settling a federal class-action lawsuit with D/HOH students who demanded the right to choose both sign interpreting and CART provision as their ADA-mandated accommodation strategy—regardless of cost.[123] But even by this time, according to NTID researcher Harry Lang, there was "a dearth of research on the effectiveness of such support services as interpreting, note taking, real-time captioning, and tutoring, particularly with regard to their impact on academic achievement" for deaf students in higher education.[124]

Paradoxically, the success of CART in college education has spawned a search for a lower-cost (meaning lower-wage) realtime translation alternative—especially at smaller colleges drawing from smaller geographic areas with a tiny incidence of D/HOH students.[125] Just as in court reporting and television captioning, high costs were linked to a labor-market shortage—not enough trained stenocaptioners to provide CART service in every geographic area.[126] As early as 1989, alternatives to CART were emerging under the name of "computer-assisted notetaking" (CAN). CAN used a sixty-word-per-minute typist and a standard QWERTY keyboard hooked up to a television or computer display, sacrificing verbatim content and speed for affordable cost: "Depending on the skills of the typist and the rate of speed with which the speaker speaks," CAN could "approximate a verbatim transcript if the notetaker types 100 words per minute and the speaker talks slowly."[127] This was the old classroom notetaker position enhanced with live-display technology.[128] Midway between CAN and CART sat a system developed at NTID called C-Print. In this hybrid arrangement, a skilled typist used a QWERTY keyboard (like CAN) but in addition, learned a limited language of shortcuts and "macros" (like CART). According to the system's innovators, Michael Stinson and Ross Stuckless, "such a system uses portable, low-cost equipment, a large pool of potential operators is available, and pay for their services is likely to be significantly less than that for stenotypists and interpreters. Training is brief, perhaps four to six weeks, and many people with reasonably fast typing speeds of approximately 65 words per minute probably have the skills to function successfully as operators in the classroom."[129] In the continuum between the three approaches, then, CAN was least expensive, the least verbatim, and required the least training, but it required the most thought by the notetaker in quickly digesting and summarizing speech; CART was the most expensive, the most verbatim, and required the most training, but it relied on the stenographic principle of writing down phonetic representations of words rather than understanding their content and context; C-Print fell somewhere in between.

College-level D/HOH students are perhaps able to judge for themselves the pros and cons of each method—especially when comparing them to traditional signed-language interpretation and student-labor notetaking. But a very different set of concerns arises when these speech-to-text methods are applied to elementary and secondary school classrooms, where there has not been much experience with CART yet. The first full-time CARTer hired in a Texas public high school started in 1999, only because in that particular school "they had a family that had requested a CART provider for their daughter."[130] Evaluating the place of CART in grade school and high school is difficult because of the uneven geography of D/HOH education in the United States. Pilot CART programs in the late 1980s involved both segregated schools designed for the deaf and public schools oriented toward the hearing. In one case this led to a classroom environment where deaf students were forced to divide their attention between a video monitor and sign-language interpreter over on one side of the room, a speaking instructor over on the other side, and a chalkboard in between.[131] Only in the 1990s have educators on both sides begun to realize that the application of CART in residential or day schools designed for large concentrations of D/HOH students might differ dramatically from the way CART should be used in regular public schools where a few D/HOH students are "mainstreamed" with hearing students.

Debates over deaf education date back to the nineteenth century, largely between manualist advocates of teaching signed languages and oralist advocates of teaching lip-reading and speaking skills (chapter 1). Historian Douglas Baynton described how "sign language enjoyed great popularity and esteem among hearing Americans through most of the nineteenth century and then, near the end of the century, fell into such disrepute that hearing educators and reformers waged a campaign to eradicate it by forbidding its use in schools for the deaf."[132] (Ironically, the oralists were led by communication-technology innovator and avowed eugenicist Alexander Graham Bell.)[133] In the twentieth century, "the number of children taught entirely without sign language was nearing 80 percent by the end of the First World War, and oralism remained orthodox until the 1970s."[134] But whether D/HOH children were taught under one philosophy or the other, they tended to be segregated from hearing children in residential and day schools.[135] This uneven geography largely escaped public scrutiny until 1965, when a report by a special advisory committee chaired by Homer Babbidge Jr. entitled "Education of the deaf" was submitted to the Department of Health, Education, and Welfare.[136] The Babbidge report reopened the deaf-education debate at the federal level, noting that "individuals who are termed deaf may vary widely in degree of hearing loss, in age at onset of hearing loss, in methods of communication used, in their attitudes toward their deafness, and in many other factors."[137] The modern mainstreaming debate that resulted brought both the question of oralism versus manualism and the question of segregation versus integration together in a new way in the 1970s.

This debate began just as television captioning was gaining momentum as a D/HOH justice issue, and just as the medical model of deafness was being increasingly challenged by the cultural model of deafness (chapter 2). As activist Harlan Lane has pointed out, this debate is centered on young pre-school-age children precisely because "there is no simple criterion for identifying most childhood candidates as clients of the one position or the other."[138] Given the demographic reality that 90 percent of deaf children have parents who are hearing, Lane argued, "advocates of the disability construction contend these are hearing-impaired children whose language and culture (though they may have acquired little of either) are in principle those of their parents; advocates of the linguistic minority construction contend that the children's native language, in the sense of primary language, must be manual language and that their life trajectory will bring them fully into the circle of Deaf culture."[139] Thus the question of communication skills (manual versus oral) is linked to the question of social networks (deaf versus hearing), as Katherine Jankowski observed: "For many Deaf people, school is where they meet other Deaf people, often for the first time; at school they develop socialization patterns and friendships that frequently last throughout their lifetimes; there they meet spouses, acquire a language that accommodates their visual orientation, and become a part of a culture that extends beyond the school years."[140]

The debate between socializing children into deaf culture versus hearing culture soon escalated with two key court cases, *Pennsylvania Association for Retarded Citizens v. Commonwealth of Pennsylvania* (1972) and *Mills v. Board of Education of the District of Columbia* (1972), where "it was ruled that exclusion of retarded individuals from a free public education was illegal" and "regular classroom placement was judged to be preferable to special class placement, which in turn was preferable to placement in a residential school or institution."[141] This educational norm for "retarded" children was then applied to all children with "handicaps" in the Education for All Handicapped Children Act of 1975, which, according to educator Donald Moores, was "a mandate for a free appropriate public education for all handicapped children in the least restrictive environment appropriate to an individual child's needs."[142] The effect of this law on the landscape of D/HOH education was swift and dramatic. In 1973, just before the 1975 act, "only about 10% of all hearing impaired children were even partially mainstreamed with their normal hearing peers."[143] By the early 1990s, two-thirds of D/HOH children were mainstreamed in local public schools.[144]

These numbers hide some internal differences, however. Shortly after the 1975 law, Gallaudet surveyed forty-nine thousand "hearing-impaired" students and determined that "a small percentage of profoundly deaf students (18%) are educated even partially in regular classrooms, while the majority of 'mainstreamed' students (62%) exhibit only mild to moderate hearing losses," such that "mainstream programs enroll two to three times as many post-lingually deaf students as do other programs."[145] This trend has continued today; although the number of D/HOH children in the United States still hovers around fifty thousand, Moores wrote, "the numbers of children who are classified as having mild and moderate losses are increasing substantially," which "may partially explain why there is a drop in enrollment in residential schools and why the numbers of children who are receiving instruction orally are increasing."[146] In the post-mainstreaming world, deaf students are more likely to attend residential or day schools together with other deaf, manually communicating peers, while hard of hearing students are more likely to go to public schools along with hearing, orally communicating peers.

This polarization brings great importance to the meaning of the phrase "least restrictive environment" (LRE) in the Education for All Handicapped Children Act and its amended version from 1990, the Individuals with Disabilities Education Act (IDEA).[147] Many educators initially interpreted LRE to simply mean the environment that least restricts one's travel away from home—in other words, the local public school. Under this meaning, "the more an educational environment resembles a normal public school classroom, the less restrictive it is thought to be."[148] But within the Deaf culture movement, LRE meant the environment that least restricts one's communication with peers—in other words, a residential or day school designed by and for the Deaf. In 1988 the special Commission on Education of the Deaf (COED) specifically addressed this debate when it recommended "the Department of Education should refocus the least restrictive environment concept by emphasizing appropriateness."[149] But the 1982 *Rowley* decision—the same Supreme Court case in which realtime CAT had first been used in the courtroom—established the precedent that "it is the obligation of the schools into which children are mainstreamed to give 'sufficient' education rather than to 'maximize' those students' potential."[150] Thus, determining the LRE was left up to the parents of the student, working together with the local school district, in a difficult negotiation between benefits and costs. The only consensus that emerged was the realization that the meaning of LRE could differ for each D/HOH student. As Thomas Kluwin argued after studying over two hundred "hearing impaired" public high school students and their sixty-three teachers, "Since a variety of differences exist between the populations, the real issue should be the appropriateness of the instructional environment for the individual child, not the supremacy of a method."[151]

The debate over the meaning of LRE could not be solved simply by assessing the academic success of mainstreamed D/HOH students. According to deafness researcher Susan Foster, most researchers agreed "deaf students in integrated classes have better academic achievement than their peers in special programs." But the hoped-for rich interactions between deaf and hearing students could not be found—and in an environment where "half of all deaf children in public school have either no deaf classmates at all or very few."[152] As a result, in Foster's estimation, "the personal and social adjustment of these students suffers, especially over time."[153] Or as Baynton put it, "What teachers sometimes call the 'unwritten curriculum,' the casual conversation that takes place outside the classroom and that accounts for perhaps 90 percent of all learning, is lost."[154]

For D/HOH students, mainstreaming helped produce a new, uneven geography of education. As Kluwin described, "Because deafness is a low-incidence condition, it is difficult to collect sufficient numbers of deaf children within an area to operate a program. This has resulted in some instances in isolated, disconnected, and inadequately supervised staff; incomplete service provision for children; and the absence of a deaf community or even deaf adult contact for the children." Trying to consolidate and centralize students, funds, facilities, and expertise in regional solutions pushes the problem the other way: "While a critical mass of deaf children is accumulated to form the nucleus of a deaf peer group, the long transportation times involved militate against participation by these children. Since hearing peers are outside of the deaf child's [local] community, meaningful contact with them is difficult. [And] because of the distances and time involved, such programs can erode family contact and local neighborhood involvement."[155] Claire Ramsey's mid-1990s study of this contradictory geography revealed that a student's deafness was treated quite differently depending on where that student was placed. "In the mainstreaming classroom [deaf children] were defined as children who merely needed their civil right to educational access ensured," but "the teachers in the self-contained classroom, on the other hand, were working toward educational goals. They expected the deaf students to improve their face-to-face communicative competence, acquire and practice basic literacy skills, and learn how to be students who are deaf."[156]

The 1990 IDEA law complicated the debate further by asserting that students had a right to demand "assistive technology" in their LRE.[157]

Researcher Joe Stedt found that in the late 1980s, "mainstreamed deaf or hard of hearing students used educational interpreters in 56% of their mainstreamed classes." But the quality of this interpreting varied greatly. In the 1980s, Stedt found, "most school interpreters have minimal to no training; of those who have training, only a minority are certified specifically as interpreters."[158] And the standard of certification was unclear through the 1990s; both the Registry of Interpreters for the Deaf and the National Association of the Deaf offered different credentials.[159] A study by Hoffmeister of the content of thirteen commonly used special-education textbooks meant to train educators and interpreters on the needs of D/HOH students in the classroom found that "no Deaf person is named as a resource or an author in all the chapters reviewed." He concluded that most of these texts "portray the Deaf as isolated, searching for contact with the hearing," which for Hoffmeister is a clear example of "audism."[160] Even having a certified, sympathetic, well-informed sign-language interpreter was no guarantee that a D/HOH student would learn the currency of the Deaf community, American Sign Language, as over 90 percent of all interpreters were trained in a signed English method.[161] And "because oral interpreters provide no written notes, hearing-impaired students must arrange to get notes from classmates."[162]

CART, C-Print, or CAN could offer much-needed communication options for students in mainstream situations, allowing parents, students, and educators a wider range of possibilities in creating a custom LRE. But CART may exacerbate the socialization problem. As Stedt defined it, "Optimal interpreting would permit participation in a discussion as well as adequate support to allow for the personal involvement with others in the classroom."[163] Communication researcher James Carey has similarly theorized that communication processes may be understood on a continuum from a "transmission model" of communication—just send the data from one mind to another—to a "ritual model" of communication—where the very act of communicating, and all the mental meaning-making wrapped up in it, defines what it means to participate in culture.[164] Even in the best of situations, CART tends to focus only on transmission needs, accurately reproducing lecture materials, and not on the ritual needs, such as interacting with other students informally. As Foster pointed out, "There may be several dimensions to mainstreaming, ranging from access to information in the classroom to full acceptance and participation in social activities," and "access to information is easier to accomplish than social interaction with non-disabled peers."[165]

In the late 1990s, the NCRA consciously rebranded CART to stand for "communication access realtime translation" rather than the earlier "computer-aided realtime translation" because, the association claimed, "the CART provider possesses additional expertise in conveying the intent and spirit of the speaker's message."[166] In bringing CART into the world of D/HOH education, reporters needed to learn which communication functions they could assist in and which they could not. A persistent danger remains that their efforts will be seen as providing a "technological fix" to the "problem" of deafness. One *Newsweek* reporter, discovering in 2003 that "some public schools have begun offering a real-time captioning service," leapt to the conclusion that "for the first time, deaf children are on equal footing with their hearing classmates."[167] But "equal footing" has not yet been granted by CART, or by any other assistive technology, as long as information inequalities in both formal and informal communication persist.

According to one university disability coordinator, demand for CART at the university level will likely increase because D/HOH students will increasingly have exposure to CART services in the public schools. At one large midwestern state university system, having eight full-time students use CART services for a single academic year cost an average of $16,000 per student; however, this was still $4,000 cheaper per student, per year, than using ASL.[168] In an era of state budget crises and declining public funding for higher education all across the nation, this trend may be unsustainable. Yet the right of students to choose CART as part of an accessibility package—accommodating communication both with teachers and with peers—was recognized when Congress reauthorized the IDEA act in 1997 and mandated that a D/HOH child's "individualized education program" (IEP) take into account the "direct communication needs" of the child, including "direct communication with peers and professional personnel in the child's language and communication mode" and "direct instruction in the child's language and communication mode."[169] CART, like captioning, can fit into such a progressive model of communication access, but only if its benefits are understood (and its costs are legitimized) as a tool, not a fix. As educator of the deaf Robert Stepp wrote in 1994, "Our goal must be to graduate [deaf] students who have been brought up to take for granted the use of computers and other electronic marvels so that media will be literally *infused* into their daily lives—into their schooling, their careers, their citizenship. Not only will technology have ensured them a broader background of information, it also will have indoctrinated them with confidence that they have the means to make themselves understood and to share their ideas and skills with the world."[170]

## The Courtroom of the Future

Although realtime CAT had helped court reporters move into the captioning booth and the college classroom, in the early 1980s the system had still not affected practice in the courtroom. As described above, even before NCI began true live captioning of the *ABC World News Tonight*, the InstaText system had received its first real test in the halls of the Supreme Court, assisting a deaf litigator in pleading his case for a deaf client. Similar experiments followed in the next few years. The judge in a Los Angeles murder trial involving a late-deafened defendant who could neither lip-read nor sign gave up on attempts to use a "speed typist" and instead contracted with CAT vendor Xscribe to bring in both a court reporter and a scopist from San Diego.[171] A Wisconsin Court of Appeals judge who was deafened himself in 1983 employed a court reporter with a TSI CAT system working in realtime so that he could continue to hear oral arguments.[172] But realtime reporting in the courtroom generated little interest outside of its application as an assistive tool for D/HOH lawyers, clients, witnesses, and judges—especially when the typical custom-built, realtime-ready CAT system cost around $30,000.[173] As late as 1989, even closed-captioning expert Jeff Hutchins assumed that "except on rare occasions involving deaf participants in the court, court reporters are not required to both transcribe and translate the proceedings."[174]

But the NSRA, still fighting against efforts to replace court reporters with electronic recording (ER) in the early 1980s (see chapter 3), took a different view of realtime CAT. In June 1983 the chair of NSRA's ER Task Force, Bill McNutt, organized a congressional reception at which the NSRA demonstrated the potential of realtime captioning in the courtroom, a demonstration later characterized as having "dramatic" consequences. By November of that year the NSRA had resolved to spend $100,000 annually "to hire a public relations firm to put forward an accurate image of the court reporter" and develop a "model courtroom" project.[175] The result, a humble Wayne County, MI, courtroom (serving

Detroit) transformed into "a laboratory for realtime reporting," was unveiled two years later, in 1985, and christened the "Courtroom of the Future."[176] Further pilot installations followed in 1986: Chicago, Phoenix, and Denver all received their own Courtrooms of the Future as well, each equipped with about $25,000 in microcomputer, display, and networking hardware (although software and training costs could push the cost up to $60,000 each).[177]

What made such installations possible was not only the development of realtime CAT but also the movement of CAT to relatively inexpensive, general-purpose office microcomputers in the mid-1980s.[178] By the end of the decade, a single-user, PC-based CAT system with realtime capability, available from any of a dozen vendors, would cost between $5,000 and $15,000.[179] During the 1980s, increasing numbers of reporters experienced CAT for the first time: in 1985, only about a quarter of all official reporters had used CAT, but by 1990 nearly three quarters had.[180] Beyond hardware costs and labor availability, the final innovation necessary for the Courtroom of the Future was local-area computer networking: some way to link the court reporter's CAT machine with similar machines (or at least displays) on the counsels' tables and the judge's bench.[181] With these three components, "the information entered by a court reporter . . . can be stored, immediately displayed, organized, immediately accessed, and shared by varied persons and functions."[182]

Information sharing sounds nice in the abstract, but how did this really add value in the courtroom—especially in cases where no D/HOH participants were involved? Even the provision of an electronic transcript after the fact was such a change from previous practice that it shook up the division of labor within the legal office. As one lawyer put it, "Just think, no more dictating deposition summaries or yellow-lining key testimony for your secretary to type all over again."[183] In 1991 Idaho became the first state to compel court reporters to make their transcripts available on floppy disk if attorneys requested them.[184] Now with realtime reporting in the courtroom, just as in the CART classroom, having a transcript immediately available—even a "dirty transcript," which had not undergone post-editing by a scopist—removed the need for lawyers and judges to take notes during proceedings. Instead of waiting until the end of a trial, lawyers could take home the day's rough transcript on a floppy disk each night.[185] Some lawyers found that they could even sequester upcoming witnesses in a different chamber but still let them follow along with the trial proceedings, setting up a closed-circuit captioning system of their own.[186] Private bench conferences were also expanded in space and time: "Real time permits second-chair lawyers, paralegals, and the parties in civil and criminal cases to 'be present' at bench conferences by remaining at counsels' tables and reading the content of the sidebar conference on the monitor."[187] One observer of realtime transcription in the courtroom was convinced that "several witnesses have been more forthright and honest because they actually saw their own testimony and knew that it was being reported verbatim" (the same claim made earlier by advocates of ER).[188] But perhaps the best argument for the realtime courtroom was that, in speeding up the process of litigation, it saved the court money. According to one judge, "It normally costs approximately $10,000 a day to run a criminal courtroom . . . If you can save 30 minutes in the course of the day, you're saving $1,000."[189]

Of course, the original idea of the realtime courtroom serving D/HOH participants remained. Ever since the Bilingual, Hearing, and Speech-Impaired Court Interpreter Act of 1979, US courts had been compelled "to appoint qualified interpreters for Deaf people in any criminal or civil action initiated by the federal government." Such practices were strengthened with the passage of the ADA in 1990 (see above), which "require[d] all state and local courts to be accessible to Deaf individuals" by the time it went into effect in January 1992. As in the classroom, realtime stenography helped alleviate a shortage of qualified sign interpreters: "In Massachusetts, for example, the courts filed over twelve hundred requests for signed-language interpreters in the 1995 fiscal year, but only half of those could be filled by the state referral agency."[190] Just as with captioned television, because most late-deafened adults did not understand any signed language they could not benefit from interpreter services. Court-reporting advocates cited the case of Alton Curtis Adams, a nonsigning deaf man convicted of sexual assault in 1986 and sentenced to a jail term of twenty-five years. Retried in 1990 in a realtime-ready courtroom, Adams "was able to follow the events at trial and participate in his defense" and was found innocent.[191]

In promoting realtime CAT, the NCRA attempted to disentangle the duties of a reporter acting as a CART provider for a deaf litigant, lawyer, juror, or judge, and those of a reporter acting as a keeper of the verbatim transcript for the court record. In the first case, just-in-time intelligibility was the goal; in the second case, verbatim accuracy. Because a single person could not fulfill both goals simultaneously, the needs of the Courtroom of the Future demanded at least *two* realtime reporters if D/HOH individuals were to be involved.[192] Such distinctions could be tricky to enforce in practice, especially if the stenographer taking the official transcript was not working in realtime but the CARTer providing communication access was. Court reporter and CART advocate Catherine Bauer recalled that more than once, while she was CARTing in the courtroom for D/HOH clients, judges have asked her to read back testimony because the official court reporter was unable to do so accurately.[193] Such moments highlight the tensions in the shift from the commodity production of the printed transcript to the service production of the realtime translation. If the CARTer's translation appeared "more accurate" than the reporter's untranscribed notes, it could undermine the claims to authority and accuracy upon which stenography had been built.

The NCRA promoted the Courtroom of the Future to court managers as a way to both meet new accessibility requirements and streamline the process of litigation. However, just as with regular CAT a decade earlier, court reporters had to come to their own decisions about realtime. One prominent reporter warned, "Do not expect real-time writing to double or triple your earnings."[194] But there had to be some monetary benefit or else reporters would not undertake the substantial equipment and skill investment to retool for realtime. Key to their calculation was the question of whether, in the end, they were selling a service or a product. Early realtimers were able to command a higher in-court price for their realtime-display reporting, but at the same time, they feared that if lawyers got their hands on dirty-copy transcript for free at the end of each day, they would have less incentive to order expensive official transcripts later.[195] As one report put it, "There appear to be no clear reasons of self-interest for court reporters to routinely practice real-time reporting so long as policies governing transcript production and pricing remain unchanged."[196] One solution was to involve the scopist in a realtime mode as well: while the reporter's computer sent the dirty copy for the screen, a scopist could access that copy from an attached computer and actually post-edit it during testimony.[197] Or sometimes two realtime reporters would trade off duties as reporter and scopist, with the goal of increasing their pooled income while enabling them to cover an entire day's worth of testimony (as in the O. J. Simpson trial described above).[198] By the late 1990s, realtime reporters would even connect with realtime scopists working offsite via telecommunications.[199] Such arrangements allowed realtime reporters to charge their normal prices for finished, daily transcript, never letting the dirty transcript out of their hands.

Keeping control of the dirty transcript was important to court reporters because the sight of a raw, unedited, conflict-ridden, context-dependent first-pass translation might call into question the cornerstone of the reporter's claim that only reporters (and not tape recorders) produced a verbatim transcript. One reporter explained, "Many of us fear exposing our raw steno for what it really is: a hastily written phonetic code that is edited and proofread before becoming a final transcript."[200] Reporters had long dreaded in-court "readbacks," those times of stress when a judge would ask a reporter to stop stenotyping and instantly translate by hand the last few minutes of testimony—or as one reporter put it, "suffer the torments of the damned."[201] Now realtimers talked about the experience of exposing their stenographic skills on the screen in a similar way, describing the experience as "writing naked": "Your writing is exposed, with all the warts and moles you've grown accustomed to living with since you began court reporting."[202] One reporter observed that realtime CAT work required not only skill but "a great degree of poise under fire, since counsel and the court are reading every word moments after they are typed."[203] Many reporters who started realtiming hoped that the association with "high technology" would offset any loss of status through revealing their dirty copy: "The value-added court reporter is a more highly skilled, technologically advanced court reporter," argued one.[204] A commonly held sentiment was that, warts and all, "real-time writers are regarded as the reporting elite" (just as stenocaptioners were regarded as the captioning elite).[205]

In response to these concerns, the NCRA began to subtly change the branding of the Courtroom of the Future idea. By 1989 a writer in the *ABA Journal* wrote, "The computer technology installed three years ago under a pilot program has been copied so often it could be renamed 'courtroom of the present.'"[206] In the 1990s, the NCRA began to refer to it as the computer-integrated courtroom (CIC) instead.[207] Putting the "computer" up front reminded lawyers and judges that realtime reporters who could produce instant transcript fit into the more general legal automation and "litigation support services" from WordPerfect

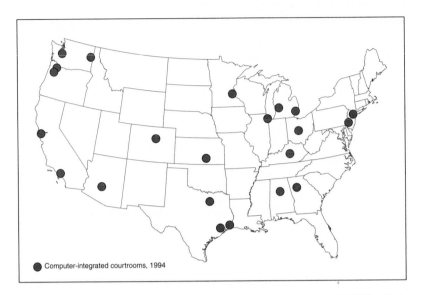

Fig. 4.2. Computer-integrated courtrooms in the United States, mid-1990s. Only about two dozen jurisdictions around the country had experimented with bringing realtime CAT and captioning into the courtroom by the mid-1990s. The majority of realtime work in the courtroom was taking place unofficially at the instigation of individual court reporters and freelance agencies. Source: Map created from data in Frank Andrews, "Computer-integrated courtrooms: Enhancing advocacy," *Trial* (Sep. 1992), 37–39, and William E. Hewitt and Jill Berman Levy, *Computer aided transcription: Current technology and court applications* (Williamsburg, VA: NCSC, 1994)

and Westlaw to Lexis and Nexis.[208] When the New York State Supreme Court installed its own CIC in 1997, named Courtroom 2000, it even added a "digital evidence presentation system" which displayed not only the realtime transcript, but all other sorts of digital video and photographic evidence.[209] By the mid-1990s, there were some two dozen CICs in existence around the nation (see figure 4.2).[210]

Even though court managers, judges, and lawyers were clearly enthusiastic about getting their courtrooms computer-integrated, it was the court reporters—either individually, or through their associations—who ended up instigating, wiring, and paying for most of the early CIC installations (although they often received donations from vendors as well).[211] For example, one reporter revealed that in the New York Supreme Court, "real-time is available only if the court reporter invests in the cabling, digital sharing device, computer and the like to which counsel and the Court may connect their own personal [laptop] computers."[212] The Massachusetts superior courts, employing twenty-five stenographers, were in much the same state a few years later, as "court reporters, nearly all of whom are government employees, have had to pay for new equipment themselves. The Superior Court did buy 25 computer-assisted transcription systems about 10 years ago, but once those machines became obsolete the experiment ended."[213] When the NSRA surveyed the sixty CICs in existence by 1996, it found that 23 percent were funded by donations, 27 percent by the CIC reporter, 15 percent by court-reporter associations, 23 percent through courts themselves, and 12 percent from other nonpublic funds (likely meaning vendors).[214] Yet even though the CIC, like the original CAT-enabled courtroom, was not an expense borne by many public courts, this hardware investment became less of a problem as laptop computers and lightweight flat-screen monitors fell in price in the 1990s. By 1992 an installed system that cost $30,000 a few years earlier could be set up as a portable system for only $15,000.[215] As reporters Olson and Moxham demonstrated in the Simpson trial in 1994, with a few pieces of portable hardware, court reporters (or freelance-reporting agencies) could now create a CIC on demand.

Despite the growing ease of creating a CIC, the NCRA had a tough time convincing its members to take their CAT skills to the realtime level—there were still reporters around who had started their careers as pen writers, unsure about making yet another technology and skill switch.[216] Up until the 1990s there was "no formally designated education process for captioning reporters"; in fact, it had taken the NCI a decade of realtime captioning before it released, in 1991, its own in-house best-practices guide to conflict-free, realtime stenographic theory for the court-reporting community at large.[217] That same year the NCRA began to draft guidelines for its proposed "certified realtime reporter" (CRR) standard.[218] The resulting standard—"180–200 words per minute for a period of 5 minutes with at least 96% accuracy"— meant slightly fewer errors, at a slightly slower speed, than the standard for courtroom and deposition reporting of 225 words per minute for a period of five minutes with at least 95 percent accuracy.[219] The difference in standards might not seem like much, but in realtime reporting—especially for captioning or CART—the universe of possible words was greater and reporters had to compile (and remember) much larger dictionaries. Thus the definition of "allowable errors" was slightly different in the two standards.[220] In April 1992, the association held a training session (including such well-known realtimers as Joe Karlovits and NCI's Tammie Shedd) in Seattle for 120 reporters to jumpstart the certification process. According to one attendee, "We reporters who were working when computer-aided transcription came about remember well the frustrations and the time necessary before a computer could really help us in our work. At the [realtime] workshop, there was an obvious resurfacing of those old frustrations when the participants were jolted into an awareness that a writing style that allows an accurate transcript to be quickly produced usually isn't good enough to be genuinely real-time compatible."[221] Even though the training and test were administered to three hundred reporters, only seventy-two passed.[222]

The push to get a significant number of NCRA members to earn their CRR, thereby making realtime reporting the new standard in the field, met with limited success. In 1993, when the Stenograph Corporation donated equipment to create a CIC at the College of William and Mary Law School, dubbed Courtroom 21, unfortunately "there were a limited number of qualified realtime court reporters in the immediate area" so the courtroom could not be used as intended.[223] A 1994 NCSC report estimated that out of some fifteen thousand official and freelance court reporters working in the United States, only 5 percent had "some experience with real-time reporting" (roughly the same percentage of reporters who still used pen-written shorthand.) In fact, out of the estimated three-quarters of all reporters who used CAT, only about half owned systems capable of realtime performance.[224] Even by 2000, there were only 1,300 CRRs working in the field, less than 10 percent of all reporters.[225] However, this dearth of CRRs did not stop freelance agencies from offering realtime service—even in 1995, in a survey of 144 responding court-reporting firms, 63 percent offered realtime for sale.[226] This may have been because uncertified reporters were doing realtime work, or it may have been that each agency had a single CRR and a low volume of realtime demand. In 1998, two court-reporting instructors surveyed 197 responding CRRs and found that 87 percent were female, 56 percent were freelance reporters, and 66 percent used realtime skills in over three-quarters of their jobs.[227] As with many other aspects of court reporting (described in chapter 3), the demographic differences in court reporters and the workload differences between official and freelance settings translated into geographic differences as well, enabling realtime work in selected, lucrative sites outside of the map of official CICs.

As the data on realtime reporting emerged over the 1990s, it appeared that perhaps a new social and spatial polarization in court reporting was emerging. Even in the original Courtroom of the Future, early realtime reporters felt they had achieved "higher status, higher skill level, and more respect from their peers, supervisors, and coworkers."[228] In part, this may have been because CIC equipment and techniques were often used only on the most complex and/or high-profile cases, lasting long weeks and involving multiple attorneys. CAT pioneer Doris Wong later argued, "Anybody who can write realtime has a master's degree in court reporting, because not only have you spent the two or three years in school, you then have had to, in effect, apprentice yourself for two, or three, or four, or five years."[229] Unlike the rest of the field, where freelance reporters could replace official reporters at a moment's notice, realtime reporters were uniquely skilled, and essential to the working of a CIC: "There is nothing more devastating to the operation of a CIC than to have an important case assigned to that courtroom, have the primary reporter call in sick and have no one able to step in and operate the system, resulting in delay of the case and/or inability to use the CIC."[230] As a result, realtime reporters gained both a higher wage rate and a higher transcript rate from the official courts. Since 1996, "federal courts have permitted us to charge the parties $2.50 per page more for realtime."[231] By 2002, the Judicial Conference had approved "a 10 percent salary increase for federal official court reporters who have been certified to provide realtime services."[232]

Salary and wages are important, but the biggest difference between "offline" CAT reporters (either official or freelance) and "online" realtime reporters (using their own equipment or in a CIC) is the association with digitally networked information technology and the new status and duties that association confers. As one reporter put it, "The CIC changes how the courts do business, and it alters the position of the court reporter in that system. In addition to the traditional function of providing an accurate and complete record of the proceedings, the court reporter becomes the gatekeeper to a fully computerized courtroom."[233] Such a gatekeeper might be responsible for setup, maintenance, and training on all the courtroom's networked computer applications.[234] After all, realtime transcription was not the only new technology to enter the courtroom in the 1990s. PowerPoint and digital evidence both became more common. "In some cases, the courtroom itself is becoming outdated. More courtroom proceedings occur through videoconferences in which a camera transmits a judge's image to lawyers in offices elsewhere."[235] In such an environment, a new ideal role for the court reporter emerges: "The 21st century court reporter is more akin to a data management specialist, facilitating the capture, storage, management, retrieval and communication of information through every step of the legal process."[236] Various important-sounding terms for this new job—information manager, CIC coordinator, litigation manager—floated through the field during the 1990s.[237] But not everyone agreed with the new language. One reporter argued, "Anybody can manage. Anybody. Only talented, highly trained individuals can do the court reporter's vital task: producing the record."[238]

The question of how to position court reporters for the twenty-first century was an important one because the threat of replacement through ER only escalated in the 1990s. A 1994 NCSC survey, which sampled local courts in 27 states, found that in over half the courts surveyed, audio recording was employed "regularly for some case types or by some judges."[239] This new ER threat came not from Soundscriber audiotaping, but from computer-mediated, digital video recording systems. In the early 1970s, as affordable videotape systems were emerging, reporters had rehashed the same arguments that they made against audiotaping in the 1960s: the sound quality was insufficient for good transcription, the audio track could be accidentally or maliciously erased, and expensive labor was needed to tend the camera anyway.[240] Proponents claimed that videotapes preserved crucial visual information in a way that both stenographic and audio methods could not, but reporters replied that the time and space requirements for reviewing videotapes were prohibitive: "Let us say a trial has been taken on videotape exclusively . . . Where does [an appeals lawyer] go through the time-consuming procedure of viewing the trial tapes? Will special viewing rooms be available at the courthouse with more playback machines and monitors? . . . The courthouse might be closed. So eventually he will have to purchase his own playback machine and monitor for $1,000."[241] In a 1973 pilot test of videotaped proceedings in Columbus, OH, one group found that "it took about seven times longer to review a videotape record than a typed transcript, and that the videotape record was boring and often filled with moments when recorded speech was inaudible."[242] Not surprisingly, the NSRA position on video-taping was the same as with audiotaping: "A shorthand reporter shall not arrange for or participate in a videotape recording of a deposition unless a stenographic record is simultaneously made."[243] But some court reporters feared the dreaded day "when proponents of the video-taping of depositions joined forces with the CAT purveyors."[244]

This day came during a 1984 pilot project in Kentucky—a state that, much like Alaska in the 1960s with audio recording, became the focus of nationwide attention from the legal community. In the mid-1980s in Kentucky, court reporting looked much like it had thirty years before in other areas of the United States: out of eighty-nine official court re-porters employed by the state, fully two-thirds of them still used man-ual, pen-based shorthand.[245] Although the Fayette County Circuit Court in Lexington, KY, had spent $100,000 on a pilot CAT system, the system had, according to one official, "fallen apart." Even Kentucky's low-tech reporting did not come cheap. By the 1983–1984 fiscal year, the "direct costs to Kentucky courts for court reporting services" as reported by the Kentucky Administrative Office of the Courts had risen to just over $2 million per year, a nearly 50 percent increase over the figure for 1979.[246] State court managers sought a technological fix for the situation, just as Alaska had a quarter-century before, and found it through the outside vendor Jefferson Audio Video Systems (JAVS).[247] JAVS won a chance to pilot a new computer-controlled videotaping system in the Jefferson County Circuit Court (Louisville, KY)—"four fixed, wall-mounted color cameras in the courtroom and one such camera in the judge's cham-bers, along with six voice-activated microphones in the courtroom and one microphone in chambers"—where a computer automatically sensed speech and switched from one camera/microphone combina-tion to another during a trial. The key benefit to the $30,000 system, be-sides the savings in labor costs, was that the resulting video transcript was supposedly free of the "dead times" and inaudible mumblings of fixed-video systems. After a year the court managers concluded, "The Kentucky courts should commit themselves no further to CAT than they already have"; instead, "statewide use of videotape should be the goal of the court system, and firm steps should be taken to realize the potential benefits of having appellate review on the videotape record, without resort to transcripts, in appropriate cases."[248] Not only would court reporters be eliminated but transcriptionists as well. As one writer

in the *ABA Journal* later noted, with reporters, "litigants pay upwards of
$450 to obtain a written transcript of one day in court, and that may
take weeks or even months to receive. With the recording system, a dub
of the video can be made the next day and costs $15—the cost of the
tape."[249]

The legal world was abuzz. When the JAVS solution "won finalist
status in the 1988 Innovations in State and Local Government Awards
program cosponsored by the Ford Foundation and the John F.
Kennedy School of Government at Harvard University," it was poised
to be adopted by state courts around the nation.[250] A 1991 NCSC study
reported that the Kentucky courts saved a quarter-of-a-million dollars
the first year by eliminating sixteen court reporters. Similar savings
were reported on a smaller scale in pilot video projects in Seattle, WA,
Pontiac, MI, and Raleigh, NC. By 1993 JAVS had systems in "190
courtrooms in 21 states."[251] The NSRA tried to counter the arguments
coming out of Kentucky with reports of its own. Analog video was a
storage nightmare, they claimed: "If a court is storing videotapes for
five years and there are approximately 500 videotapes per courtroom
per year and there are five video courtrooms, there will be 12,500
videotapes by the end of five years. All these tapes have to be logged
and stored in such a way that they can be retrieved in a timely fashion
whenever needed."[252] By 1992 an NCRA-sponsored study conducted
by the Justice Research Institute claimed that videotaping was $50,000
more expensive per year than court reporters using a (non-video) CIC
system. NCRA president Roger Miller even claimed that in the automated JAVS system, "attorneys learn to trick the system by, for instance, dropping a book just as damaging testimony comes out."[253]
But even as association leaders made such claims against video, others
suggested that "reporters need to view ER as a tool rather than a threat.
Many companies are actively working to combine the two technologies. How much more effective is a video tape when it can be linked to
a reporter's electronic transcript?"[254]

Thus a third attempt to brand the Courtroom of the Future emerged
in the 1990s to take this potential into account. The "total access courtroom" (TAC) brought together both the "information management"
services of realtime reporters, as well as the "information justice" services of realtime CARTers.[255] But crucially for the NCRA, the TAC also
acknowledged a permanent place for the videotaping of proceedings—
as long as that videotape was closed-captioned by a court reporter.

According to courtroom realtime pioneer Karlovits, "The TAC shall integrate real-time text with simultaneous videotaping of the proceedings" such that the videotape may be indexed by the text, and "the videotape shall contain captions of all proceedings open to the public or permitted by the Court."[256] This compromise was attractive to lawyers who disdained wading through hours of cheaply videotaped proceedings to find a particular moment, phrase, or fact.[257] In the TAC, "once the user has performed a search for a word or phrase on the transcript, the computer can then use the signals to locate that testimony quickly on the videotape."[258] One writer in the *ABA Journal* described such a TAC in the Los Angeles superior court as "a combination of the best features of audio, video and CAT in an integrated system that eliminates the individual weaknesses of the three alone."[259]

One of the first TAC projects came about in September 1992 through a partnership between the Certified Shorthand Reporters Association of New Jersey and the Stenograph Corporation (with the help of some high-profile realtime reporters, including Karlovits).[260] The timing of this demonstration was crucial—in 1991 New Jersey had just begun the process of "winnowing out court stenographers and replacing them with audiovisual tape recorders."[261] Reporters in the Coalition to Preserve the Record had been waging a publicity campaign against ER, and losing (see figure 4.3). But the TAC project provided a unique compromise. Because the NCRA's TAC, like the JAVS video system, involved physically integrated video and audio equipment—inevitably tied to the visual and acoustic qualities of a particular courtroom space—this new kind of CIC was a return to the model where the court would have to invest in capital equipment rather than the court reporter: "The cost could range from $5,000 for a system displaying text only on monitors at the judge's bench and attorney tables to more than $60,000 for a court where video and textual records are synchronized." The NCRA made sure to add that even in the TAC, "The initial computer system is already purchased by each individual court reporter—no cost to the State of New Jersey."[262] A decade later, the TAC idea had not stemmed the loss of court-reporting jobs completely—from a high of 205 official New Jersey reporting jobs in 1991, only about a third had survived—but at least three dozen realtime court reporters had found employment with the state courts, alongside another three dozen non-realtimers. The realtimers earned from $6,000 to $8,000 more than their more traditional counterparts,

# The Tapes Of Wrath

Fig. 4.3. Anti-electronic recording flyer distributed in New Jersey, early 1990s (detail). The rebranding of the realtime CAT and captioning as the "total access courtroom" coincided with renewed efforts to replace court-reporter labor with computer-controlled videotape systems, recalling the audiotaping debates of the 1960s. Source: NCSC and NCRA, *Technologies in court reporting* (Denver: NCSC; Vienna, VA: NCRA, 1991)

perhaps a factor in the state's goal "to have at least one real-time reporter per vicinage" but not to completely transform its court reporting labor force.[263]

Although videotaping has by no means left the scene, the new computerization of courtroom procedures—whether under the banner of the Courtroom of the Future, the computer-integrated courtroom, or the total-access courtroom—granted realtime reporters a clear occupational advantage by the late 1990s. One reporter described how her area "was one of the first Florida counties to install a complete video system in a circuit courtroom, leading me to believe that my position as one of seven officials working for the Eighth Circuit was on the way out." But after convincing the judge that she could report in realtime and complement the video record, "not only did I not lose my work in the courts, but the administrators took me on as the county's only full-time salaried court reporter."[264] Reporters who could make this transition found that their work took on a new dimension, both in the taking down and writing up phases; they were the managers of audio-video technology at the time of recording and the indexers of the outputs of this technology after the fact. The court-reporting industry has now come to accept, if not embrace, video—as long as it goes hand-in-hand with realtime captioning. Court-reporter textbooks proclaim, "Video recording enhances what is called litigation support—those extra services offered to lawyers by court reporting firms." Even the NCRA now offers a "certified legal video specialist" (CLVS) credential.[265]

Like the change from machine stenography to computer-aided transcription in the 1970s and 1980s, the switch from "batch" CAT to realtime CAT in the 1990s—even when made together with video—was neither complete nor conclusive. As of 2003, according to the NCSC, "while federal courts provide real-time reporting in nearly all courtrooms, most state courts use a mixture of tape recording, real-time reporting and conventional reporting."[266] At the same time, only "one-quarter of the courts in the nation's 94 federal districts have at least one high-tech courtroom."[267] In such computerized spaces, the role of the court reporter as "information manager" is far from assured. In the layouts and diagrams included with the 1999 Administrative Office of the US Courts "Courtroom technology manual," the video processing, audio recording and amplification, and network computer systems are to be managed by dedicated professionals working in custom-built spaces.[268] Reporters, despite their new technology, are still relegated to their traditional position off to the side of the bench.

In 1993, after realtime reporting had emerged in the captioning context, matured in the classroom context, and returned to the courtroom context, one reporter predicted the future might see all of the professional stenographer's technological tools folded into a single, digital, portable, and ergonomic artifact dubbed the Stenochair: "computer and double-sided screen for viewing from both sides, CD WORM with scanner and player, a phone, a fax, a modem, and video with video-text synchronization," conveniently "encased in a portable, foldable chair" with touch sensitive stenotype keys on each armrest, all of which "weighs less than the former steno machine case when fully loaded." The Stenochair represented an optimistic future for court reporters in which the threat from videotaping would be eliminated through ubiquitous live-captioning; it would "allow simultaneous digital videotaping/recording of the proceedings in synchronization with real-time text via the computer."[269] This vision also carved out a literal place for the court reporter in any number of public and private situations where a speech-to-text translation might be needed. Have Stenochair, will travel.

However, as this chapter has shown, the development of realtime CAT technology and practice in the three contexts of television captioning, college education, and court reporting left a more complicated and contradictory legacy. New national-scale opportunities opened up for realtime reporters to work as closed-captioners on network newscasts, but at the same time, prewritten text fed through teleprompters at the local level kept the bulk of the profession from tapping this new potential market. New demands for disability accommodations in higher education put realtime reporters in direct competition with signed-language interpreters and paid notetakers, but across the landscape of public and private elementary and secondary schooling, the role of CART in the "mainstreamed" classroom remains controversial. And even though realtime reporting seems to be a useful component of the "courtroom of the future," there is no guarantee that another recording technology—like automatic speech recognition—might not someday threaten the reporters' core market once again. Captioning expert Jeff Hutchins encapsulated the profession's perennial dilemma in 1989 when he argued, "If there's a weak link in today's real-time captioning method, it is the human one. The use of people instead of machines to input the sounds and words to the captioning computer does not, as some people have believed, hurt the accuracy or lag times of captioning. To the contrary, the human factor almost certainly enhances those two measures of captioning quality because of the human's infinite ability to adapt and be flexible. But people weaken the chain because they are fallible—they have bad days; they get sick or take vacations or resign, leaving the captioning company to spend weeks or months preparing replacements; and they demand to be paid."[270]

That question of payment still loomed as a key issue in television captioning, educational CARTing, and realtime reporting. Shortly after realtime captioning was inaugurated on television by NCI in 1982, television producer and entrepreneur Sheldon Altfeld (a hearing person) decided to start a sign-language-based cable television network for the deaf called the Silent Network.[271] Finding that he needed to open-caption certain programs, he turned not to the postproduction offline captioning community represented by most of the writers at NCI and WGBH but to the new online captioning community growing out of court reporting. Why? "When one considers that it needs between 30 and 40 hours to caption a one hour program in comparison to only one hour with 'Real Time', it is quite obvious that if time is the critical factor, then the system [the Silent Network] uses must be seen as a clear winner."[272] Using realtime captioning techniques for postproduction captioning needs might have saved costs in some situations, but it also undermined all of the skill and artistry that previous rounds of captioners and subtitlers had developed ever since the end of silent film. Careful timing, editing, and proofreading in the video production process were replaced with dictionary-building, "just-in-time" performance, and, if viewers were lucky, post-editing for all showings after the first.

No matter how skilled the realtime stenocaptioners were, the nature of computer-aided stenography and live human performance ensured that some errors would occur. At NCI in 1985, during a sober discussion about the Supreme Court, the phrase "under Warren Burger" had gone out live nationwide as "underwear even burger."[273] During the O. J. Simpson trial a decade later, defense lawyer Howard Weitzman's name appeared on the realtime screen as "Mr. Whites Man."[274] For the first time, the errors evident in closed-captioned text—on television, in the classroom, and in the courtroom—became the stuff of urban legend. Even on the space and time of the television screen itself, the newer and more chaotic "roll up" method of displaying live captions as continuously scrolling text on four lines at the bottom of the screen began to take attention away from the more carefully timed and positioned "pop-on" captions that appeared on the screen next to the person speaking.[275]

By the end of the 1990s, even the top for-profit captioning company acknowledged in its official training materials that intelligibility with the presence of error, rather than the lack of error itself, should be the realtime captioner's goal: "Realistically, since translation cannot always be perfect, it should be at least phonetically readable to the caption viewer, and never offensive."[276] But even properly displayed realtime captioning could be misunderstood. At the public memorial service marking the death of Senator Paul Wellstone (D-MN) in 2002, attended by an estimated twenty thousand supporters at the University of Minnesota Williams Arena, one "Republican political analyst and lobbyist" misinterpreted the live, big-screen CARTing of the event—which included the standard audio description terms indicating when the audience was clapping or laughing—as some sort of "outrageous" direction *telling* the audience to clap and laugh on cue. What was perhaps an honest mistake was then repeated on C-SPAN's *Washington Journal* by a pundit who claimed "the people who were in attendance were told by screen when to cheer and when to jeer."[277] As this unfortunate episode illustrates, even though realtime tools and techniques have brought captioned content into more contexts, to bear on more social problems, and in view of more varied audiences than ever before, an understanding of how captions are produced behind the screen is necessary to fully judge their meaning on the screen.

# Chapter 5

# Public Interest, Market Failure, and Captioning Regulation

Closed-captioned television just might be the most underrated technology of the past decade.

*Article in* The Reading Teacher, *1991*

In January 1986, *The Cosby Show*—a half-hour, family-oriented sitcom featuring comedian Bill Cosby and a talented cast of fellow African American actors—had captured the imagination of audiences to become NBC's highest-rated program. It was also a financial windfall for the previously last-place network. When this prime-time staple was first closed-captioned at the start of its 1985–1986 season, it was raking in advertising revenues of $1.5 million per week; the captioning labor cost of $1,250 per week was a trivial expense by comparison. Nevertheless, NBC refused to pay for this cost, arguing that its only responsibility to its viewers was to pay for "encoding the signal that the network transmits over the air," at roughly $500 per episode. Instead, NBC said, Casey-Werner Productions and part-owner Bill Cosby should foot the captioning bill, as NBC already paid them $500,000 per episode for the right to screen the show each week. Besides, the production company stood to reap a syndication price of $1 million per episode when the series entered reruns. In the end, *Cosby* retained its captions, but its producers paid for only half the cost. The rest of the bill was picked up by the captioning agency, the WGBH Caption Center—a nonprofit, publicly funded, PBS affiliate.[1]

This was not the first time NBC had balked at paying captioning costs. Just two years after closed captioning had begun, in February 1982, NBC Vice Chairman Irwin Segelstein sent a letter to the National Captioning Institute (NCI) president John Ball informing him that, because of disappointing decoder sales, the network had decided "to withdraw from the project," saving some $520,000 per year.[2] A hastily prepared counterproposal by Ball, as well as the fact that "200 people from two schools for the deaf picketed outside the New York headquarters of NBC's parent company, RCA," led NBC to postpone its decision.[3] But by the start of the 1982–1983 season in August, the network—"in last place in the ratings and cutting costs across the board"—announced it was indeed abandoning NCI.[4] In a congressional hearing a year later, an NBC executive explained that his network's participation "was expressly conditioned on the representation that the hearing impaired public would obtain at least 100,000 decoders . . . a year," but "today, almost 4 years after the project commenced, we understand that there are fewer than 75,000 decoders—or less than 20 percent of expected demand." As a result, NBC remained "skeptical . . . of the extent of interest by the hearing impaired in closed captioning."[5] Yet even as the network pleaded poverty, it was pushing forward with another Line 21 innovation that brought text to television screens in a different way. In spring of 1983 "NBC became the first broadcast network to produce and transmit for demonstration purposes a full high-resolution teletext service," some eighty "pages" (screens) of text and graphic information delivered over the TV, containing magazine-like sections for news, weather, sports, money, people, "your body," "living," horoscope, kids, travel, movie reviews, and soap opera reports. Unfortunately, because no affordable decoders for this teletext service were available for purchase, "the viewership for teletext on the networks was virtually nil."[6]

As the NBC saga illustrates, even with NCI's new ability to realtime-caption news programs—as well as other popular live events, such as sports—the chicken-and-egg problem in closed captioning persisted. From the networks' point of view, not enough deaf and hard-of-hearing (D/HOH) households were purchasing caption decoders; from the D/HOH point of view, not enough programming was captioned to justify such an expensive purchase. The networks bristled at the suggestion that the government might mandate the captioning of all television content, however. ABC argued such regulation was "unnecessary in light of the success of the voluntary approach," and CBS said that "mandatory captioning would infringe upon First Amendment rights of broadcasters and producers, placing an economic burden upon them that often would not be justified."[7] Yet, at the same time, the networks were more than willing to invest in risky information-technology ventures, like teletext, that might reach a larger, more affluent, and more evenly distributed mainstream audience. Through the 1980s, as both techno-logical and legal challenges kept the fate of closed captioning in doubt, the federal government refused to step in, either to mandate the free distribution of caption decoders or to force the universal cap-tioning of all television broadcasts. However, as the 1990s opened, something changed. Suddenly Congress and the Federal Communica-tions Commission found the will to re-regulate closed captioning on both the demand side (decoder purchases) and the supply side (cap-tioned content). But this only happened after the "public-interest" beneficiaries of closed captioning were redefined as not only a rela-tively small (and economically powerless) D/HOH audience but also a larger (and potentially lucrative) community of students learning to read and immigrants learning to speak English. This solution drew on the long history of captioning for language translation and literacy training pioneered by cinematic subtitlers and deaf educators.

### The Resistance to Government Regulation

Everything that the D/HOH education community had accom-plished with captioning in the 1970s—from the non-exclusive FCC authorization to use Line 21 of the vertical blanking interval (VBI) for closed-captioning signals, to the nonbinding agreement by the net-works that they would contribute money to closed-caption television shows through NCI—had been a compromise between public-interest goals and private-industry profitability. Even before the rise of neolib-eral economic policy with the Reagan administration in 1980, the benefits of captioning to the D/HOH community were always calcu-lated against the potential loss of audience (and advertising revenue) to the broadcast lobby. Yet since the Nixon administration in the early 1970s, with its hostility toward state-funded media, the Public Broad-casting System (PBS) had advocated for captioning under the rhetoric of taxpayer-supported broadcast justice. For this reason, it is rather sur-prising that the main attempt to force the federal government to legis-late a solution to the captioning chicken-and-egg problem of the 1980s was rooted in a 1970s D/HOH lawsuit *against* PBS.

What would eventually grow into a decade-long legal battle to force public and private TV broadcasters to provide captioning—involving the FCC, the Supreme Court, the Department of Health, Education, and Welfare (HEW), and the Corporation for Public Broadcasting

(CPB)—began as a single complaint against Los Angeles PBS station KCET for not airing the open-captioned and time-delayed WGBH version of the *ABC Evening News* in the mid-1970s. When Sue Gottfried, a hearing-impaired member of the Greater Los Angeles Council on Deafness (GLAD), requested that KCET carry the program, the station's refusal touched off a round of picketing by the D/HOH community, which, in May 1977, resulted in a victory as KCET (and, presumably, the local ABC affiliate) agreed to carry the news broadcast.[8] This might have ended the matter had it not been for two preexisting laws that GLAD now decided to use. The first was the recently enacted Rehabilitation Act of 1973, called "the first civil rights law for the protection of the handicapped" in the United States (see chapter 4).[9] The crucial portion of the act was Title V, Section 504, otherwise known as "the handicapped person's 'Bill of Rights,'" which "prohibits discrimination against qualified handicapped individuals in any federally assisted program or activity."[10] As a writer in the *New England Law Review* explained it, "Section 504 was designed to prevent employment discrimination against handicapped individuals. In 1974, however, the Act was broadened to prohibit all forms of discrimination against handicapped individuals."[11] Originally, no agency was designated to enforce the rule, but in 1976, an executive order designated HEW as having enforcement responsibility.[12] Gottfried and GLAD reasoned that if television stations— both public and private—could be considered "federally assisted," then they should have to caption all programming to avoid discriminating against the deaf. The second law that GLAD used was a provision of the original Communications Act of 1934 that established the FCC and granted it the power to issue and renew broadcast licenses, as long as those licensees served the "public interest, convenience, and necessity."[13] If a substantive public-interest complaint could be brought against a broadcaster—again, public or private—then the FCC was empowered to hold a license-renewal hearing and potentially revoke the license. Such happenings were rare—between 1968 and 1978, "out of the approximately three thousand renewal applications considered each year" from both radio and TV broadcasters, only sixty-four stations lost their licenses over that decade—but the possibility was there.[14] Gottfried and GLAD decided to try to beat the odds. In October 1977, they filed petitions with the FCC to deny the license renewals of KCET plus seven commercial Los Angeles TV stations for failing to open-caption their programming (closed captioning had not yet begun).[15]

After nearly a year of consideration, the FCC ruled that it would not even hold public hearings on the license-renewal question. Regarding the private stations, the FCC disagreed both with the argument that their license to use the public airwaves amounted to "federal assistance" under Section 504 and with the claim that these stations had failed in their duty toward the "public interest" in serving the deaf, as there was no specific FCC rule they had violated (these stations had met the FCC's existing emergency-information guidelines for the deaf, as described in chapter 2).[16] Regarding the public station, while the FCC agreed that KCET received "federal assistance" for up to 30 percent of its budget, it ruled that the funding and monitoring agency—HEW—needed to find a violation with Section 504 for the FCC to take any action.[17] The case was appealed to the District of Columbia Court of Appeals in 1979, resulting in a partial victory for the petitioners in April 1981: although the FCC's renewal of the private station licenses was upheld, the renewal decision on KCET was reversed. Two of the three judges believed that Section 504 did impose a legal obligation on the public station.[18]

Both sides appealed to the US Supreme Court.[19] Even though the justices agreed that "there was no question that the public interest would be served by making television broadcasting more available to the hearing impaired," the Court argued that "it was unfair to criticize any licensee, whether public or commercial, for failing to satisfy requirements of which it had no notice."[20] Thus the Court refused to hear arguments concerning the private stations (letting their license renewals stand) but agreed to consider the case of the public station.[21] In February 1983, by a 7-2 vote, the Supreme Court reversed the Appellate Court decision and granted license renewal to KCET, ruling on the question of who had authority to enforce Section 504 regulations but sidestepping the debate over the substance of the regulations themselves.[22] In effect, "the Court found no congressional intent to have the FCC enforce the Rehabilitation Act" as "enforcement comes from the funding agencies, and the FCC is not a funding agency."[23] However, "the Court did concede that if a licensee was found guilty of violating the Rehabilitation Act, the FCC would clearly be obligated to consider its possible relevance in ascertaining whether to renew the violator's license."[24] In other words, Section 504 may have been violated, but it was not the FCC's job to make that finding; thus until someone *else* declared a Section 504 violation, the FCC could not use captioning as a criteria in license renewal.

In this chain of legal decisions, the issue for the courts was not whether the deaf deserved broadcast justice but rather who in the federal government had been assigned the responsibility to make and enforce rules about broadcast justice. Back in 1977, when HEW had issued its first regulations on implementing Section 504 (four years after the act was passed), it used language mandating "reasonable accommodations" for the handicapped unless those modifications constituted an "undue hardship" on the affected organization.[25] Both terms were subject to debate, but that debate could hardly begin if no agency would take further responsibility for the application of Section 504 to broadcasting. In 1978 when Gottfried petitioned the HEW to declare KCET out of compliance with Section 504 so the FCC could then make a license-renewal decision, HEW refused to make any ruling without "a policy clarification"—from the FCC!—"as to how the Rehabilitation Act related to public broadcasters."[26] In response, GLAD brought a second suit against KCET, PBS, CPB, and HEW seeking someone who would issue specific regulations on how Section 504 applied to broadcasters (PBS and CPB were dismissed from the case, but the Department of Justice was added).[27] The case dragged on so long that HEW was actually taken off the hook by both administrative changes in the Reagan administration and technological changes in the world of captioning. In May 1980, HEW was split into the Department of Health and Human Services and the Department of Education (DOE), with Section 504 handed to Education. Six months later, responsibility for drafting Section 504 regulations was transferred to the Attorney General's office.[28] By the time DOE got around to presenting its firm opinion on the case in August 1981, its officials felt that NCI's new closed-captioning system had rendered the entire case moot: "KCET [now] satisfied the requirements of Section 504 by airing the 'closed captioned' versions of programs supplied by the Department of Education, [and] the station did not need to add its own captioning to other programs in view of the substantial financial burden such a requirement would impose."[29]

This was all too much for the Ninth Circuit District Court, which was hearing the case, because in November 1981 it surprisingly ruled that "the government was discriminating against hearing-impaired viewers by failing to promulgate regulations." The court froze all $14 million in funding for public broadcasting and ordered not only DOE but the FCC and the Department of Justice to issue specific regulations on the matter.[30] The court ruled that because of the high cost of closed-captioning decoders, and the (statistical) lower economic resources of the hearing impaired, only open captions met the definition of "equal access."[31] This striking judgment did not last, however, and was overturned entirely exactly two years later (with further appeals declined by the Supreme Court).[32] As in the previous case, the higher courts dodged the central issue of what broadcast justice might actually mean, wrestling instead with the "legal paradox" that "until there is a specific governmental standard as to what constitutes compliance with the nondiscrimination dictates of the Rehabilitation Act, the act will not be enforced against public or other broadcasters."[33] As one legal observer concluded in 1984, "Broadcasters are unwilling to provide captioning without a mandate and clear guidelines from the FCC, and the FCC, in the present period of deregulation, is not apt to formulate such guidelines absent an order from Congress or the courts to do so."[34] The FCC did, however, amend its procedures in two ways that may have had something to do with the Gottfried and GLAD cases. First, in 1980, the commission specifically rejected proposals to include the "handicapped" in its list of relevant members of the population whose needs had to be "ascertained" by broadcasters in order to prove they were serving the public interest, "finding that the handicapped are not a significant group in all or most communities."[35] (Here again, the spatial dispersion of D/HOH populations, not just their overall numbers, worked against their interests.) Second, in 1981, the commission adopted new rules making it harder for activists to affect station license renewals by allowing "ninety-five percent of television and noncommercial radio licensees" to file for renewal using a simple, five-question form.[36] With two legal loopholes closed, and a case precedent established all the way to the Supreme Court, D/HOH advocates for government regulation to increase access to captioning decoders or provision of captioned content were stymied.

## Trying to Sell the Caption Decoder

Removed from such legal challenges, NCI had one primary strategy for its own survival in the 1980s: increase the sales of caption decoders. NCI president John Ball recalled that in 1972, National Bureau of Standards (NBS) engineers had bragged that "a five dollar decoding chip was only a year away."[37] A decade later, though, that chip was nowhere in sight, and instead the bulky add-on TeleCaption decoder still sold at Sears for just under $300 (the overly expensive decoder-within-a-TV had been discontinued after its first manufacturing run of eighteen thousand).[38] "We have barely scratched the surface in reaching the hearing impaired people of this country," lamented Senator Tom Harkin (D-IA) in 1983.[39] Only about 73,000 decoders had been sold in all by the end of that year, and only 20 percent of the general public was even aware of the term "closed captioning" according to one NCI survey.[40] NCI, having long abandoned its original goal of selling a hundred thousand decoders per *year*, was desperate to finally exhaust its original run of a hundred thousand first-generation decoder circuit boards in order to fend off claims that it was pushing obsolete technology.[41] Finally, in late 1985, NCI announced that it had decoders in a hundred thousand homes—representing 400,000 total viewers, 250,000 of whom were deaf or hard of hearing—and that the D/HOH community had invested some $30 million in the effort (at $300 apiece).[42] John Ball set a new goal for NCI, a target of 500,000 total decoder households by 1990, at which point he claimed NCI would "relieve the federal burden" and "become a commercially attractive service," representing "a sufficiently large audience that commercial networks, advertisers and the like will be willing to spend money to have programs captioned to attract these particular viewers."[43]

But who and where were these viewers? And why did some buy decoders while others did not? As described in chapter 2, NCI tried to find out by surveying decoder-owners who returned the customer-response cards included with almost every unit. By the early 1980s it was clear that "households with decoders tend to consist of young, well-educated people who have a relatively high income": 48 percent were under thirty-five years of age; 40 percent of had attended college, and 63 percent had a household income over $20,000 per year.[44] Not only were these the elite of the D/HOH population, but they were unevenly distributed across the country. Counting nearly all the TeleCaption I sales, the top five states ranked by raw decoder households were California (11.6% of all decoder households), New York (9.1%), Illinois (5.6%), Pennsylvania (5.5%), and Texas (4.7%).[45] But considering decoder households per capita results in a different ranking. Using 1980 census figures and computing decoders per ten thousand persons, the top five states were Iowa (7.3 decoders per 10,000 persons), Nebraska (7.2), Minnesota (6.1), Maryland (6.1), and Connecticut (5.1)—with Washington DC (5.2) in this range as well (see figure 5.1).[46]

Clearly, closed captioning had reached deaf individuals much more readily than hard-of-hearing (HOH) individuals; even though "hard-of-hearing people represent about 85% of the hearing-impaired population of the United States," NCI found that "83% of decoder owners consider themselves deaf rather than hard-of-hearing." In order to survey this HOH population, NCI targeted households with members who were affiliated with the group Self-Help for the Hard of Hearing (SHHH) or who had recently purchased hearing aids. Of the 468 respondents they found, 59 percent were over sixty-five years of age, and 70 percent had a household income under $25,000. They subscribed to cable and owned VCRs at about the same rate as the general population. Some 84 percent of these HOH respondents reported that they "have trouble hearing television," but turning up the volume and using hearing aids were their preferred solutions. "More than half (52%) of the [HOH] respondents had never heard of closed captioning, 79% had never seen it, and 92% did not own a decoder."[47] This helped explain why, even though hearing-impaired respondents reported "nearly one hour more of daily viewing than the hearing," decoder sales were not meeting expectations. As a 1984 study in the *Journal of Communication* put it, "Whether or not a given program is closed captioned does not seem to fully explain the viewing behavior of the hearing-impaired."[48] Such research led NCI away from its earlier, more-ambitious estimate of a decoder audience numbering up to 13 million, and to a more conservative estimate of 4.75 million persons "who could benefit from captions"—this time including only 25,000 of the 5.6 million Americans with "minor loss" of hearing.[49] In this way, NCI finally embraced, a decade later, Donald Torr's original 1974 estimate of a closed-captioning consumer market of around 4 million D/HOH viewers (see chapter 2).[50] At least Ball's new goal of reaching a total of 500,000 decoder households by 1990 seemed a bit more attainable—NCI only had to convince 10 percent of the likely market for decoders to purchase one.

NCI still needed assistance to reach this smaller market, however. In May 1983 Congress heard that "NCI's research department has estimated that only 14 percent of the households across this nation who

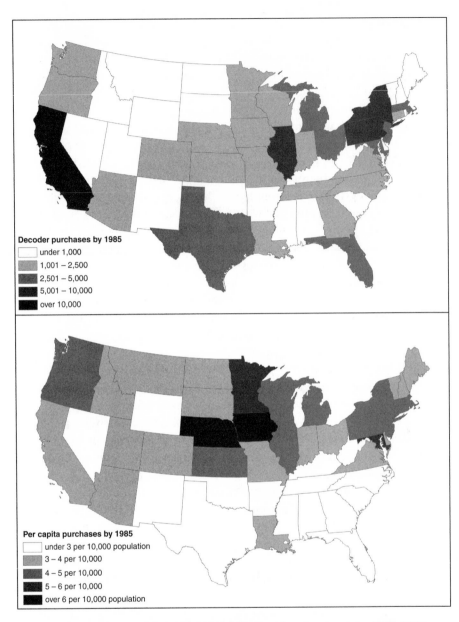

**Decoder purchases by 1985**

- under 1,000
- 1,001 – 2,500
- 2,501 – 5,000
- 5,001 – 10,000
- over 10,000

**Per capita purchases by 1985**

- under 3 per 10,000 population
- 3 – 4 per 10,000
- 4 – 5 per 10,000
- 5 – 6 per 10,000
- over 6 per 10,000 population

Fig. 5.1. Purchasers of first 100,000 TeleCaption decoders by state, 1980–1985. While the most purchases were made in the same states having the greatest total population of deaf and hard-of-hearing individuals, per capita purchases revealed a different pattern, with proportionately lower rates in the Sunbelt. NCI found that the geography of broadcast-captioning consumption was unevenly concentrated in deaf households rather than in hard-of-hearing households. Source: Map created from data in NCI, *Characteristics of the audience for closed captioned television on Dec. 31, 1984,* report 85-1 (Falls Church, VA: NCI, 1985), and US Census Bureau, *Decennial census of population, vol. 1: General social and economic characteristics, part 1: United States summary* (Washington, DC: GPO, 1980), Table 230

want and need closed caption TV will ever be able to afford to buy [the decoder] at present prices." Key groups needing "incentives" to purchase decoders were "under-employed heads of households," "parents of a whole family with only one hearing impaired child," and "elderly on fixed incomes." Therefore, in FY 1984, NCI asked the Department of Education for $1 million to direct-buy decoders at half the retail price from the manufacturer and place them in ten thousand new homes, at a special cost of $100 each.[51] NCI also looked to "foundations and corporations" for such subsidies, eventually "running programs in New Jersey, Delaware, North Carolina, Georgia, Louisiana, and in D.C., Detroit and Chicago to help people get the decoder for as low as $35."[52]

Besides securing decoder subsidies, various captioning agencies and D/HOH groups made a greater effort to build captioning awareness. In 1984 the Caption Center released *The Caption Workbook,* a twenty-eight-page children's activity book filled with cartoon-like illustrations, intended to train children to interpret captions on television—four years after closed captioning had begun.[53] By 1986 WGBH had created an eighty-dollar "Caption Kit," sold through the National Association of the Deaf, including "a 15 minute color videotape and three booklets" designed to introduce kids to captioning. The video, "Caption Marvelous," featured "Bernard Bragg as a deaf superhero who teaches two deaf children how captions can open a new world of understanding."[54] For the adults, the hard-of-hearing group SHHH released a "decision tree" in 1985, meant to help users sort through eleven different combinations for connecting a caption decoder, a television, a VCR, and a cable box together in the right way. Even five years after closed captioning had begun, this was still not a simple task.[55]

Enter the TeleCaption II decoder in 1986, complete with cable-ready TV tuner and remote control.[56] Ball reported that NCI had "spent close to two years and over $300,000 to develop this second generation device," and in its first few months the new decoder sold at "more than five times the sales level of the preceding years."[57] Although NCI had been targeting the elusive the $100 price point with this model, it ended up retailing at double that—even after a "$1.5 million subsidy grant from the US Department of Education."[58] After the initial rush of sales, NCI only sold about eighty thousand units in all, bringing the total number of caption-enabled households up to 180,000 by 1988 (an optimistic estimate, assuming that all of the original decoders were

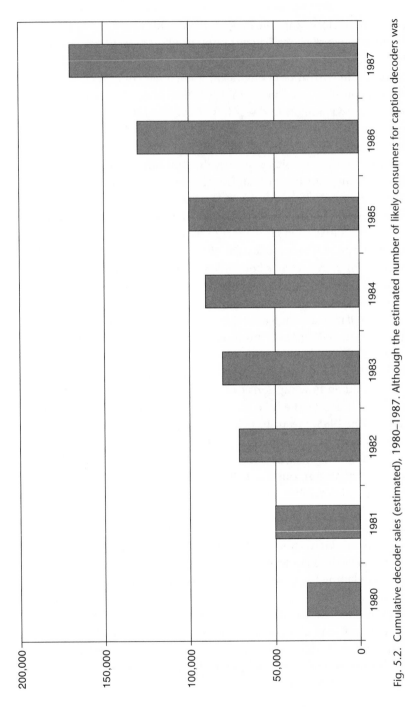

Fig. 5.2. Cumulative decoder sales (estimated), 1980–1987. Although the estimated number of likely consumers for caption decoders was repeatedly lowered through the 1980s, cumulative caption decoder sales never met expectations. Source: Graph redrawn from Commission on Education of the Deaf (COED), *Toward equality: Education of the deaf* (Washington, DC: GPO, 1988), 119

still working and that none of the original decoder owners had "traded up" to the new model).[59] Clearly Ball's goal of 500,000 decoder households by 1990 was unattainable—yet an important report by the Council on the Education for the Deaf (COED) released in 1987 upped the stakes for closed captioning, arguing that "to make the captioned service economically viable and self-sustaining, the service must reach into at least 500,000 homes and ideally 1,000,000 homes by 1990."[60] A third model, the TeleCaption 3000, was rushed to market in 1988—this time breaking the $200 price barrier by abandoning the cable-compatible tuner—but it only sold about twenty-eight thousand units, with a federal subsidy of thirty dollars per unit.[61] Market research showed that nonadopters either did not think their hearing had degraded enough to need a decoder or still judged the decoder to be too expensive.[62] Despite new estimates by the White House Office of Technological Assessment that "by the year 2000 . . . there will be more than eleven million senior citizens with a significant hearing loss who could benefit from closed-captioning," the hard-of-hearing market simply was not responding to NCI marketing efforts.[63] Although NCI pressed on with its plans for a TeleCaption 4000, it looked like the closed-captioning installed base would not even reach 250,000 households— half of Ball's stated break-even minimum—by 1990 (see figure 5.2).[64]

## Trying to Fund Captioned Content

Because legislators blamed NCI for the lackluster decoder sales in the 1980s, one might imagine that legislators would blame the networks for the small amount of captioned content on the airwaves. NCI had argued before Congress in May 1983 that its caption viewers "have access to less than ten percent of the television programming today."[65] As one deaf viewer pointed out in hearings later that year, "To make it even more dismal, sometimes three captioned programs appear at the same time," and "you can only watch one."[66] The amount of cable TV programming increased faster than the rate of cable TV captioning, and the total percentage of captioned programming had dropped even further by the following year, when Ball lamented: "The volume of programming from which the hearing impaired, hearing handicapped can select, is . . . about five percent of what you and I have available."[67] By May 1985, NCI had nearly doubled its 1981 level of captioned programming to seventy hours per week, but the service still captioned "*no* daytime programs."[68]

In Congressional hearings dealing with this problem, however, some lawmakers displayed a striking ignorance of the labor process—including costs in both time and money—which made captioned content possible. In May 1984 NCI's John Ball and Rep. Silvio Conte (R-MA) had this exchange:

> Conte: "How does that captioning machine work? If I were on TV, my words would be printed out through that device?"
>
> Ball: "Yes, sir. As the photograph indicates, you can sit this on top of your TV set, and tune in just as you would on the regular TV set, and if the program is captioned, the captions appear as subtitles."
>
> Conte: "So, the program has to be captioned by somebody?"
>
> Ball: "Oh, yes."
>
> Conte: "It isn't an automatic translation of words into the caption?"
>
> Ball: "No. That is the thing. It is not just a question of getting hold of one of these [decoders]. The network or advertisers has to pay about $2,200 per program [hour] to us to have each program captioned."
>
> Conte: "That's what I was wondering—an automatic captioning machine would be a heck of a breakthrough . . . What you had me thinking at first was that as you uttered the words they appeared automatically through that machine."
>
> Ball: "I wish they did."
>
> Conte: "We have got to work on that."
>
> Ball: "Yes . . ."[69]

But NCI could hardly afford to work on the problem of automatic machine translation of speech—a problem that continued to vex computational linguistics researchers, the inheritors of the cold war machine translation project (see chapter 3). Nine out of every ten captioned hours of television flowed through NCI by 1987.[70] Each captioned program-hour cost $2,500 to produce, taking between twenty and thirty hours of labor to do it (resulting in a labor rate of about $100 per working hour).[71] That $100 per hour purchased skilled and trained workers, such as Joy Ritchie-Butland, "one of the original group of caption editors at NCI," hired in June 1979 with a master's degree in Spanish. In 1980 the NCI newsletter explained the "five-week formal training program" necessary to bring "editors" like Joy up to speed: "Their first challenge is to edit captions for *Nova*, a program with a high degree of technical language." Then they move on to a situation comedy, where they "learn to work around different camera shots and scene changes." Finally, they edit an "interview show," where "editing haphazard conversation and uncompleted thoughts that accompany interviews is a true final exam for the caption editor."[72] NCI was arguing that high captioning costs resulted in high-quality captions.

The responsibility for paying for such high-quality captions was supposed to fall on the program networks and producers; however, this was a shaky promise. CBS still had not agreed to participate in the program when NBC pulled out of captioning temporarily in 1983. ABC was able to increase its captioning bill by a half-hour in the 1983–1984 season to help out (captioning a total of 5.5 hours weekly cost the network $629,200/year), but even PBS, under fire from the Reagan administration, cut its annual captioning outlay by $100,000 due to budget cuts.[73] The events of 1983 had "forced [NCI] to lay off more than 20 employees, reducing its staff to about 75." By 1987 NCI had bounced back to 150 captioning jobs between its three locations in Virginia, Hollywood, and New York—a fourfold increase over its 1983 size—but it still faced the tricky challenge of keeping its own captioning rates low enough to entice the networks to caption programming but high enough to attract and keep quality staff.[74]

Although NCI's original government subsidy from the Bureau of Education for the Handicapped had been intended to expire after three years, in FY 1982 the new Department of Education contracted for $1.2 million of captioned programs from NCI.[75] Over the next few years, this figure remained relatively constant, subsidizing both household decoder purchases and network captioning of programming. Ball and his staff requested appropriations each year for the overall Media Services and Captioned Films effort (still including the Captioned Films for the Deaf distribution system from chapter 1), with a portion always going to NCI for television captioning. In May 1983 NCI's Doris Caldwell testified that "NCI's singular service for handicapped people is not yet commercially attractive in the fiercely competitive marketplace of television, but with continued strong support from this Congress, it can become self-supporting and thus self-sustaining."[76] The following year Ball proudly testified, "We are about fifty percent federally funded. The rest comes from the corporate contribution."[77] But NCI could not go much beyond this fifty-fifty split. Far from weaning itself from federal support, NCI requested a larger appropriation each fiscal year through 1987 (see figure 5.3). By FY 1989 Ball was

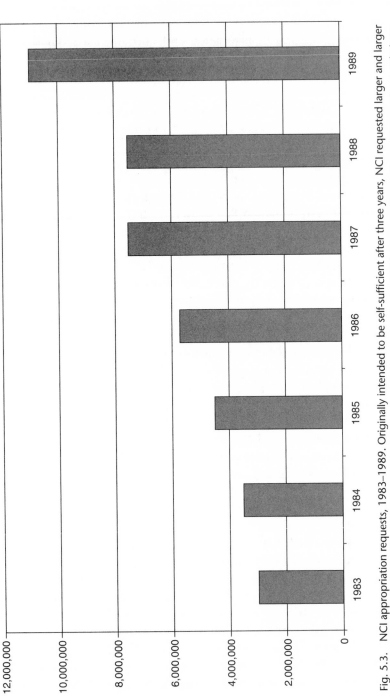

Fig. 5.3. NCI appropriation requests, 1983–1989. Originally intended to be self-sufficient after three years, NCI requested larger and larger federal appropriations for captioning labor, research, and development through the 1980s. Source: House Committee on Appropriations, "Departments of Labor, Health and Human Services, Education, and Related Agencies Appropriations" for 1984: part 8, CIS-NO 83-H181-75 (5, 9–12, 16 May 1983), 363; 1985: part 9, CIS-NO 84-H181-87 (1–3, 7 May 1984), 803; 1986: part 10, CIS-NO 85-H181-94 (22, 29, 30 May 1985), 247; 1987: part 10, CIS-NO 86-H181-57 (6, 7 May 1986), 97; 1988: part 8, CIS-NO 87-H181-61 (21, 22 Apr. 1987), 435

requesting $11 million for NCI alone: $6.5 million for network captioned programming (especially daytime); $1 million for local captioned
programming (news, as described in chapter 4); $1.5 million for the
purchase of decoders by low-income families; $500,000 for further de-
coder development; $1 million for advertising; and $500,000 for "basic and applied research," especially in captioning as an educational
tool.[78]

NCI's response to its persistent government dependency—rather
than, say, arguing that such dependency resulted from a market failure
and represented the collective will of the state acting in the public interest to safeguard the media access of an underprivileged few—was to
try to help networks and producers find additional ways of paying for
NCI's own captioning. NCI turned to the very community of private
capital that balked at high captioning fees to try to figure out how to
do this, creating a Corporate Advisory Council in 1983 made up of
firms like Kellogg, Prudential Insurance, Equitable Life Insurance,
Nabisco, Mobil, Xerox, and AT&T.[79] One of the first ideas NCI and this
council came up with was to solicit charitable contributions—from
the deaf and hard of hearing themselves. An NCI Caption Club was
formed in October 1983 for "annual contributions from individuals,
families, and organizations . . . toward expanding the volume of closed
captioned programming," donations ideally "matched by contributions from corporations, foundations, and various groups in the television industry."[80] The D/HOH activist T. J. O'Rourke had started the
idea when he became "the first deaf person ever to underwrite the captioning of a television show" by subsidizing captioning for the *Perry
Como Christmas Special* himself in 1982.[81] Following this example, in
the club's first six months, "over 2,000 families nationwide . . . contributed over $70,000 towards us providing more captioned program-
ming" according to Ball.[82] Eventually NCI specified that funds
donated from D/HOH households (who were already paying for
decoders at $300 each) could not be used for "more than 35% of the
program's captioning costs."[83] Volunteer efforts from the D/HOH
community helped bring the level of captioned content up to "180
hours of programs a week, including 35 hours of cable TV" by 1987,
but even then only fourteen programs had their captioning fees paid
entirely by their producers.[84]

The final attempt that NCI made to raise funds during these years
did target private business, but in a novel way. NCI proposed that large

advertisers caption their commercials to reach the D/HOH audience. The idea was not a new one. As early as 1978, PBS and HEW had talked to major television advertisers—"companies that symbolize wholesome, family virtues" such as General Foods, Hallmark, IBM, Eastman Kodak, Quaker Oats, AT&T, Bristol Myers, J. C. Penney, and McDonalds—about underwriting the closed captioning of programs.[85] Interestingly, they were more successful in convincing corporations to caption government events—Xerox underwrote the captioning of all of President Reagan's televised speeches in 1981—than private programming.[86] But in terms of captioning commercial spots themselves, as *AdAge* had feared in 1978, "there seems to be little interest among advertisers."[87] Part of this disinterest rested on a misconception by advertisers that the D/HOH did not watch as much TV as their hearing peers. In 1980 *Business Week* claimed, "Surveys show that the segment of the TV audience with hearing problems typically watches only 3 hours a day vs. a national viewing average of 6½ hours."[88] Academic research in the early 1980s contradicted these findings, however, suggesting that D/HOH audiences watched *more* television than hearing audiences.[89] NCI's own surveys found that "most respondents wanted commercials to be closed captioned even if the program was not, and were more likely to watch a commercial if it was closed captioned."[90] Some advertisers were interested in viewers with hearing difficulty. But they targeted older, late-deafened, HOH viewers—"a target audience particularly ready for commercials selling laxatives, antacids, pain relievers, and other items used by senior citizens" according to J. Walter Thompson—rather than prelingually deafened children who might use signed languages.[91]

By offering a bargain-basement rate of $165 per minute, however, NCI was able to convince a handful of brand-name advertisers to experiment with captioned commercials during its first year in 1980, including Sears, IBM, AT&T, Bristol Myers and Seiko.[92] The experiment met with some success. A year later, about 115 television advertisers had closed-captioned at least one commercial.[93] By 1988, NCI had captioned an estimated 9,300 commercials.[94] In the end, the main deterrent to convincing advertisers to caption commercials was not cost (by 1990 an average "no-frills commercial" cost $200,000 anyway), but time. Explained *AdWeek*, "the only ads that are not closed-captioned are quick-turnaround spots that corporations want to get on the air right away."[95] In response, NCI added a surcharge to its fee for forty-eight-hour service and even dedicated its New York office, opened

August 1985, to "overnight service for commercials (in by 5 p.m., ready by 10 a.m.)."[96] But NCI could never prove that captioned advertisements increased sales for vendors.[97] Even though these advertising-captioning services did not bring in enough revenue to wean NCI from federal funding (an estimated ten thousand commercials through the 1980s at $200 each would only have grossed NCI $2 million over the whole decade), these efforts did at least contribute to the legitimacy of captioning as a standard step in the television-production process.

## Technological Obsolescence and the Threat of Teletext

The legitimacy of captioning was at stake in the 1980s. The networks could only plead poverty so many times, and even the Reagan administration was willing to funnel money to make television accessible through its Department of Education, so NCI knew it might hang on until 1990 if it could just reach an important technological milestone: the supposedly inevitable drop in decoder prices to trivial levels. However, this strategy framed NCI's existing base of decoders as technologically obsolete. This was a risky move for the advocates of Line 21 captioning because ever since the early 1970s there had been another claimant for the precious bandwidth of the VBI, an alternative technology that promised to do all that Line 21 closed captioning could do, and more. This technological competitor promised to target not only an ever-shrinking (and ever-contested) number of deaf and hard-of-hearing TV viewers but the ever-growing (and ever-coveted) number of hearing TV viewers. The technology was called teletext.

Like Line 21 closed captioning, teletext was a method of hiding digital characters within an analog television signal in the VBI. Developed in the late 1960s by the British Broadcasting Corporation (BBC), it too was originally intended as "a means of transmitting captions that would help deaf people enjoy TV programs."[98] The television signal was to function as a carrier for character data that would be output on a small, quiet, and inexpensive home printer for the deaf, in much the same way as US TTY/TDD systems used recycled teletype printers hooked up to phone lines (see chapter 2). But by 1968, the BBC had turned to "semiconductor technology to generate characters on the screen" as a better solution. Around 1970 the BBC realized that the same digital character technology might support two entirely separate applications: subtitled television programs for deaf households, which

they called "teletitles," and full-screen textual information for all
households, which they called "teledata."[99]

Development of teletext in the UK proceeded much more swiftly
than the development of captioning in the United States. In 1972 the
BBC was testing and prototyping its new CEEFAX system (for "see
facts"), which would combine the teletitle and teledata ideas into a
single teletext standard.[100] By March 1974 the BBC, the Independent
Television service (ITV), and the British Radio Equipment Manufacturers' Association announced a unified teletext standard for CEEFAX,
known as the teletext *White Book*.[101] After a two-year trial period, the
British government gave its approval, and in November 1976 the BBC
began public broadcasting of its nonprofit, advertising-free CEEFAX
service.[102] ITV followed a few weeks later with its for-profit, advertising-
based ORACLE (Optical Reception of Announcements by Coded Line
Electronics) teletext system, using the same standard.[103] As in the
United States, getting decoders into the marketplace was a challenge;
until 1978 most CEEFAX and ORACLE users were "home electronics
enthusiasts who had built their own decoder units as add-ons to their
television receivers using plans published in a popular electronics and
wireless hobbyists' magazine."[104] But in a society where many rented
televisions instead of purchasing them—a long-standing system that
supported BBC programming costs—the new technology spread rapidly. About 75 percent of UK viewers rented their televisions in the late
1970s, and many spent the equivalent of an extra six dollars per
month to rent a new teletext-capable set.[105] BBC and ITV broadcast to
about 100,000 teletext receivers by 1980, the same year that closed
captioning began in the United States.[106] It would take NCI five years
to reach the same number of captioning decoders. Amazingly, "the
whole annual budget for a nationwide, two-network teletext service
was less than the cost of a single television program," the equivalent
of only $1 million each year.[107] However, this did not include much, if
any, deaf captioning.

Observers in the United States enthusiastically described teletext as
a new hybrid medium combining television, printed magazines, local
newspapers, and radio.[108] One US journalism professor wrote, "You
dial a three-digit number on your teletext keypad for the latest news
flashes or for an index of information. Brilliant six-color 'print' displays flash the latest headlines."[109] The magazine metaphor was a
common one.[110] This was partially a result of technological limita-

tions: CEEFAX only transmitted one hundred "pages" (screens) using its four lines of the VBI, with access time per page ranging from one-eighth of a second up to fifteen seconds (each home decoder had to sit and wait for the correct page to be delivered in a continuous cycle of updating).[111] A writer in *Telecommunications Policy* described how CEEFAX was actually two differently targeted magazines: "Ceefax 1 is more popular, news and sports-oriented. Ceefax 2 gives more documentary and less perishable information."[112] But other metaphors seeped in as well. One observer argued, "Teletext is something halfway between radio and television. It could be seen as a kind of *printed radio* displayed on the ordinary home television screen."[113] In the early 1980s, analysts in the *New York Times* predicted that "more advanced applications of the technology will include local traffic conditions, in-depth news stories, maps, charts and other information that could be gleaned from libraries of information."[114] Moreover, teletext would "allow television stations to offer some of the services now offered by newspapers, such as television listings and some classified ads, as well as news."[115] A decade before the World Wide Web would take the media industries by storm, teletext offered a tantalizing glimpse at a new electronic information industry—and a new electronic information market.[116] Computer-networking expert Paul Baran had already forecasted in a 1971 report that by the 1980s, the average US household might spend a total of twenty dollars per month on new information services, yielding a $22 billion market among the projected 90 million US households.[117] Here was a business venture that dwarfed the scale of closed captioning.

The loudest proponent of teletext in the United States was CBS. While traveling abroad in 1976, a CBS-affiliate station manager had seen CEEFAX in operation.[118] Soon CBS was conducting teletext pilot tests of its own, comparing the British "World System" standard to the newer French "Antiope" system in St. Louis in 1979.[119] CBS also organized a formal study group on teletext under the Electronic Industry Association, a twenty-three-person committee chaired by a CBS executive.[120] The plan, drawn directly from the closed-captioning example, was to come up with a standard acceptable to CBS, gain FCC approval for that standard, and then convince consumer-electronics makers to produce decoders that could understand the new standard.[121] But one thing stood in CBS's way. To make teletext a richer experience, in both access speed and page counts, CBS needed to control a large swath of the VBI. However, in its St. Louis experiments, CBS felt hampered because "two lines have been allotted to the teletext experiment and one line to the closed-caption system"; if CBS could have Line 21 too, "teletext would be greatly enhanced and more efficient."[122] CBS's ambitious plans for teletext signals in the VBI and teletext decoders in every household conflicted with NCI's agenda for closed-captioning signals in the VBI and caption decoders on every D/HOH television. In March 1979, the network's *CBS Reports* series had aired a special program entitled "How much for the handicapped?" in which their news team had soberly considered "a mounting conflict between conscience and cost—the cost of making public facilities available and accessible to the handicapped."[123] The network provided an answer to its own question a year later, in February 1980, on the eve of the introduction of NCI's closed captioning. CBS head of engineering Joseph A. Flaherty declared, "The rest of the world is going to teletext, and I think it's wrong to support a closed-caption system that is already obsolete and destined to be replaced."[124]

NCI's John Ball responded quickly, attempting both to bring CBS into the closed-captioning partnership and to defend Line 21 from the CBS attacks. "I can assure you," wrote Ball to CBS president Gene Jankowski in July 1980, "that NCI would never wish to perpetuate an 'obsolete' system or to perpetrate such a system on a large population which has historically been denied access to television, the most pervasive communications force in this country today." Further, offered Ball, "NCI wishes to cooperate with CBS in creating a set of conditions under which CBS would participate in the closed-captioning service without thwarting progress toward its longer-term teletext goals."[125] Others within the D/HOH community tried to broker a compromise. Noting that "Teletext would be a viable alternative to a radio for the deaf," Gallaudet professor E. Marshall Wick sent a report to the FCC in October 1980 recommending that it "provide some means by which network Line 21 signals could be converted to Teletext Line signals and network Teletext captions with SMPTE time codes could be converted to Line 21 signals so that deaf persons would have a choice of either decoder."[126] For its part, CBS held a demonstration of teletext for the D/HOH community at Gallaudet on November 20, 1980.[127] The network argued that "updated news, weather reports, traffic conditions, school closings, local events, home education and other information, previously unavailable to the hearing-impaired, will fill an important need." As for captioning, CBS bragged that "teletext technology provides the opportunity for captioning at different speeds (for different audience reading levels), in different colors and different places on the screen (to aid in distinguishing who is speaking) and in different sizes (as a service to those with impaired vision)."[128] However, the network failed to recognize the amount of captioning labor that would be required to encode and enable such variations.

By 1981 the time for compromise was over. NCI researcher Carl Jensema argued, "CBS says their teletext system puts them in conflict with the line 21 system. How?? NCI has offered to (a) caption for CBS on line 21 until teletext is available, and (b) help CBS make teletext decoders electronically compatible with line 21 so that a teletext decoder could receive both types of signals. CBS has refused to cooperate in any way." Jensema also pointed out, "Line 21 offers about 25 hours of captioning per week after being on the market nine months. British teletext offers about two hours of captioning per week after being on the market seven years."[129] Although some work had been done in the UK at Southampton University in establishing subtitling guidelines, just because CEEFAX and ORACLE had the *potential* to offer deaf subtitles did not guarantee that this potential would be realized unless both funding and labor were organized first.[130] In order to demonstrate that its teletext system could caption programs at least as well as NCI's could, CBS contracted with the WGBH Caption Center (shut out by NCI from Line 21 captioning as described in chapter 2) to open a Los Angeles office and pilot teletext captions starting in March 1981.[131] (NCI had refused to do the captioning unless CBS agreed to broadcast those captions simultaneously on Line 21 as well.)[132] The mustering of the Caption Center to the teletext side split the D/HOH community further on the issue. Joseph Blatt, former producer of the *Captioned ABC News* at WGBH, wrote an article in the *Volta Review* arguing that the economic benefits of teletext could provide an answer to the chicken-and-egg problem: "Unlike TDDs, for example, hearing-impaired consumers will not face the burden of financing a product with limited appeal. Instead, teletext decoders can become a mass-market commodity, with the cost of transmissions and of manufacturing spread across the general population."[133] Nevertheless, as of May 1982 D/HOH activists were still picketing CBS headquarters in New York in "silent protest."[134]

In the end, the CBS effort to enshrine teletext as a new broadcast data standard was thwarted by the same market-based FCC policy that had left Line 21 captioning open to attack. Back in July 1980, even before the Electronic Industries Alliance committee that CBS headed had made its recommendation, CBS filed a petition with the FCC to establish its modified version of the French Antiope system as the "national standard for a broadcast teletext system in the United States."[135] The committee was unable to come to an agreement that satisfied the required three-quarters of its members, although press reports indicated that the CEEFAX standard was favored "almost 2 to 1" over Antiope.[136] Lacking industry consensus, the FCC was unwilling to make a choice among these technologies, and in October 1981 chose instead "an open environment (or deregulated) approach to Teletext standards."[137] The FCC argued that "the forces of competition and the open market are well suited to obtaining the kinds and amounts of service that are most desirable in terms of the public interest."[138] The FCC heard comments on the issue for over a year but had not changed its position by March 1983 when it officially authorized teletext transmissions without a standard (and simultaneously reaffirmed the use of Line 21 of the VBI for closed captioning for at least five years).[139] The following month, CBS began its national teletext transmission, called ExtraVision, despite that "the required decoding equipment is still not available to consumers."[140] By the start of the fall 1983 television season CBS would claim that 85 percent of its affiliates were carrying the transmission.[141] Nevertheless, D/HOH activists continued their pickets of CBS through the end of the year.[142] NCI requested $500,000 in extra technical funding from Congress, linked to teletext threat, testifying that "we are not the National Line 21 Institute. We are the National Captioning Institute and our captioning can go on any line of the spectrum, but we must have funding in order to keep the Line 21 present technology in step with emerging teletext and videotext technologies."[143]

By late 1983, government officials began to wonder if they should continue funding NCI at all. In November the House Committee on Science and Technology held a hearing in which they considered NCI's chicken-and-egg problem in light of the teletext euphoria. CBS's Flaherty was on hand to argue that teletext would appeal to a mass audience of TV households and thus subsidize decoder sales and content development for the smaller D/HOH audience, to "bring hearing-

impaired viewers into the mainstream of visual communications."
CBS acknowledged that only custom-made teletext decoders costing
over $1,000 were currently available, but predicted that mass-market
teletext decoders would arrive "in roughly 18 to 24 months" for about
$150. But a representative from the National Association of the Deaf
responded that CBS had itself contributed to this problem by not captioning its own programs: "It is well-known in this country that CBS
has the best programming. It is a little bit like a person buying a new
car and not being able to travel over the best highways. A car's value is
substantially decreased if you can't use it for its full purpose." In the
end Congress was perplexed by CBS's willingness to offer captioning
through teletext later but resistance to offer captioning through Line
21 now—especially considering the network's "tremendous profits"
through the early 1980s. Echoing E. Marshall Wick's original advice,
the committee noted the technological fix where "a simple and relatively inexpensive device called a caption transcoder" could translate
captions from Line 21 to teletext and vice-versa.[144] If Congress could
fund captioning to the tune of millions of dollars each year, certainly
CBS could play along.

By the time CBS was ready to introduce deaf captioning on Extra-
Vision in early 1984, a compromise had been reached. Although its
first captioning effort, color-coded subtitles on the evening soap opera
*Dallas,* highlighted the supposed benefits of teletext captioning, the
network also agreed "to caption programs in both modes so that deaf
and hearing-impaired viewers who had purchased the decoders
needed for the closed-captioned system would have access to some
CBS programming."[145] Both the WGBH Caption Center and Joe
Karlovits of realtime CART fame helped CBS develop its "dual mode"
captioning system.[146] CBS even proudly proclaimed that "the network
will accept captioned programs and commercials supplied by others,
providing they are dual mode captioned"—neglecting to mention that
NCI did not dual-mode caption any of its programming.[147]

With the entrance of CBS into the captioning club (even at a paltry
three hours of captioned programming per week), Line 21 captioning
was momentarily safe from the teletext threat.[148] But that threat soon
withered away on its own. The first mass-produced ExtraVision decoders, from Panasonic, only reached market in January 1984.[149] By
then, a competing teletext venture sponsored by Time Inc. had already
come and gone—leaving that company some $15 to $30 million

poorer.[150] Despite their rhetoric, both CBS and NBC had scaled back their teletext experiments by the end of 1983.[151] NBC's system would not survive past 1986, though CBS's ExtraVision held on till about 1988. So-called videotex services for accessing information not on the home television but on the increasingly affordable home computer stole some of teletext's thunder.[152] Dial-up offerings like CompuServe and Prodigy made teletext seem obsolete by comparison, just as teletext did for Line 21 captioning.[153] These "wired teletext" services were billed as more "interactive" than broadcast teletext and relied on a subscription model for revenue rather than an advertising model.[154] In the 1990s, new Web TV systems to link the television with the Internet would return to hiding data in the VBI and tempting viewers to seek online information through their television sets.[155] But even then, an important objection first articulated in the early 1980s with teletext still persisted: "viewers might switch to teletext during television commercials," an advertising-avoidance behavior which would provide little incentive for either marketers or networks to support text on TV.[156] In the end, those who studied teletext consumers found that these systems simply did not perform well enough for the quality of information they provided. A 1982 NYU study of one teletext system in "upscale" households in Washington, DC, found that "at the levels at which usage stabilized, a typical household used the service three or four times a week, accessing approximately 10 to 15 pages or 'frames' at each session." This very low level of use might have been because "users found the service unacceptably slow."[157]

The supply side was equally important in considering the fate of teletext—especially since teletext shared more with closed captioning than its advocates liked to admit. The NYU researchers noted in 1982, "Both the design of sophisticated graphics and the frequent updating of frames are highly labor intensive activities" in teletext production.[158] CBS was unwilling to commit to such labor on a national level, assuming that local affiliates would undertake the production of teletext content and teletext advertising. As early as 1980 one CBS vice president called teletext "primarily a new and different local information and advertising service," saying "We see very limited use from the point of view of the network."[159] CBS expected its local affiliates to invest $100,000 each, both to retransmit nationally produced teletext and to "insert local news, which is seen as critical to the success of teletext."[160] The fate of the Charlotte, NC, CBS affiliate, touted by CBS in

November 1983 as the first local station to install its own teletext production equipment, is instructive.[161] According to teletext entrepreneur Leonard Graziplene, out of a hundred "pages" on the system,
eighty were "generated twenty-four hours a day by a staff of twelve
based in [CBS headquarters in] Los Angeles," but the local station had
to generate the other twenty. The local station reported that "it would
cost them approximately two hundred thousand dollars a year" to
maintain this local teletext operation. In general, "In order to offer
local pages it would have been necessary for each station to invest
approximately $200,000 for equipment and another $100,00[0] to
$125,000 each year for operating expenses." A year later, in November
1984, the high costs of equipment and labor had discouraged all but
three local affiliates across the nation—Los Angeles, CA, Buffalo, NY,
and Charlotte, NC—from producing local teletext pages.[162] This dy-
namic was strikingly similar to the resistance of local stations in funding live news stenocaptioning on the Line 21 system (described in
chapter 4). It was a lesson the teletext advocates learned too late.

## Widening the Audience for Captioning: The Education Argument

Advocates for closed captioning did learn an important lesson from
the teletext years, however. The wider the audience they could claim
for their service, the more likely that service would be to find continued government funding, and possibly even renewed government regulation. This was especially important after 1989 when a Department
of Education study recomputed the market for closed captioning
downward from the previous estimate of 4.5 million to only 1 million
persons, "eliminating people under five who may be too young to read
the captions and people with low levels of educational attainment who
may not read well enough to benefit from captioning."[163] The urge to
rhetorically expand this market hearkened back to the original enthusiasm of the 1971 Southern Regional Media Center for the Deaf conference, where open- and closed-captioning were first demonstrated for
the D/HOH educators and activists. WGBH had claimed that their
open-captioned *French Chef* experiment targeted, besides the D/HOH
population, "the approximately 10 million bilingual and 4 million educationally retarded persons who could benefit from captioning as a
tool for language learning."[164] The creation of NCI in 1980 largely ignored this wider possibility in favor of rhetoric claiming "the hearing

impaired will be the only audience seeing these closed captions."[165] But as NCI struggled in the 1980s, returning to a wider audience for captioning meant drawing upon the long history of captioning in deaf education to claim that a universal closed-captioning system—with decoders in every household and captions on every program—would serve not only the needs of deaf communication but also the needs of literacy education and immigrant Americanization.

First, however, the D/HOH community had to prove that captioned media actually contributed to learning in the deaf classroom. Even after decades of captioning by D/HOH educators through the 1950s and 1960s, researchers in the early 1970s found "no record . . . of a systematic investigation of the most effective levels and modes for presenting visual material to deaf children."[166] As one researcher explained, "For many of the initial projects of [Media Services and Captioned Films] there was no time to do study or research and many decisions were based on assumptions," especially the assumption that "the basic learning processes of deaf children were the same as for normal children."[167] New research performed at schools for the deaf in the early 1970s comparing captioned video to uncaptioned video generally concluded that captions helped convey information to students, although in at least one study the captions seemed to have little effect.[168] Residential schools for the deaf in particular adopted video display, production, and captioning equipment at a greater rate than other deaf schools and programs; by 1977 most residential deaf schools had televisions and video cameras, with 24 percent owning video character generators (essential for caption production).[169] By the time NCI inaugurated closed captioning in 1980, the WGBH Caption Center had already spent three years researching "multi-level captioning" for children, "a system which generates captions at three reading levels" corresponding to second, third, and fourth grades.[170]

NCI extended this research in its own efforts to market closed captioning. In 1980 the institute captioned "23 hours of educational programming" for use in schools, part of a three-year contract (which incidentally brought in much-needed revenue).[171] In 1981 it reported that 97 percent of decoder-owning parents thought captioning "helped improve the reading and language skills of hearing-impaired children."[172] Surveying the top seventy D/HOH programs in the United States, representing "a total enrollment of 15,949 students" or "roughly a quarter of all students known to be enrolled in special education programs for the hearing impaired in the United States," NCI found that "by 1984 99% had at least one decoder."[173] An NCI-funded study in a Chicago school argued that even though "deaf students do not have the opportunity to hear the sounds and patterns of English," making learning to read "a difficult task," in general "deaf students are highly motivated to read print when it is in the form of captions on television."[174] Such research recalled the claims that television was a "vast wasteland" (fears that had led to the creation of PBS), but turned these fears around: "Television, which is sometimes charged with contributing to reading problems, may now be used as one of the solutions to these problems."[175] By 1987, John Ball had rewritten history in his own public statements, claiming that NCI had been created not to bring broadcast justice to deaf viewers but "to open new educational horizons for all hearing-impaired Americans through access to television."[176]

Through the 1980s NCI developed this educational argument for captioning by widening the target audience for such captioning. In May 1983, as part of NCI's budget request for FY 1984, Doris Caldwell claimed that "retarded children in Florida are learning to speak through use of captions, normal hearing students in Fairfax County, Virginia public schools, who are slow readers, have markedly improved their reading and comprehension skills through use of captions[,] and Harvard University is experimenting with English language captions with their students who are non-native speakers of English."[177] In its FY 1985 appropriation request, NCI specifically asked for $200,000 for research on "the potential educational benefits of closed-captioned television for prospective beneficiaries beyond the deaf population."[178] NCI-funded researchers studied captioned video with "normal" hearing children, ESL students, and learning-disabled students—and found encouraging results to promote in the media each time.[179] Of course, D/HOH educators had long argued that their pioneering captioning efforts deserved wider recognition from mainstream reading educators. In 1974 one author in the *American Annals of the Deaf* argued, "Hearing children are benefiting greatly from *Sesame Street* and *The Electric Company*, programs whose educational value derives largely from captioning of key words in entertainment material."[180] Now that NCI was promoting this idea a decade later, non-D/HOH educators were finally taking up the question themselves, often reaching more skeptical or limited findings. One 1982 study argued, "Adding redundant print information" for hearing students at the university level "exerts a negative effect on learning."[181] Another 1987 study of eighty mixed hearing and nonhearing undergraduates concluded that "contrary to the supposition that captions are distracting [to the hearing students] and hence would interfere with learning, there was no significant difference between groups in the mean score on the test of knowledge."[182] But for the most part, NCI was buoyed by the increased focus on captioning's educational value, and by the time John Ball made his FY 1987 budget request to Congress, he was citing "at least 1,500,000 learning disabled children" as part of the audience for captioning (and the market for decoder purchases).[183]

NCI did not restrict its educational arguments to the plight of children, however. In 1988, NCI deflected attention away from its still-weak decoder sales by bragging that "over 10,000 [captioning] units last year [out of a total of 45,000] were sold to folks with no hearing impairment whatsoever."[184] In the marketing materials produced for its TeleCaption 4000 decoder in 1989, NCI claimed "people who speak English as a second language . . . accounted for over 40 percent of all TeleCaption decoders sold in 1988!" The sales pitch reported that this market of 20 million "and growing daily" had a "combined disposable income" of $169 billion.[185] According to June 1990 Senate testimony by Ball, *most* of the 1989 decoder sales "were to people learning English as a second language."[186] And as a 1990 NCI report by Susan B. Neuman put it, "35 million of our nation's children come from homes where English is not spoken."[187] Even First Lady Barbara Bush was quoted in 1990 as musing, "I really think literacy organizations might want to investigate how captioned television might be used in their programs."[188] Counting up all of these literacy students—native-born and immigrant, adult and child—finally provided a combined captioning audience with both the numbers and the wealth to make the networks sit up and take notice (or so NCI hoped).

The effort to recast captioning as an educational force had two main results. First, in D/HOH education, mandatory captioning was now cast as an essential ingredient in deaf educational reform, receiving prominent attention in the *Toward Equality* report commissioned under the Education of the Deaf Act of 1986 and published in 1988. The authors of this report declared, "The present status of education for persons who are deaf in the United States, is unsatisfactory. Unacceptably so." One solution they proposed was that "Congress should require the Federal Communications Commission to issue regulations as it deems necessary to require that broadcasters and cable-TV programmers caption their programming" and "to make new TV sets capable of decoding closed captions," because for the deaf, "the attainment of literacy and a wider acquaintance with the world at large" were processes that "can be most effectively enhanced by the accelerated use of captioned TV."[189] Second, in mainstream education, a longstanding assumption about captioning going all the way back to the early 1970s was finally questioned: the idea that hearing viewers could gain no benefit from, and thus would not stand for, text on their screens. Education specialists who looked at the record and realized "there was no research that substantiated this belief" opened the door to a new set of arguments that captioning was not only a broadcast right for the few but a broadcast enhancement for the many.[190]

### Re-regulating Caption Decoders

Besides its new strategies of research and rhetoric about the educational value of captioning, NCI was also counting on a technological innovation that had been predicted ever since the first closed-captioning demonstration in 1971: the "decoder on a chip." In the mid-1980s, the heart of the TeleCaption decoder, the so-called data slicer that extracted digital information from the analog VBI, was a dozen-chip circuit board relying on a 68A03 microprocessor with 2K of RAM and 12K of ROM.[191] In April 1988, John Ball asked Congress to fund research on shrinking this decoder module to a single chip, reducing both its size and its cost.[192] The Department of Education contracted the task to EEG Enterprises of Farmingdale, NY, which recommended that a single application specific integrated circuit (ASIC) could replace the entire existing decoder module for a cost of only five dollars per unit if quantities of a hundred thousand units per year were achieved.[193] Unfortunately, it would cost about a million dollars to design and prototype such a circuit.[194] Congress decided that the investment was worth it and awarded NCI the money; in October 1989 NCI announced a million-dollar contract with the ITT Corporation "to develop and produce the decoder chip" by the first quarter of 1991.[195] Not only would the new chip be smaller, it would allow new features such as "pop-on" captions, which could appear anywhere upon the screen (rather than on four designated rows at the bottom of the picture).[196] ITT was required to sell the device to TV manufacturers "for no more than $5."[197] The new chip would make or break the captioning system.

Unfortunately, from the consumer's point of view, the new chip would only bring down the cost of captioning if it were integrated into televisions rather than sold in a stand-alone decoder. As the 1980s ended, only 300,000 decoders had ever been sold.[198] If NCI could convince Congress to finally mandate the inclusion of the new chip in all new television sets, the survival of the closed-captioning system (and perhaps even its financial viability) would be assured. Congress, for its part, was becoming more and more accustomed to regulating captioning, at least on the margins. In late 1988 both the House and the Senate proposed rule changes to caption the C-SPAN coverage of congressional proceedings, which had begun in the mid-1980s. Although the Congressional Budget Office estimated that the annual cost of closed-captioning 1,200 hours of proceedings would be $1 million, Congress seemed prepared to pay.[199] Congress gave increasing attention to captions in other venues as well. Hospitals and nursing homes receiving federal funds would have to supply caption decoders under a provision of HR 5178 (August 1988); hotels where federally funded conferences were held would have to offer in-room decoders to guests on demand under HR 2968 (July 1989); and public service announcements funded by the federal government had to be captioned according to HR 2273 (May 1989).[200] All of these efforts meshed with the intent of the 1990 Americans with Disabilities Act (ADA), which mandated accessibility measures as a condition of federal funding in education and the workplace (see chapter 4). But interestingly, no mandate for universal closed captioning made it into the ADA. According to one report, "The three major captioning services . . . through their daily contact with broadcasters and producers, believed that the broadcast industry would be more favorably disposed to legislation increasing viewership through requiring built-in decoder circuitry." Thus when officials from "the major national deaf and hard-of-hearing organizations" met at Gallaudet University to discuss the issue, "all agreed that a provision for mandatory captioning in the ADA would create major broadcast industry opposition to the ADA that might weaken its chances of passage."[201]

Captioning advocates did not give up on legislation, however. There were clear precedents for congressional intervention in the telecommunication marketplace. The All Channel Receiver Act of 1962 had mandated the inclusion of UHF tuners in all televisions sold after April 30, 1964; more recently, the Hearing Aid Compatibility Act of 1988 had required all telephones sold in the United States after August 1989 "to be compatible with hearing aids."[202] Thus it was not a total surprise when, in November 1989, Senators Tom Harkin (D-IA), John McCain (R-AZ), Daniel Inouye (D-HI), and Paul Simon (D-IL) introduced the Television Decoder Circuitry Act (SR 1974), which would "require that any television with at least a 13-inch screen which is manufactured, or imported for use, in the United States be equipped with built-in decoder circuitry designed to display closed-captioned TV transmissions."[203] A House version soon followed in March 1990.[204] As with previous legislation dealing with broadcast justice for D/HOH viewers, congressional hearings in May and June 1990 saw a star-studded panel of advocates for universal captioning, including actress Linda Bove of *Sesame Street* and actor Geoffrey Owens of (ironically) *The Cosby Show*.[205] As in previous debates over captioning, the size of the D/HOH audience (this time estimated at "more than 15 million Americans [with] some degree of hearing loss"), the expense of the stand-alone decoder (around $200), and the paucity of captioned programming (this time focused on the fact that "only 90 out of the 1400 television stations in the United States provide closed captioned local news programs") figured prominently in the discussion.[206] But this time around, with the technological innovation of the five-dollar captioning chip and the rhetorical construction of a captioning audience made up of more than just D/HOH viewers, the debate proceeded differently.

Captioning advocates now described stand-alone decoders as troublesome, regardless of their price. The tale of one elderly member of Self Help for Hard of Hearing People was used to illustrate how hard they were to use: "I wasted $98 on two servicemen trying to get my decoder number 3000 hooked up. The first one couldn't figure out the hook-up with all the TV attachments. The second one got the VCR in backwards, so I could not record the captions. Finally, an electronics expert who is also my neighbor, got everything straightened out."[207] A representative from the WGBH Caption Center added that "many deaf and hard-of-hearing people—primarily the elderly—choose not to purchase a separate decoder because of the stigma attached."[208] The new NCI decoder chip, if embedded in all televisions, would eliminate such frustrations and stigmas, while still preserving a 5 percent royalty to NCI.[209] This royalty was meant to fund "further development of captioning technology, particularly that for captioning live events—news and sports, for instance—in real time."[210] With some 20 million new TV sets sold in the United States each year, universal decoder availability, it was hoped, would finally fix a broken market mechanism and "create market incentives for broadcasters to invest in and provide more closed captioned programming."[211]

The networks were strangely silent during the hearings, but television manufacturers objected to government interference in their consumer-products market. After complaints that ITT would have a monopoly on caption-chip production, "the bill was revised to let companies buy their chips from anyone or develop them on their own."[212] Negotiations with "Japanese and Korean TV manufacturers" apparently led to an agreement "not to fight the act if the effective date was moved forward at least 18 months."[213] And the Electronic Industries Association (EIA) demanded the legislation be limited to "one model for each screen size beginning with television sets with screens twenty inches and larger," a compromise that would have brought decoders only to the most expensive 40 percent of all television sets.[214] What held these complaints together was an overall argument, articulated by an economist at the Brookings Institution, that mandating decoder chips in all TVs was a "tax": "It doesn't look like a tax, but it is nonetheless—people are paying. The goal is laudable, but perhaps it's not the most efficient mechanism."[215] Argued the EIA, "We do not . . . believe it is fair to impose what amounts to a regressive excise tax on all purchasers of television sets in order to benefit a small, albeit deserving minority."[216]

But this time the captioning advocates had an answer to the "small, albeit deserving minority" argument—and Congress agreed with them. The House report on the hearings cited 24 million "deaf or hearing impaired" people in the United States and claimed 38 percent of "older Americans" who had "some loss of hearing" would benefit from the act. They also cited "23–27 million functionally illiterate adults, 3–4 million immigrants learning English as a second language, and 18 million children in grades kindergarten through three, learning to read." Finally they claimed that, according to the Department of Education, "13 percent of the adult population is unable to read."[217] Captioning no longer pitted a profitable television audience against a minority viewing community; instead, it offered a low-cost technology that would end up in every single American household, reminding citizens that Congress was watching out for their interests. When the Congressional Budget Office estimated that the new law would only cost the federal government a maximum of $100,000 per year for the five fiscal years 1991–1995, based on costs that the FCC would incur "developing the rule and responding to requests," the deal was sealed.[218] The Television Decoder Circuitry Act (TDCA) was signed by President Bush in October 1990, only three months after the ADA.[219] The following spring, Bush told the attendees of the annual meeting of PBS, "You've pioneered new broadcasting techniques, including closed captioning for deaf students," casting his signature on the bill as a triumph for minority education.[220] But NCI saw the legislation as a triumph for the majority as well, predicting that, due to the rapid rate of TV replacement in the United States, "virtually every home in the country" would have a caption-capable TV by the turn of the century.[221] For once, NCI's prediction was correct.

## Re-regulating Captioned Content

Although the TDCA did not mandate the sale of televisions with the captioning chip until 1993, Zenith got a head start on the market and delivered the first TVs with built-in decoders in October 1991, using a chipset of its own design.[222] Other television manufacturers, as well as network affiliates, were gearing up for an unprecedented level of public scrutiny of captioning. One expert wrote, "Many people unfamiliar with the purpose or technology of captioning will suddenly be exposed to it with little or no preparation. TV programmers and manufacturers can expect a huge wave of questions and complaints about the hardware and the service."[223] But after a 1991 national forum on closed captioning, communications lawyer Stuart Brotman reported that "although 68 percent of major commercial network programming is captioned today, only 20 percent of local news programming is captioned, and 16 percent of syndicated television programming is captioned."[224] In fact, only as recently as fall 1989 had the entire prime-time slate of shows on the Big Three networks all been closed captioned.[225] Cable television was not performing much better: "HBO and other pay cable services now caption 36 percent of their fare, but basic cable programming—Turner Network Television, ESPN, and the like—woefully lag behind at 1.2 percent."[226] Even during the crucial weeks of the first Gulf War that year, it was revealed that "CNN, the indispensable 24-hour mainstay of the gulf coverage, has no captioning."[227] Brotman argued, "The federal government has supplied roughly one-third of all funding for captioning, with the balance supported by networks, program producers, advertisers, and foundation support, all in the spirit of 'doing the right thing.' But we must move to a system where captioning can be cost-justified, even profit-motivated."[228] The networks feared that if the TDCA were as successful as everyone expected, new pressure would be put on them to caption all of their programming.

More frightening changes than captioning were brewing in the telecommunications industry, however. At the end of the 1980s, television-based teletext and its computer-based cousin videotex had been largely laid to rest as failed consumer-information utilities—then suddenly, seemingly out of nowhere, a free, global information system emerged from a particle-physics lab in Switzerland and a supercomputing lab in Illinois. The HTTP hypertext protocol from CERN's Tim Berners-Lee and the Mosaic "browser" software from the NCSA's Mark Andreesen combined to make the World Wide Web the hottest new real estate in cyberspace during the early 1990s.[229] So-called digital convergence questions over which industry would come to dominate the "information superhighway"—the telecommunications "baby Bells," the cable companies, or the computer hardware and software giants—forced Congress under the Clinton administration to consider an overhaul of the Communications Act of 1934.[230] As early as the first draft of the legislation in 1994, it was clear that mandated captioning for the new world of audiovisual "content" would be part of the eventual law. The House Committee reporting on the bill for the National Communications Competition and Information Infrastructure Act of 1994 expected that "most new programming will be closed captioned, and that preexisting programming will be captioned to the maximum extent possible"; however, it specified that "the Committee does not intend that the requirement for captioning should result in a previously produced programming not being aired due to the cost of the captions."[231] The eventual law would provide a governmental mandate for captioning and allow private capital certain loopholes to avoid captioning. Both sets of rules would likely become the responsibility of the Federal Communications Commission.

In a surprising change of pace, the FCC decided to act even before the telecommunications legislation had been finalized. In December

1995, FCC chair Reed Hunt asked at a symposium cosponsored by the WGBH National Center for Accessible Media (the new name for the Caption Center), "If we require buildings to be wheelchair-accessible, and phones to be hearing-aid compatible, should broadcasters be required to make programming accessible to people who are deaf and hard of hearing?"[232] That month the commission formally requested comments on closed captioning—as well as a complementary service for the blind, audio-based "video description"—in order to "assess the current availability, cost and uses" of these technologies.[233] The captioning questions that the FCC was wrestling with centered around audience composition, program content, and labor process. With regard to audience composition, the FCC wanted the latest numbers on the proportion of D/HOH individuals who required captioning "to enjoy television programming"; the number of D/HOH children who could "benefit" from captioning; and the number of people who might benefit from "teaching literacy skills to children and illiterate adults." In terms of program content, the FCC asked for figures on how much captioned programming was currently available based on source (network, cable, local), genre (news, documentary, children's, sports), and provider size. Finally, with respect to the labor process, the FCC asked for statistics on capital costs, prerecorded and live labor costs, and "the role free-market forces have played and can play," especially whether there were finally "a sufficient number of decoder-equipped television receivers in the market to provide the hoped-for incentive for the television industry to provide closed captioning."[234] It took the FCC fifteen years, but it was finally asking the right questions of the closed-captioning system as a whole, in preparation for pragmatic legislation to make the system work better.

Sure enough, the final draft of the Telecommunications Act of 1996 required the FCC to complete "an inquiry to ascertain the level at which video programming is closed captioned" within 180 days, leading to regulations eighteen months later. But the law included an important, and expected, loophole, allowing the FCC "to exempt specific programs, or classes of programs, or entire services" from captioning if doing so were "economically burdensome to the provider or owner of such programs."[235] The FCC report would thus provide a baseline to judge what "economically burdensome" meant in any appeal. Submitting its findings to Congress in July 1996, the FCC reported that three years after all new TVs were able to decode captions, "between 50 and 60 million U.S. homes can receive closed captioning"—a hundred-fold increase from the 400,000 or so before the TDCA. Predictably, though, captions were "less likely to be included in programming intended to serve smaller or specialized audience markets." The Department of Education still funded "40% of the total amount spent on captioning," and although the FCC originally estimated the cost of that captioning at $2,000 to $5,000 per hour of programming, it found that the actual hourly costs were only $800–$2,500 for offline work and $150–$1,200 for online—debunking the notion that realtime stenographic captioning was more costly than carefully edited offline captioning.[236] Additional studies from the National Association of Broadcasters (NAB) and NCI appearing at the same time rounded out the picture of mid-1990s captioning. According to the NAB, live local-news captioning was still far from universal, with only 57 percent of local stations providing it (although two-thirds of these "had a sponsor for this service").[237] And as reported by NCI researcher Carl Jensema, "Counting both broadcast and cable, about 100 hours of captioned television programs are shown on national television in the United States each day."[238]

In turning these reports into rules, the commission was once again heavily lobbied by a reluctant broadcasting industry.[239] Not only over-air networks such as CBS and NBC but also cable networks such as HBO argued that programming producers, providers, or distributors—not broadcasters—should be charged with captioning "legacy" (rerun) programming. Similarly, the Association of Local Television Stations argued against requiring local stations to caption programming that they broadcast but did not produce. The National Cable Television Association claimed, "The costs of captioning the hundreds of thousands of hours of basic cable programming alone could range from $500 million to $900 million per year."[240] (NCI responded that such estimates still used outdated and inflated estimates of captioning labor costs.)[241] Broadcasters requested blanket exemptions for "commercials, regional sports and promos," as well as any "programs where viewership is low."[242] News and information channels like C-SPAN, Bloomberg TV, and the Weather Channel argued that they already showed enough text on their screens to exempt them from captioning.[243] And there was an interesting debate over whether music videos should be captioned, with Viacom (parent company of MTV and VH1), BET, and the Recording Industry Association of America all objecting. One industry lobbyist exclaimed, "It's crazy," fearing that "a music video captioning requirement would allow young viewers easy access to pop music's most suggestive lyrics."[244] As one analyst put it, "Virtually every sector of the industry wants an exemption from the plan."[245]

The FCC's final ruling, delivered in August 1997, mandated a ten-year graduated phase-in period where "95% of new shows must have closed captioning by first quarter 2006." Despite the NCI's best efforts to convince advertisers to caption their commercials, the FCC exempted "advertisements of less than five minutes, promos and public service announcements"—but not music videos. Certain "locally produced and distributed programming" was also exempted, and live news programming could be captioned through teleprompter-based "electronic newsroom" techniques (see chapter 4). Finally, there was an important "economic burden" exemption granted without any need for special appeal: "Companies with annual gross revenue less than $3 million will not be required to deliver captions with their shows," and "companies will not need to devote more than 2% of their gross revenue to captioning."[246] The FCC decided (over the objection of the Caption Center) that "program distributors, such as TV stations, cable systems and DBS operators, rather than program producers, will be responsible for assuring captioning quotas are met." The FCC's reasoning was that "it would be easier for consumers to identify program distributors for purpose of filing complaints," and that distributors would quickly become accustomed to "including captioning requirements in program contracts," shifting the labor cost and time of captioning to producers anyway.[247]

All in all, the National Association of Broadcasters was pleased with the rules, but the National Association for the Deaf was not.[248] NAD executive director Nancy Bloch argued that the exemptions would allow "huge gaps in programming" and that, in particular, the regulation allowing local electronic newsroom captioning in place of stenocaptioning would result in "denial of video access by deaf and hard-of-hearing people to live interviews, sports and weather updates and other late-breaking stories."[249] The FCC explained that this exception was driven by a labor shortage, "to allow captioning companies sufficient time to recruit and train more captioners." But after considering the ability of realtime stenocaptioners to work remotely over telecommunications networks, the FCC bowed to its critics and came back a year later to toughen its own rules.[250] Modified regulations issued September 1998 mandated that all programming, not simply 95 percent of all programming, be captioned by 2006. Live news captioning could no longer be ignored: "The four major networks, broadcasters in the top 25 markets and nonbroadcast networks serving 50% or more of households subscribing to multichannel services also will be required to caption all live newsroom reports."[251] These big-market broadcasters were directed to use only human stenocaptioners for live news, not electronic newsroom captioning.[252] After nearly thirty years of petitioning, the D/HOH community finally had both a technological infrastructure and a regulatory regime that would deliver universal television accessibility at both national and local scales.

## The Educational Captioning Backlash

The long-awaited captioning successes of the 1990s had one negative side effect: they exposed captioning to public scrutiny like never before. In March 1998, Senators Daniel Coats (R-IN) and Joseph Lieberman (D-CT) publicly attacked *The Jerry Springer Show* as "the closest thing to pornography on broadcast television," and called upon Secretary of Education Richard Riley to end federal funding for closed-captioning of the show—some $50,000 in subsidies per year.[253] "We are confident," the senators wrote, "you will share in our outrage that the federal government is not only using taxpayer funds to subsidize their degrading and prurient program at all, but it is also judging it to be of some educational and cultural value."[254] But the Department of Education responded that the total of $7 million or so it spent on captioning that year was the only way to "provide access to the shared cultural experiences of television" to D/HOH viewers, and that not only taxpayers but also the television industry and private charitable donations funded captioning costs.[255] Two peer-review panels drawn from the D/HOH community, one at NCI and one at the Department of Education, selected the shows to be captioned, to ensure that the selections reflected the needs and desires of D/HOH viewers.[256] Secretary Riley argued, "As distasteful as *The Jerry Springer Show* may be to you and me, I do not believe it should be the role of this Department in administering the captioning program to single out particular television programs and make a cultural judgment that individuals who are deaf or hard of hearing will be denied the same access to those programs that are watched by America's hearing community."[257] The head of captioning at DOE, Judith Human, pointed out that "as long as a show passes FCC muster, which Springer has, then it can be captioned. And, since Springer just nosed out Oprah to become the most popular program, it's an obvious candidate." In an argument reminiscent of the *Cosby* debate a decade earlier, Senator Lieberman responded, "I think the least that *The Jerry Springer Show* could do, based on the enormous amounts of money it appears to making from the garbage it's putting on the air, is to pay for the closed captioning itself."[258]

The attack on closed captions for *Jerry Springer* quickly became a symbol for other causes against government "waste" and "abuse" that spring. In April 1998 Representatives Joseph Pitts and Roy Blunt used the flap to promote their Dollars to Classroom Act, which would have affected many more DOE programs than just captioning, arguing "while American school children lag behind the rest of the developed world in basic academic skills, our federal education dollars are paying for children to lean about prostitution, racism, polygamy, and other values which are completely contrary to traditional family values."[259] Senator Don Nickles (R-OK), in his opposition to a Clinton administration child-care initiative, brought "Oklahoma stay-at-home mom" Susie Dutcher to testify before the Senate. She said: "There are many things I would like to do with my husband's earnings, but . . . you seem to believe you have the moral authority and the superior judgment to make those choices for us . . . I'd like to buy more books for Lincoln, Elizabeth, and Mary Margaret, and put more money in their college fund, but you've already seen fit to use that money funding closed-captioning for the Jerry Springer show."[260] Representative Pete Hoekstra (R-MI) argued against increasing funding to the Department of Education, saying "the federal bureaucracy is out of control": "I mean, we're struggling with reading, writing and math—where do some of these dollars go? Federal dollars pay for closed-captioning of *Baywatch* and *Jerry Springer*—is that going to help us teach our kids how to read, write and do math?"[261]

The irony in these accusations was that a law had already been passed to phase out DOE funding of noneducational programming by 2001. As early as May 1996, in the wake of the Telecommunications Act, the Senate had reported that "once the FCC regulations are in place, the committee would expect the Department of Education to direct its captioning resources toward programming in which there is a clear public interest in ensuring its accessibility."[262] So it happened that in the Individuals with Disabilities Education Act amendments of 1997 (PL 105-17), DOE was instructed that its captioning funds could be used only for "educational, news, and informational television, videos, or materials." But after the *Springer* episode, DOE found itself in the odd position of soliciting "comments and recommendations from the public on what the term 'educational, news, and informational' encompasses in reference to the description and captioning of television, videos, and materials."[263] In future years, recipients of DOE captioning funds would be expected to "identify and support a consumer advisory group, including parents and educators, that would meet at least annually" in order to "certify that each program captioned or described with project funds is educational, news, or informational programming."[264]

Nevertheless, captioning remained an easy target for those eager to score "family values" points against the perceived excesses of "big government." Despite its own previous rules, the Department of Education under the Bush administration "cut off its closed captioning for nearly 200 TV shows" in late 2003.[265] Bypassing the D/HOH parent committees and professional educator recommendations, DOE based its decision on "an external panel of five unnamed individuals," complained the National Association of the Deaf.[266] A DOE spokesperson "declined to name the five experts who volunteered their time . . . or to offer the parameters indicating that *Andy Hardy* and *Inside Edition* are educational and informational but *Discovery Jones* and Lifetime's *Biographies of Women* are not."[267] The NAD argued that this amounted to "censorship," excluding D/HOH children from "shows that help them learn about the trends, culture, and society around them."[268] But from the federal government's point of view, broadcast justice was no longer the purpose of federal funding for closed captioning; education was. The rhetoric that had helped universalize the captioning infrastructure had thus also helped to minimize the power of captioning's core audience of deaf and hard-of-hearing Americans.

In January 1972, the *Deaf American* enthusiastically reported that "NBS engineers estimate that the simplest caption display model, which may be available as early as 1973, would cost less than $20 and would have numerous options to assist the hearing handicapped."[269] But creating a universal broadcast closed-captioning system was not to be so simple. Caption decoders, not even available until nearly a decade later, would cost over ten times this original estimate and would not fall to a negligible cost until after the government subsidized their development on a chip and mandated their inclusion in all TV sets. Captioned content, already pioneered by PBS in 1972, would not really hit critical mass until after the government mandated full captioning. And the "hearing handicapped," although instrumental in developing and promoting the captioning system in the 1970s and 1980s, would become just one of many captioning audiences, behind hearing children and immigrant adults learning to read English. The struggle for broadcast justice took so long that several instrumental actors, including captioner Doris Caldwell and captioned films chief Malcolm Norwood, would not live to see the success that they helped bring about. Caldwell passed away in 1985, at the tail end of the teletext debate, and Norwood died in 1989, just before the Television Decoder Circuitry Act was passed.[270]

Nevertheless, by the time corporate captioning advocate Julius Barnathan of ABC passed away at the age of seventy in 1997, closed captioning—by his own admission, "his proudest accomplishment"—was finally an accepted feature of the time and space of the television screen.[271] Late-night television comedians Conan O'Brien and Andy Richter even had a regular skit where they would pretend to turn on the closed captioning for home viewers (and the studio audience) during a conversation, with the realtime captioner surreptitiously making fun of the two hosts as the bit unfolded—having the audience sing "Row Your Boat" or making fun of O'Brien's famous hair.[272] Closed captioning had finally displaced the shouting of Garrett Morris as the proper way to poke fun at TV accessibility.

But the success of closed captioning also held some bittersweet irony. It came at a time when the television screen was already littered with text—station logos, sponsor slogans, and sports, news and weather "crawls." Even before the horrific, nationally televised events of September 11, 2001, put text-based news on nearly all broadcast and cable channels, CNN had made over its *Headline News* channel to look more like a Web page in hopes of attracting younger, more computer-savvy viewers.[273] A few years later, CNN's market research confirmed that 70 percent of its viewers liked the ever-present crawl.[274] By this time, the bottom-of-the-screen text had become so ubiquitous that it generated its own marketing scams, such as the TV Shield, a "space-age specially formulated film which quickly adheres to any TV set" so that home viewers could supposedly cover up the crawl (otherwise known as masking tape).[275] Not only did all of these textual enhancements compete with closed captions for precious space on the television screen, but they also revealed the faulty assumptions behind the decades-old foundational belief of television captioning—the idea that, as one PBS executive explained it in 1980, "you really can't do much open captioning because we've found the rest of the audience finds it disruptive."[276]

This was not the only conventional wisdom the history of closed captioning overturned. Neither the hands-off, voluntary suggestions of the FCC in the 1970s nor the neoliberal "public-private partnership" of NCI in the 1980s proved effective enough to sustain the closed-captioning system; instead, a decidedly progressive round of re-regulation on both the demand side (decoder distribution) and the supply side (program captioning) was necessary to finally bring the promise of broadcast equality to the deaf and hard of hearing. Yet many alternative models for media access did not enjoy anything near the same success; late 1970s projects dealing with "Morse Code for the Deaf" and "captioned radio" died a quiet death even before teletext appeared in the 1980s.[277] In the end, the decision to adopt any kind of same-language text-captioning system—even hidden "closed" captions, which would not bother hearing viewers—was not only a technical question of finding the "best" standard (Line 21 captioning versus teletext, for example) but also a normative question of how much the government should compel privately owned equipment manufacturers, program producers, and network distributors to participate in the creation of a public information system for a minority of the population.

As time passed, the official D/HOH audience for captioning (and the official target market for decoder sales) kept shrinking, from 14 million in 1979 to 1 million in 1989. But at the same time, more and more publics were (rhetorically, if not actively) mobilized as potential audiences for captioning, as Gallaudet University's Mark Goldfarb argued before Congress in 1994: "Closed captioning is like the sidewalk curb cut in that it helps many groups of people, not only individuals who are deaf or hard of hearing. Nearly 100 million Americans can benefit from watching captioned television."[278] Soon this "curb cut" was appearing in more and more public spaces, bringing television to hearing viewers in situations where either the audio track had to be turned down (such as in a hospital or a health club) or where the noise was too great to hear an audio track anyway (such as in a tavern or on a busy street). Quipped one writer in the *New Yorker* in 2000, "If it weren't for closed captioning, all those Times Square tourists with their necks craned backward wouldn't be able to 'hear' the *Today* show on the Panasonic Astrovision Screen every weekday morning."[279]

This spatial and temporal ubiquity of closed captioning has had some contradictory effects on its original D/HOH constituency. The fight for television accessibility brought together the different factions of the D/HOH community—oralist versus manualist, non-hearing versus hard-of-hearing, culturally Deaf versus educationally mainstreamed—for the first time in a long time.[280] But along with the social unity and program success came a new set of "disability" labels and associations, especially troublesome for Deaf-culture advocates who saw themselves as a distinct language community to be cherished, not a population of handicapped individuals needing to be "cured."[281] Hopeful D/HOH activists had long assumed that television would be the first in a series of consciousness-raising battles to "make deaf people all the more alert to what can be done, not only on television but also in other media."[282] But the legitimization of captioning had only come when captions had been seen as tools for the hearing: "a way to follow the action on TV while you answer the phone and to watch a late movie without disturbing your dozing significant other."[283] On the eve of the 1993 effective date of the Television Decoder Circuitry Act, two Gallaudet University researchers cautioned, "With more people watching captions, the number of programs with captions should increase and full access may become a reality. The larger potential audience, however, also means that television manufacturers and caption providers will be attending to the needs and wishes of more special interest groups, and they may tend to address the desires of the larger groups or those with greater spending potential."[284] Especially for the newest caption providers to be brought into the mix—the realtime reporters—the "spending potential" of all audiences, coupled with the new political-economic environment of mandatory captioning laws, would produce a new geography of for-profit captioning over the next decade, further distancing the technology from the world of D/HOH education.

# Chapter 6

# Privatized Geographies of Captioning
# and Court Reporting

Who is the person who sits in front of the courtroom, usually between the
judge and witness, furiously tapping his or her fingers on a strange looking keyboard? Who is the person who walks into a crowded room filled
with lawyers and potential witnesses and calmly takes control of the situation? Who is the person responsible for having closed captioning appear
at the bottom of millions of television screens across the nation?

*From a court-reporting training manual, 1999*

In 1993, the same year that the Television Decoder Circuitry Act
(TDCA) went into effect, ensuring that anyone who bought a new television set in the United States would have access to closed captioning,
the market for captioning labor behind the television screen was undergoing a dramatic change. The National Captioning Institute (NCI)
still performed about 80 percent of all captioning for the nation's television production and distribution companies. But the nonprofit monopoly it once held was under pressure from a slew of for-profit
startup firms creating new competition for skilled captioning labor. In
Hollywood, nearly three dozen NCI captioners and technicians who
had organized with Local 53 of the National Association of Broadcast
Employees & Technicians (NABET) threatened to strike in August, re-
jecting NCI's contract offer of a three-year wage freeze and a reduced
number of vacation days and holidays. Captioners picketed with signs
charging NCI was "deaf to TV closed-captioning labor" in early September. Interim NCI president Edward Merrill Jr. responded, "We're
caught like everybody else with rising prices and competition." But
the labor tactics worked, as the two sides were eventually able to reach
an agreement calling for a pair of 2 percent wage increases rather than
a wage freeze.[1] These were the only unionized captioners among NCI's

three sites in New York, Virginia, and California, and they performed only prerecorded captioning, not live stenocaptioning.[2] But nonunion and live captioners were exerting pressure on NCI in other ways.

That same year, on the other side of the country in Washington, DC, former court reporter and NCI-trained stenocaptioner Lorraine Carter entered into direct competition with her previous employer after she started her own realtime captioning firm, Caption Reporters Inc. (CRI). Carter got started with $30,000 in capital secured from her first client, local television station WUSA, which had hired her to caption its evening newscasts. To compete with NCI, Carter and her captioners worked out of an office in the WUSA studios, just as a court reporter, or someone performing computer-aided realtime translation (CART), would bring equipment to a client's site (see chapter 4). By 1997 her company grew to three captioners and a quarter-of-a-million dollars in yearly revenue, counting Gallaudet University among its clients. Instead of the $1,200 per hour that NCI had charged for live captioning when Carter started, by the late 1990s CRI charged only $450/hour.[3] One observer summed up the shift in the landscape in 2000: "Years ago, the challenge was to convince potential clients that captioning was affordable, and that they couldn't afford *not* to do it. Now that live captioning is mandatory in some markets, the challenge is the price competition."[4]

This new landscape developed slowly. In the late 1980s, back when closed captioning's future was still very much in doubt, only a few private closed-captioning firms competed with NCI and WGBH— two in California and one in Pittsburgh, PA.[5] But the dual legislative victories for captioning in the 1990s—the 1990 TDCA, which mandated universal decoder provision to households, and the Telecommunications Act of 1996 (TCA), which mandated universal captioning of program content—provided the incentive to transform both offline and online closed captioning from a centralized, nonprofit, government-subsidized institution to a decentralized, profit-oriented, market-based industry. As one media analyst wrote of the closed-captioning audience in 1993, "Would anyone be interested in a national market segment that is scheduled to increase in size by more than 3,000 percent over the coming year?"[6] By the late 1990s, there were over a hundred private, competing captioners across the United States.[7] However, privatization did not end the government subsidy to closed captioning; rather, it benefited from such subsidy. Enabled by the new telecommunications innovations of the Internet and direct-broadcast satellite television, privatization helped change the spatial, temporal, and social division of labor in captioning. By the end of the 1990s, caption production had been transformed into a side business of increasingly concentrated video-production and court-reporting firms, and caption labor had moved out from the full-time broadcast booth and into the part-time home office. This uneven geography of captioning training and production emerged in the 1990s at the intersection of government regulation, technological innovation, and professional survival.

### A For-Profit Captioning Industry

The 1988 report of the Commission on Education of the Deaf (COED) was instrumental in making a case for government re-regulation of closed-captioning on educational grounds (see chapter 5). But besides faulting network broadcasters for failing to fully fund captioned programming, the COED faulted the nonprofit captioning agencies for failing to nurture a for-profit captioning industry. By subsidizing NCI and WGBH, the commission claimed, the federal government was helping to keep the labor costs for captioning artificially high. Because for-profit captioners could not successfully compete for federal grant money to caption programming, they were at an artificial economic disadvantage, even though "some offer captioning rates under $1,000 per hour, 50 percent lower than what [NCI and WGBH] charge." Thus the COED argued that "federal funding for captioning, instead of being distributed to captioners, should be distributed to producers and broadcasters," who would then spend this subsidy competitively, choosing whichever captioning provider could do the work for the lowest cost.[8] The 1990 TDCA failed to include such a provision, however, and the private captioners continued to protest. The president of Real Time Captioning Inc. (Van Nuys, CA) argued in the *Wall Street Journal* that "they take my tax dollars and go in and undercut me"; similarly, the head of Captions Inc. (Los Angeles, CA) related that "his company spent $3,000 last year preparing a grant application, but National Captioning Institute won the grant."[9] The more than $7 million allocated by the US Department of Education (DOE) in FY 1992 for captioning costs was all allocated to NCI and WGBH, except for the $1.16 million allocation to offline captioning of daytime programming and online captioning of daytime CNN, which was split between

NCI and the private firm Media Captioning Services (Carlsbad, CA).[10] Public comments on these DOE contracts sometimes "expressed a concern that the priority should be defined in such a way as to not give undue competitive advantage to nonprofit organizations," but the response from department in the early 1990s was that "the priority as written gives no competitive advantage to nonprofit organizations."[11]

Despite the vocal complaints of unfair competition, the top-level grants from DOE tell only part of the captioning story. True, NCI was the biggest recipient of federal funds, but it was also the biggest and oldest captioning provider, employing about 114 offline captioners and twenty-one online captioners at its Virginia site alone in the mid-1990s.[12] Its main rival, the Caption Center in Boston—which in the early 1990s employed some ninety captioners of its own—often subcontracted out work that it received through the federal government to private captioning firms. In 1992 the Caption Center awarded work to Captions Inc., Real-Time Captions Inc., CaptionAmerica—even to NCI itself.[13] Similarly, high-profile captioning sponsorship for large events, such as BellSouth's sponsorship of the 1996 Olympic Games, was often divided between for-profit and nonprofit providers.[14] A new geography of captioning was emerging which encompassed both economic forms (see figure 6.1).

There also emerged in the 1990s a new market for captioning where the nonprofits held no advantage over the startups: live and legacy television, which were not permitted to be subsidized by federal funds. For example, the head of Captions Inc. "found he could compete successfully with the nonprofits" for contracts to caption religious programming.[15] Captions Inc. also won a contract to caption Disney video releases in 1988, as federal grant money did not cover home-video captioning at the time.[16] NCI captioned many videos as well but had to solicit contributions for such work from "individual film buffs who develop hearing difficulties as they grow older."[17] By 1993 *Gopen's Guide to Closed Captioned Video* listed four companies besides NCI and WGBH who captioned most of the five thousand videos it listed.[18] That year the Computer Prompting and Captioning Co. (Rockville, MD) advertised, "For $250 we will open or close caption your 30 minute video with our premium service. Or, for just $150 our economy service will open or close caption the same video with basic captions." Clients could even buy captioning software direct from this firm for about $1,500 if they wanted to do the work themselves.[19] By

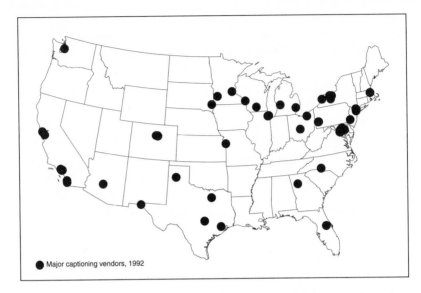

Fig. 6.1. Location of major US captioning agencies, 1992. While nonprofit captioning under NCI and WGBH had been conducted exclusively on the East and West
Coasts through the 1980s, the growth of for-profit captioning vendors expanded
the geography of captioning labor across the nation. Source: Map created from data
in Gallaudet University, National Information Center on Deafness, *Providers of captioning services for video productions* (Washington, DC: Gallaudet University, 1992)

1994 NCI was captioning some eight hundred new and existing videos
annually, but another two hundred were captioned by other firms
each year.[20] The market still had much room to grow, since nearly five
thousand videos were released each year, only 20 percent of which
were captioned at all.[21] Even the venerable Captioned Media Program
(formerly Captioned Films for the Deaf, chapter 1), which now used
closed-captioning techniques to open-caption VHS-based films for its
subscribers, increasingly contracted out to private-captioning companies.[22] For these vendors, the lines between traditional cinematic subtitling and television closed captioning were increasingly blurred.

Online captioning also flowed more easily to the startups in some
cases. NTID CART pioneer and local-news stenocaptioner Linda Miller
(chapter 4) was by the late 1980s president of her own firm, Captioning
Resources of Western New York, providing both postproduction and
live-captioning services.[23] Lacking federal funding, firms like Miller's
pioneered the practice of securing corporate sponsorship for their

realtime captioning work, "plugs at the end of the programs" (and within the caption text).[24] Local firms like these might grow by expanding geographically, by branching out from captioning to CART, or by doing both. Caption Colorado started in Greenwood Village, CO, in 1991 with just "three captioners and one customer—Denver's KCNC-TV." A decade later it had cut its original realtime captioning costs from a high of $800 per hour to "about $120 an hour." Besides captioning local news in their hometown, they used telecommunications connections to caption local news in "96 stations in 40 markets in more than 30 states." Meanwhile, back at home, they also realtimed the "live commentary to deaf or hard-of-hearing Broncos fans at Invesco Field at Mile High [Stadium]."[25] Success came through a diverse geographical mix of realtime work—local and national, in-person and remote.

What some free-market advocates forgot, however, was that the national nonprofit captioning agencies had a hand in the success of the local for-profit startups. Many new realtime-captioning business owners in the late 1980s and early 1990s were originally trained by NCI and WGBH, especially former court reporters doing local-news captioning in their home cities.[26] In 1988 NCI described how it assisted local television stations by "training local court reporters to caption, fundraising to underwrite the captioning, advising the station on promoting the captioned newscasts and working with area decoder retailers on cross promotional activities."[27] By 1990 the WGBH Caption Center had established an External Captioning Projects department to consult with local-news directors and station managers, train potential realtime news captioners, and even "work with the community of court reporters and their schools to develop new curricula and certifications" for realtime captioning.[28] By the late 1990s, the WGBH Closed Captioning University was offering both free training and free software for public-television-station employees engaged in offline captioning—and building a ready labor market for freelance work at the same time.[29]

Not only were NCI and WGBH helping to grow the private captioning industry, but the federal government began to target captioning subsidies directly to private firms. In July 1997, a House report on FY 1998 appropriations to the DOE stated that "as the captioning program transitions from an appropriated program to be mandated in association with the Telecommunications Act of 1996, the Committee recognizes the importance of very small businesses in a robust and competitive captioning market that provides the taxpayers and consumers the best value."[30] Some firms that had barely survived to this point suddenly found themselves with windfall contracts. Max Duckler had started CaptionMax Inc. (Minneapolis, MN) "in a corner of his son's bedroom" in 1992, anticipating a boom in business when the TDCA went into effect in 1993. But "there were times when I went months without an assignment," he reported, until 1997 when he was able to win "a three-year, $340,000 grant from the Department of Education to caption syndicated shows." By 2000 he had increased his staff tenfold, from three to thirty-three, and opened three new offices around the country.[31] Subsidizing such entrepreneurs was now an official goal of the government's captioning program. In the Department of Education's annual announcement of grant applications for FY 2002, bonus points were awarded for new captioning providers who were never primary grantees or subcontractors before—giving preference to new private businesses over the older nonprofits.[32]

## The Convergence of Captioning with Court Reporting: A Case Study

Neoliberal strategies of government subsidy to support private profits in captioning weren't the only processes affecting this new industry, however. As demonstrated in the cases of Captioning Resources and Caption Colorado above, the convergence of captioning with court reporting (see chapter 4) was happening just as the court-reporting industry was undergoing increasing consolidation and globalization. Individual court-reporting-agency owners and freelancers with captioning and CART skills often found that the new realtime aspect of their work began to crowd out courtroom and deposition reporting. Said one reporter in 1991, "In the past year that I've been captioning, I've done less and less traditional court reporting work. It's probably about 90 percent captioning and 10 percent reporting. I have other people who work with me in my freelance firm, and they take most of my deposition work."[33] Even as stenocaptioning rates fell due to competition, realtime reporters often found that "captioners can sometimes earn more money in fewer hours than in court reporting, without the time-consuming work of preparing written transcripts."[34] (Here again the split between taking down and writing up, described in chapter 3, was key to the court reporter's work life.) Two decades after realtime television captioning began at NCI in the early 1980s, the NCRA "gave its first exam in November [2003] for certification as a broadcast captioner." Association and industry experts estimated that out of the three to four hundred working captioners across the United States at the time, 90 percent were former court reporters.[35] Had court reporting become part of captioning, or had captioning become part of court reporting?

One way to explore this question is to follow a single freelance reporting firm over these two decades, to see how new technologies, labor markets, and competitive pressures in both captioning and court reporting came to bear on that firm's fortunes. As it happens, two well-known realtime-captioning pioneers—the first courtroom realtimer, Joe Karlovits, and the first broadcast stenocaptioner, Marty Block—founded just such a firm in the mid-1980s. Karlovits recounted how in 1982, his media exposure from realtiming for the Supreme Court, plus the CBS effort to push teletext (chapter 5), combined to offer his court-reporting firm a new opportunity: "I was living in Boston. My reporting company was hired by WGBH, the PBS giant, to help develop captioning for ExtraVision."[36] Around 1983, Karlovits and his first business partner, Ed Fulesday, began selling realtime captioning services.[37] Three years later, Karlovits, Fulesday, and now Block had moved their efforts to Pennsylvania and incorporated American Data Captioning (ADC) in Pittsburgh, with a contract to caption local affiliate KDKA's evening news.[38] In 1987 another colleague, William McNutt, opened a branch of ADC in Arizona, and together with the Pittsburgh team, captioned the pope's visit to Phoenix.[39] ADC even made connections to the captioning-hardware and software industry, helping the CAT firm Xscribe in "the development of a stand-alone captioning system" by playing a role somewhere between "user" and "innovator."[40]

ADC soon went national. In 1988 both NBC and CBS contracted with ADC to caption the Republican National Convention that nominated George H. W. Bush for president (ABC went with NCI instead).[41] A year later NBC returned to hire ADC as its contractor for the three-hour live-captioning of *The Today Show* each weekday, and by 1992 the company was captioning *The Tonight Show* for NBC as well.[42] As a result of these changes, around 1990 the company changed its name to CaptionAmerica, hired Jeff Hutchins of the WGBH Caption Center as CEO, and advertised its specialty as realtime captioning.[43] With an hourly stenocaptioning rate of $1,100, CaptionAmerica proudly claimed, "Our 'for-profit' rates are far lower than the 'nonprofit' rates of other companies." The company did not restrict its labor to live captioning, however, performing postproduction captioning as well for the slightly lower charge of $990 per program-hour. The company advertised its geographical location in the rustbelt as one of the secrets of its success, in contrast to the East and West Coast offices of NCI and WGBH. According to one CaptionAmerica executive, "With satellite receivers, we can still receive [live] programs from all over the country," but "having offices in Pittsburgh allows us to keep operating costs low."[44] CaptionAmerica kept growing and soon shed its name once again. By 1996 the company employed eighteen full-time stenocaptioners and was rechristened VITAC, for "vital access through captioning." The new name both drew upon the 1990s language of "accessibility" and pointed the way to the increased provision of CART services.[45] VITAC's growth continued through the 1990s with the FCC rules mandating universal captioned content. By the turn of the millennium, Karlovits was "president of the largest for-profit captioning company in the world."[46]

VITAC now employed nearly two hundred people—at least twenty-five of whom were full-time stenocaptioners, with another sixteen freelance "remote captioners" on call.[47] This difference between the two categories of VITAC employees mirrored a spatial and technical division of labor in the captioning industry as a whole. In-house captioners often worked on national accounts and were "paid a competitive salary, plus benefits, including health, dental and eye care; life and long-term disability insurance; pension; and paid vacation, holidays, personal days and sick days."[48] Remote captioners, by contrast, were more likely to work on "local news and sports" under the VITAC banner.[49] Signing on with VITAC remotely had both costs and benefits. New distance captioners were obliged to "incur an initial start-up cost of approximately $3,500, which includes the computer that has been ordered to VITAC specifications compatible with all necessary hardware and software" and had to pay for their own "dedicated high-speed Internet access." The company's rationale for such costs and standardization was that it streamlined technical support: "This system allows VITAC's technical staff the capability to download software remotely to your computer or troubleshoot your set-up at any time," or even "to drop-ship a replacement piece of equipment for any malfunction at the captioner's end to protect their earning capacity."[50] Even VITAC's in-house captioners worked remotely in that they captioned broadcasts far from where those programs originated. One full-time VITAC captioner, Melissa "Missy" Macri, who worked on the *Today Show*, was described as "a kind of super stenographer who sits in a small ergonomically correct room some four hundred miles away" from where the show was taped in New York City. An employee of VITAC since the CaptionAmerica days in 1992, the thirty-two-year-old Macri rose at 3:00 a.m. each day to drive to work and begin her shift at four. "She does *Today* from seven to nine, has breakfast, and captions *The Rosie O'Donnell Show* at ten. She leaves work at noon, drives home, naps, and usually turns in at around nine-thirty."[51] Quipped Macri, "It's not good for my social life."[52]

Finding talented stenocaptioners like Macri was not easy. Karlovits reported that many court reporter applicants had abundant stenographic skills but little media literacy: "More than 90 percent fail our current events test on subjects such as: Who is the Chief Justice of the Supreme Court? Name three members of the US Supreme Court. Who were the Democratic and Republican candidates for President and Vice President in 1992?"[53] Those who passed this screening entered a "broadcast realtime compatible" training program of steno theory and technique which could be grueling, according to Hutchins: "A young lady will come in here, pretty good court reporter, very confident about her abilities, excited that she's going to get into captioning, and she will begin the training process very fired up, excited." But "in two to four weeks she is going to be walking around with stooped shoulders, totally dejected, feeling like, 'I'll never get this.'"[54] Perhaps in an attempt to externalize such training costs, VITAC offered its stenocaptioner training manual for sale over the Web in 2003.[55]

Those who pass training are not put on the air immediately. One VITAC trainee, a veteran court reporter since 1976, wrote of her experiences as a "captioning newbie" in the *Journal of Court Reporting* in 2002. After much learning and practicing of the realtime steno style, she recounted how "the day finally came for my first simulation." From a remote location, "you hook up to the encoder at corporate headquarters, and you write an hour or so of programming to a television in a control room where your boss and perhaps their bosses watch and evaluate whether you are ready to go on-air." This trainee went through at least three such simulations.[56] Her first remote-captioning job for VITAC turned out to be an hour of captioning a home shopping channel—monotonous, but better than honing her self-described "novice" skills on ABC's *World News Tonight*.[57] VITAC even advertised that it offered "hundreds of weekly program hours that are 60–80% pre-scripted so that you can clock a high number of on-air hours with little to no risk of injury that may normally accompany high-end hours."[58] Although framed in terms of an employment benefit lessening the risks of repetitive-stress injuries like carpal tunnel syndrome, such "pre-scripted" hours also offered a way for lower-cost, novice captioners to learn the online captioning craft with minimal risk to the firm. As a result, the boundary between realtime and offline captioning blurred even more.

The boundary between employee and entrepreneur—as well as the boundary between reporting firm and captioning firm—could be blurry as well. Just as the first private captioning firms spun off from NCI and WGBH, many local captioning firms later spun off from the national private companies. Take the case of Judy Brentano, an accomplished freelance court-reporting-agency owner who learned captioning under Karlovits, Block, and Fulesday on the KDKA local news contract in Pittsburgh in 1987. The following year Brentano founded her own local captioning firm, The Caption Company, in Atlanta, GA, with a contract to caption the local TV news. In her spare time she headed the NCRA in the early 1990s, providing another high-profile voice in the court-reporting community to urge for a transition to captioning work. Brentano's captioning company won one of the contracts awarded by the Olympic Committee for the 1996 games in Atlanta and, as she described, not only captioned television programming for the local NBC affiliate but "provided virtually instantaneous text for all of Nike's daily Olympic press conferences for upload to their interactive site on the Internet," as well as CARTing "the opening and closing ceremonies on site at the Centennial Olympic Stadium, with the captions being projected on the Panasonic Jumbotron screen!" Like other successful private firms, Brentano's company prospered by combining national work with local work and captioning work with CART work. In 1999 her old employer VITAC purchased her spin-off captioning company, making Brentano "general manager of the VITAC MetroCaption division," which handled local-news captioning around the nation.[59]

Brentano's story illustrates that freelance stenocaptioning firms often performed offline as well as online captioning (or even language-translation film subtitling), and freelance reporting firms often hired one or two stenocaptioners for broadcast or CART work.[60] This blurring of industry boundaries was set against a backdrop of corporate consolidation. In the late 1990s "buyouts of locally owned court reporting services by national chains" began to "rock the established court-reporting boat."[61] For example, in 1996, the publicly traded Esquire Communications Inc. (San Diego, CA) "received a $20 million credit infusion from a Chicago equity investment firm" and subsequently purchased twenty-four court-reporting firms. A year later Esquire bought DepoNet, "the largest national referral service for court reporters and other support staff." By 1998 Esquire sold court-reporting services in twenty-four markets among eleven states and the District of Columbia, employing 1,400 freelance court reporters (though it had yet to turn a profit). Such consolidation had contradictory effects on the court-reporting community. Some smaller freelance-agency owners were happy to be able to cash out of their businesses at a time when "court reporting as a business [was] no longer growing at the huge rates of the 1980s." However, some high-earning "on call" freelance reporters claimed that their new corporate owners standardized and slashed page rates, leaving them with less compensation for the same work.[62]

Leaders within the field were split as to whether consolidation was a boon or a threat to their profession. Overall industry statistics, reported in the most recent economic census, reveal that between 1997 and 2002, while the number of overall court reporting and stenotype establishments actually grew by about 10 percent (to 3,322 firms), the total number of paid employees in these firms shrank by 13 percent (to 13,516). On average, owners were taking in greater per-firm revenues, and workers were earning higher per-employee salaries. However, these gains were unevenly distributed across the country, as the 1997 figures showed that the top five states for sales of court-reporting and stenotype services (CA, FL, TX, NY, and PA) together represented fully 42 percent of all firms in the business, 52 percent of all sales, and 49 percent of all paid employees.[63] The increase in wages was uneven as well. Between 2000 and 2004, the earnings of the top 25 percent of all non-self-employed court reporters rose from $51,704 to $60,760, an increase of 18 percent. But the earnings of the bottom 25 percent grew from $28,630 to $30,680, an increase of only 7 percent.[64]

The industry consolidation, and its uncertain effects, extended to captioning and CART work as well. One of Esquire's main competitors, LegalLink (Boston, MA), acquired fourteen court-reporting agencies by 1998, "focused in major cities such as Chicago, Los Angeles, New York City and Washington, D.C."[65] But LegalLink expanded into "transcriptioning and videography" besides court reporting, consolidating its stenographic services with the same electronic recording services court reporters had long feared. By the turn of the millennium LegalLink had restructured itself under a new, global parent company named WordWave to reflect its wider target market of "legal, broadcasting and entertainment, and e-commerce" industries.[66] Finally, in January 2000, WordWave purchased VITAC, bringing what had been a locally based, startup closed-captioning firm firmly into this new global consolidation.[67] Today WordWave describes its workforce as neither "court reporters" nor "closed captioners," but as "experts in voice-to-text conversion," part of "the largest court reporting and broadcast-captioning company in the world" and "the emerging leader in creating searchable digital audio and video, as well as captioning Internet webcasts and conferences."[68]

## An Electronic Cottage Industry: Captioning Careers

For individual captioners, the consolidation and globalization of the "voice-to-text conversion" industry might seem far away indeed. The majority of the estimated five hundred stenocaptioners employed in the United States in the early 2000s worked out of their homes.[69] One report from 2001 described how a "stay-at-home mother" discovered she could make "as much money as she did working full time as a court reporter, without having to prepare written transcripts" (taking down, not writing up) by working a sixteen-hour week captioning local and national news broadcasts out of her home.[70] Similarly, a 2004 article detailed how "a court reporter for 18 years" and "40-year-old suburban Cincinnati mother of three" was "enticed by the flexible hours of her new career as a broadcast captioner" because she could work part-time from her "bedroom-turned-office" twenty-two hours a week, captioning "newscasts, football games, tennis matches, cooking and shopping shows."[71] Sally Bennett, a court reporter for sixteen years, started captioning as an NCI subcontractor in 1996 because, in her words, "I burned out hearing about all those sad things going on" in the courtroom. Her captioning duties involved both local and national broadcasts, but since "she doesn't get paid when she doesn't work," not only was she forced to work nights and weekends, but she had even "taken her equipment on family vacations" and worked out of hotel rooms.[72] And Mary Kay Webster of Denver, CO, described that she became a captioner after sixteen years of reporting because, in part, "unlike court reporting, where we must remain unbiased and emotionless, I work alone and have the pleasure of smiling, laughing out loud, cheering, or crying."[73] Individual captioners like these were increasingly subcontracted by the larger nonprofit and for-profit firms at a temporal and spatial distance as "entrepreneurs" working out of their "electronic cottages"—in 2000, NCI alone worked with "independent contractors in 16 different states."[74] As part-time labor sited in the suburban home, this shift helped reproduce a new set of gender assumptions about realtime reporting—closed-captioning, CART, and court reporting became naturalized as "women's work."

If the daily labor of the nation's remote stenocaptioners shared anything besides this gendering, it was a facility with technology and a demand for technological space. Besides purchasing and maintaining one's own realtime CAT equipment, a captioner needed to be able to access a wide array of television broadcasts from across the nation, requiring multiple satellite-television systems for video feeds and, in cases where video was not available, multiple phone lines for audio feeds.[75] A typical remote realtimer might have in her home office "three computers; three steno machines; audio equipment, . . . to capture an audio line for those shows she cannot see on television; a DSS satellite dish with sports and entertainment packages; three modems; and five telephone lines."[76] Multiple components were needed both for backup in case of technical failure and to allow reporters to multitask during their labor process. For example, one realtimer mentioned, "I know several captioners who keep their second computer dialed up to the Internet while they're captioning. That way during breaks they can research names and spellings." Household offices needed to be closed off from family turmoil in order to enable the intense concentration needed for sustained stenocaptioning. And because one's stenographic dictionary represented years of carefully honed theory, practice, and experience, realtime CAT PCs could not be shared with children and spouses. Even ergonomics were essential: "You may be writing for long stretches at a time, depending on what kind of work you do. A good chair is critical. Don't scrimp on the chair!"[77] A home captioner's overall computer-technology and office-space expenses could be substantial.

Boosters of remote captioning argued, however, that theirs was a reporting niche that paid enough to compensate for the odd hours and specialized equipment costs. Although in the early 1990s it was common for captioners to "take a pay cut" compared to their freelance reporting peers, by 2000 some in the media quoted salaries of "$100,000 or more" for remote captioning.[78] Judy Brentano more realistically estimated in 2000 that salaried, experienced realtime captioners could earn $45,000–$75,000 annually plus benefits and that independent contractors could earn $36,000–$72,000 without benefits, at hourly rates from $55 to $100. But Brentano warned, "I don't think it's prudent to be on-air over 20 or so hours a week for quality and personal health reasons."[79] Balancing recovery time with earning potential might be tricky, but advocates argued that the work was its own reward: "It's unbelievably stressful, but it gives you the reward of a tangible product that serves a very important purpose."[80]

How could a person find such labor and achieve a balance between "doing well" (monetarily) and "doing good" (morally)? One model was the career of captioner Dorin Radin. Graduating from a Syracuse, NY, court-reporting school in 1982—just as CAT was moving to the personal computer—Radin was one of only four out of forty-five students in her class to find a job. She worked as a salaried official reporter in Onondaga County, NY, until 1988, when she "received a call from a local nonprofit organization" seeking a captioner for the local CBS newscast. She developed her realtime skills on this job for a year and, by 1989, had hooked up with VITAC, gaining further experience there until 1990. Returning to her hometown, she freelanced part-time for a series of clients until 1995, when she incorporated her own firm, Caption Advantage. The captioning provisions of the TCA of 1996 allowed her to grow her realtime business rapidly through the late 1990s. But she also captioned pretaped shows as well, charging a higher fee for offline work ($200–$400 per hour-long video) than for online work ($100–$250 per hour-long session).[81] By 2000, Radin had come full circle, still working out of her own basement but also employing "a staff of seven who also work from their homes in California, Arizona, Canada, Vermont, New Hampshire, Florida and Iowa."[82] This was a common pattern: many reporters who owned their own small businesses doing CART and court reporting locally also subcontracted as remote freelancers for national captioning companies like VITAC, MetroCaption, and Caption First.[83]

These hybrid arrangements—acting as both subcontractor and sub-contractee, captioning both offline and online—reveal that the market for local captioning work is rarely enough to sustain a business on its own. As court reporters have long known, "If you want to eat regularly, you have to do a little bit of all types of court reporting unless you are satisfied with just enough to get by."[84] One former VITAC contractor in Pittsburgh split her time captioning the local news and CARTing at a local church.[85] A thirty-four-year-old reporting school graduate in Kuna, ID, combined CARTing twenty hours a week for Boise State University students with doing "conventional court reporting at depositions."[86] And in St. Louis, co-owners Judi Bennett and Susan Rick started their freelance-captioning firm Hear Ink "out of frustration and the stress of being court reporters" but "were quickly disappointed"—finding it hard to get enough captioning clients, they went back to taking depositions as well.[87] Switching between the two modes of work—realtime captioning or CARTing and standard deposition writing—is difficult. One reporter who tried it found that she had to schedule different types of jobs on different days: "The differences in role, mindset and writing style were too great to go easily from one to another and back again in the same day."[88] But such transitions are necessary given the "feast or famine" cycle of work availability.[89] One court-reporting trainer recommended, "Realtime writers can write anything in realtime, including text documents for word processing and even e-mail messages."[90] In this way, realtime reporting is redefined as "fast input and transcription" (FIAT) or, more mundanely, "rapid data entry."[91]

Realtime reporters and captioners acting as data-entry workers in fields like medical transcription or audio transcription might be able to outperform their QWERTY-keyboard-using peers; however, the income from such labor is well below what court reporting, captioning, and CARTing can bring.[92] By the mid-1990s, the same telecommunications links that enabled realtimers to work on remote-broadcast-captioning projects offered yet another market for the skills of these reporters. Captioning for a "cybercast"—a live, digital video feed broadcast over the Internet—was technologically similar to regular captioning. As one advocate described it, "The reporter simply dials the Internet captioning server just like you'd dial an encoder at a TV station.

The technicians handle everything. That server receives the incoming captions, reformats them and distributes them to one or more Internet sites."[93] Court reporters pointed to Jack Boenau of the realtime startup AmeriCaption (Sarasota, FL) as the first person to "realtime to the Internet" when he cybercaptioned Al Gore's 1994 speech on "building the information superhighway."[94] But it was Section 508 of the Rehabilitation Act (discussed in chapter 4) that helped legitimize Internet captioning.[95] In April 1999, Attorney General Janet Reno "directed that all federal agencies must conduct self-evaluations of their electronic and information technology and report on the extent of accessibility for people with disabilities" under Section 508.[96] By 2001, regulations had been put in place mandating that government webcasts be "captioned for the deaf," unless "an undue burden would be imposed on the agency" in doing so.[97] Despite the economic loophole, cybercasted government meetings, distance-education courses, and corporate conferences were now all candidates for deaf and hard-of-hearing (D/HOH) accessibility—and for captioner profit.

The practice of captioning such feeds came to be known as "remote CART," since "the CART provider listens to the speaker by telephone and writes the realtime account to a Web site that the client is logged onto."[98] Broadcast-captioning providers like Caption Colorado and WordWave soon announced partnerships with Internet firms to add remote CART to their menu of client services.[99] Even plain old court-reporting firms without D/HOH clients found value in offering remote CART services, however. Remote CART allowed a freelance reporting firm to expand its market statewide or nationwide without having its reporters travel.[100] Such technology could even be combined to create "virtual conference rooms" where "those interested in 'observing' a deposition—lawyers, expert witnesses, paralegals—can do so from a distance."[101] One freelance owner in Aurora, CO reported that "when his service started [in 2000] it was doing about one Internet deposition a week," but by 2002, "it's up to 10 or 15 a day."[102] With the Internet enabling (and the government demanding) both remote television captioning and remote CART, court-reporting and captioning laborers, employers, and clients all found themselves fragmented across time and space, in a new virtual geography of contingent employment and service relationships. On one hand, this geography opened up more and more markets to independent contractors situated where there was inadequate local legal, television, or accessibility work to sustain and grow a small business. But on the other hand, this same geography enabled new competition between national and local providers. Surprisingly, some found that being local could still be a selling point in such a competitive environment, as hometown captioners were more likely to know the names of local personalities, recognize references to local landmarks, and better understand the communication needs of local audiences.[103]

## The Captioning Labor Crisis and the Geography of Realtime Training

The day-to-day difficulty in switching between courtroom transcription and realtime captioning, coupled with the increasing market opportunities opened up by captioning and CARTing over the Internet, led the realtime reporting profession to believe that it faced a dire labor crisis in the late 1990s. In 1993, the head of the NCRA had announced that there were only twenty-one certified "real-time reporters" in the United States.[104] At the time of passage of the Telecommunications Act of 1996, NCRA experts estimated that perhaps five hundred realtimers would be needed to meet the new television closed-captioning demand.[105] But by 2001 that estimate had increased sixfold: "There are about 300 broadcast captioners nationally, but about 3,000 more will be needed within four years."[106] The rising estimates were related to a more careful appreciation of the long-underestimated time-space constraints of live captioning. As one captioner explained, "What causes the biggest problem in terms of a shortage of [realtime] captioners is the fact that 90 percent of the work occurs in an overlap of 30-minute or 60-minute time periods. For example, everyone needs captioning at noon or 6pm or 11pm. So you need enough staff to cover a certain number of 6pm newscasts per day, but then it's difficult to have enough other shows in the right time slots to be able to provide enough work for those captioners."[107] At the same time, as more of the nation's 1,700 local broadcast stations turned to live stenocaptioning as opposed to "electronic newsroom" teleprompter captioning (described in chapter 4), the landscape over which captioning had to occur expanded dramatically.[108] In short, said NCRA president Carl Sauceda in 2000, "Never before has our profession seen such an urgent need for our skills."[109]

Yet NCRA membership had been flat for a decade, topping out at about 32,000 reporters in the early 1990s.[110] This might seem like a lot (about a hundred reporters for each of the top three hundred urban markets, for example) but the problem was that the majority of these reporters were not realtime capable.[111] The transition to standard CAT had largely been completed by the early 1990s, with over three-quarters of all reporters using computers by 1993–1994, but a similar transition to *realtime* CAT had never taken place.[112] One NCRA official optimistically estimated that 40 percent of reporters used realtime by 2002; however, an NCRA survey from just three years earlier showed that "only about eight percent of NCRA-certified court reporters" had obtained realtime certification.[113] More focused research on 254 responding reporters who had passed the 1998 NCRA basic, non-realtime certification test found that only 1 percent listed their job as a "captionist."[114] Many located the source of the problem in the stenotype-training schools, which had decreased dramatically in both numbers and enrollments through the 1990s, even as they standardized on realtime-compatible, "conflict free" training methods.[115] Although some 160 "postsecondary vocational and technical schools and colleges" around the nation offered court-reporting programs in the early 2000s, only about half of these were NCRA-approved, and even in those, enrollments were down by 50 percent.[116] In terms of targeted captioning training, the statistics were even worse. One estimate put the number of realtime training programs at only twenty-five (see figure 6.2).[117] VITAC founder Karlovits lamented in 2000, "The reporting education system is at a standstill."[118]

Echoing their discussions of CAT technology and training in the 1970s (chapter 3), reporters offered various reasons for the decline of stenographic instruction for realtime captioning in the late 1990s. Karlovits argued that "top high school students want a college education and a baccalaureate degree," but "we have very few baccalaureate reporting programs."[119] In 1998 only 13 percent of the students who passed the NCRA's nationwide "registered professional reporter" (RPR) exam had earned a four-year college degree.[120] Non-college training programs were a deterrent, it was feared, because "the highly specific court reporter training does not transfer readily into other occupations."[121] Yet many court-reporting programs took nearly as long as earning a bachelor's did. Of the 1998 newly minted RPRs, even though the vast majority (82%) had been enrolled full-time, about a third had taken two years to finish their training, another third had taken three years, and the final third had taken four years or more.[122] And court reporting was not necessarily an inexpensive alternative to a four-year college degree;

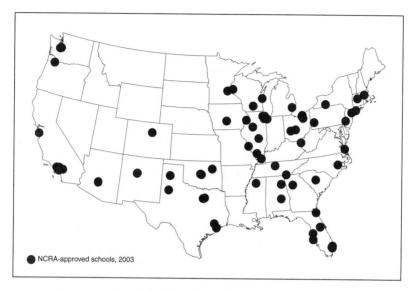

Fig. 6.2. Schools with court reporting and/or captioning programs, 2003. Though over 150 sites of court-reporting training could be found around the United States, only about two dozen programs offered NCRA-approved training in realtime CAT and captioning. Source: Map created from data in "Closed Captions-List," *AP* (16 May 2002), and NCRA, www.ncraonline.org/educcertification/schools/certified/ (visited 4 Nov. 2003)

it could "saddle a student with large debts."[123] As a result, by the turn of the century even the court-reporting programs that survived suffered with "dropout rates of up to 90%."[124] Just as it had decades before, the NCRA feared that "without the infusion of new blood into this profession, we will continue to shrink in size and power until a tape recorder is doing everyone's job."[125]

For reporters wishing to do more than deposition work, graduating from a formal training program was only the beginning. By 2001 some twenty-eight states demanded that official reporters earn a state-specific "certified shorthand reporter" (CSR) qualification; yet in California, for example, "fewer than 30 percent of those graduating from a three-year court reporting program pass the state's exam each year."[126] At the national level, in 1998, out of 3,143 candidates who took the NCRA's registered professional reporter exam, only 386 (12%) passed.[127] Even those who made it through the shrinking stenotype-reporting education system and achieved basic state or national certification were not necessarily prepared or suited for the speed, stress, and style of broadcast captioning. NCI realtime manager Karen Finkelstein estimated in 2000 that "only one in 10 who we screen has the skills to make the transition, which is a huge problem."[128] In 2004 the president of NCI, Jack Gates, similarly estimated that only "a couple of percent" of court-reporting graduates had "both the desire and the ability needed to do real-time captioning."[129] Why go into captioning anyway, if the market for other corporate and legal reporting services was still expanding through consolidation and globalization? Captioning advocates feared that with "pay for captioners . . . sliding, despite the shortage," even those few graduates with the skills to take up captioning's challenge might be dissuaded from the career.[130]

This captioning labor crisis threatened to undermine the long-promised broadcast accessibility of live news, sports, and entertainment programming to deaf and hard-of-hearing viewers; however, it was not the D/HOH community that pushed the government to address the problem but the court-reporting community. Bringing to bear congressional lobbying skills honed in their earlier efforts to avoid replacement by electronic-recording systems, the NCRA began "talking about the need to obtain federal funds for reporter training with various Representatives, Senators and their staffs in May 2000" and soon "set a target of $3 million for pilot programs at three schools across the nation," according to the NCRA's Mark Golden. Rather than basing their choice of which schools to fund on any local need for captioning or court-reporting labor, in November 2000 the NCRA sent a letter to all its officially approved reporting schools soliciting their interest. The association advised the schools that "if a school's congressional representatives did not sit on any of the key Hill committees that will appropriate funds or authorize their expenditure," the school would "face an uphill battle, at least in the short term." By December, eighteen schools had responded.[131] That month the NCRA convinced the House Conference Committee that "there is an urgent need for pilot programs to increase the availability of trained closed captioners."[132] That first year, however, only one school received funding. The University of Mississippi, thanks to the efforts of Senator Thad Cochran and Rep. Roger Wicker, was awarded a $500,000 grant through twenty words added to a 1,300-page appropriations bill.[133]

The court reporters had only just begun to lobby. Just two months later, in February 2001, representatives of the eighteen targeted reporting schools were sent to a "boot camp" run by the NCRA, where they met with their state representatives on Capitol Hill and pushed for action on a more comprehensive stenocaptioning training bill targeted for April 2001 (the deadline for congressional appropriations requests). This time their goal was not half a million dollars, but $90 million—$1 million a year over five years for each of the eighteen participating schools.[134] In June, Rep. Ron Kind (D-WI) introduced just such a bill in the House, "to provide grants for training of realtime court reporters and closed captioners to meet the requirements for closed captioning set forth in the Telecommunications Act of 1996."[135] The stated justification of these subsidies swung back to focus solely on the D/HOH population (not early readers or ESL learners as in chapter 5), citing the estimate that "28 million Americans, or 8 percent, are considered deaf or hard of hearing."[136] In the final appropriation put together in December 2001, the NCRA failed to attain its $90 million goal, but it did manage to increase its training subsidy tenfold from the previous year, securing $5.75 million for fourteen NCRA-approved court-reporting programs to train realtime captioners, "one of its greatest legislative victories to date."[137]

NCRA held a second boot camp in early 2002, this time seeking $75 million in total training subsidies over five years. Now they supplemented their face-to-face lobbying with "more than 5,000 letters" from member court reporters, D/HOH interest groups, and even the National Association of Broadcasters.[138] In 2002 Senators Tom Harkin (D-IA) and Charles Grassley (R-IA) were the key sponsors of the bill— unsurprising because Harkin had long been active in D/HOH accessibility legislation and because the AIB College of Business in Des Moines, IA, was one of the targeted grantees.[139] The NCRA had shifted its rhetorical strategy, pushing for these subsidies by arguing that the TCA of 1996 was an "unfunded mandate" and that without the stenocaptioning training money, the goal of 100 percent broadcast captioning by 2006 was "unrealistic."[140] But this was a bit of a bait and switch. Not all the program captioning required to occur by 2006 was live, realtime captioning. Yet the NCRA, with the apparent blessing of key D/HOH organizations, was redefining all closed captioning labor as realtime, stenographic captioning labor. Due to "key issues, such as homeland security" the 2002 bills were not acted upon.[141] However, individual appropriations were still inserted into key bills for over a dozen schools that fiscal year.[142]

By the time the 2003 lobbying season started, once again Ron Kind in the House and Tom Harkin in the Senate had introduced real-time training bills, seeking "$20,000,000 for each of fiscal years 2004, 2005, and 2006."[143] By now the NCRA boot camp team had honed its arguments for the subsidies even further. First, the audience for captioning was again expanded beyond the D/HOH community to encompass over 100 million persons: 28 million "deaf and hard-of-hearing"; 16 million "young children learning to read"; 44 million "illiterate adults"; 31 million people "for whom English is a second language"; and "millions of older Americans who have some level of hearing loss" (who seem to have been counted twice).[144] The rest of the population was thrown in as well, as Harkin stated, "I see people using closed captioning to stay informed everywhere—from the gym to the airport." The second strategy constructed stenocaptioner training as a solution to the "digital divide." The Senate bill suggested that funded training programs would contribute to "the creation of educational opportunities for individuals who are from economically disadvantaged backgrounds or are displaced workers." And Senator Grassley asserted that "distance learning opportunities in particular will have an enormous impact by making training accessible to individuals who want to become realtime writers but do not live in metropolitan areas."[145]

But the most prominent new strategy was to link realtime captioning to the post–9/11 "War on Terror." The Senate Committee on Appropriations reported that it was "deeply concerned about the ability of the 28 million Americans who are deaf or hard-of-hearing to be informed of critical news and information in the post–9/11/01 environment."[146] Harkin himself described how "the morning of September 11 was a perfect example of the need for captioners," citing the experiences of captioner Holli Miller (described in the introduction).[147] But the same "post–9/11/01 environment" that suggested the need for a captioner-training subsidy was a wartime political economy where purse strings for such initiatives remained tight. Once again the legislation failed to pass, and once again a dozen or so individual stenocaptioning programs resorted to line-item appropriations in the Department of Labor and Department of Education budgets in order to subsidize their realtime reporting programs (see figure 6.3).[148]

Even as the pork-barrel geography of Congress favored certain stenocaptioning schools in particular states, however, the same space- and

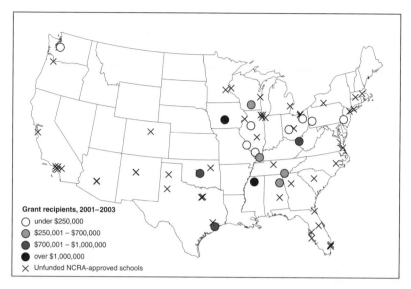

Grant recipients, 2001–2003
○ under $250,000
◑ $250,001 – $700,000
◕ $700,001 – $1,000,000
● over $1,000,000
✕ Unfunded NCRA-approved schools

Fig. 6.3. Appropriations for captioning training programs, FY 2001–FY 2003. Although major funding for captioning education repeatedly failed to pass Congress, targeted NCRA lobbying yielded significant line-item appropriations for over a dozen schools mostly in the Midwest and South. Source: Map created from data in Jeffrey Brainard, Ron Southwick, and Jennifer Yachnin, "Congressional Earmarks for Higher Education, 2001," *Chronicle of Higher Education* (10 Aug. 2001); Anne Marie Borrego, Jeffrey Brainard, Richard Morgan, and Ron Southwick, "Congressional earmarks for higher education, 2002," *Chronicle of Higher Education* (27 Sep. 2002); and Anne Marie Borrego, Jeffrey Brainard, and Will Potter, "Congressional Earmarks for Higher Education, 2003," *Chronicle of Higher Education* (26 Sep. 2003)

time-conquering technologies of digital video and data transfer that allowed captioning labor and CART provision to decentralize and fragment across the country allowed captioning- and CART-training programs to do the same. One instructor described a 1997 distance-learning collaboration between Johnson's Community College of Allegheny County (Pittsburgh, PA) and DuBois Business College (DuBois, PA) where, over a video link, "I dictated sentences, students wrote them on their machines in realtime, edited them and printed the transcript, and then we corrected the documents in class." Distance students sent transcripts via the mail for final grading.[149] Two years later, an instructor at Alfred Technical College (Alfred, NY) set up "one of the earliest online court reporting programs," an asynchronous email-based program with students enrolled from Utah, Florida, North Carolina, and New York. Federal funds helped expand such efforts. At Lakeshore Technical College (Sheboygan, WI) classes were "broadcast to 27 students at seven sites around the state" to reach interested trainees.[150] One thirty-nine-year-old woman attended remotely "from her home in West Bend [WI] because she [was] the only court reporting student in her area," avoiding "the two hour round trip to and from the college five days a week." Such students had few other options in Wisconsin; programs at two other technical colleges had recently closed due to lack of students.[151] Remote training could even cross state lines. In Senator Harkin's state of Iowa, AIB College of Business in Des Moines, "the only institution in Iowa offering a court-reporting program" also served Kansas, Nebraska, Minnesota, and South Dakota, since those states had "no schools offering such education."[152] Distance education in realtime work was meant to conquer the uneven geography of regular court-reporter training and employment, inserting high-tech captioners into the "space of flows" of the new economy with "no need to move for training or jobs."[153]

This latest crisis of court-reporter availability—not a "transcript crisis" of writing up (as described in chapter 3) but a "captioning crisis" of taking down—revealed the contradictions inherent in the presumed linear and progressive transition of stenographic reporting from computer-assisted transcription in the 1980s to realtime translation in the 1990s. On one hand, advocates of realtime reporting, captioning, and CARTing argued that these were the new services, new skills, and new standards that would not only preserve the professional careers of court reporters but also deliver the promise of universal captioning to deaf and hard-of-hearing citizens—all through an increasingly privatized, decentralized, and contingent market of freelance firms and subcontracted workers. But at the same time, these firms were globalizing, consolidating, and restructuring in ways that seemed to diminish the competitive power of individual reporters to demand higher wages and higher page rates for their "taking down" and "writing up." Similarly, while advocates of the new, combined, private reporting and captioning industry initially lamented the "unfair competition" resulting from government subsidies to NCI and WGBH, they later petitioned the federal government for subsidies of their own, pointing to the failure of their own infrastructure for realtime-reporter training, credentialing, and employment as evidence of a labor reproduction need in the public interest. Yet if realtime reporting represented the future of the profession, why was that future unable to inspire a new generation of reporting students on its own? If realtime reporting was the new standard by which a profession should be measured, why did so few within that profession aspire to reach that standard? Despite the rhetoric of uplift through technological advance, the effects of realtime CAT technology and re-regulated television captioning on the economic landscape of court reporting—whether from the point of view of the student, the subcontractor, or the agency-owner—were uneven and unpredictable.

### The Threat of Electronic Replacement—Again

Despite the recent focus on the "captioning crisis," the original "transcript crisis" never really went away. Even today the alarms are sounded that "nationwide, court administrators are canceling depositions, even postponing criminal jury trials, because they can't find enough limber-fingered folk to man the little black instruments with the weird keys."[154] In Massachusetts, "trials have been delayed or even canceled because a reporter was unavailable."[155] Was the problem one of low wages in public courtrooms, "most of which pay about half of what a freelancer can make working for private lawyers and TV companies"?[156] Or were reporting salaries "substantial for a career that requires two or three years of education"?[157] The key was perhaps the low level of realtime court reporting, described above; in Massachusetts, "only a few of the state's 63 court reporters have the equipment and training to produce transcripts instantaneously."[158] But just as the NCRA was convincing Congress to shore up its training industry in order to ensure a larger labor pool of realtime reporters, another technology and labor practice had spun off from both realtime CAT and the old electronic-recording threat of voice recording and was moving into the courtroom in answer to the perennial transcript crisis. If these "voice-to-print" systems succeeded, all the NCRA's efforts at preserving their century-old skills of machine stenography would be for nothing.

The lure of creating a machine to automatically transcribe the spoken word into written text had captured the imagination of inventors for more than a century. Alexander Graham Bell himself had spoken of his desire to create a "phonoautograph" in the 1870s as a tool for the deaf.[159] A century later, a writer in the *American Annals of the Deaf* agreed that "instantaneous automatic captioning would make it possible to receive communication from all of the hearing population. Nothing else short of hearing would do more to normalize deaf communications."[160] In the 1950s and 1960s, experiments in training computers to recognize spoken words—about the same time that the machine translation (MT) effort had been mobilized to train computers to translate printed text (chapter 3)—met with limited success at first.[161] The field of "speech understanding research" (SUR) ramped up with a $3 million DARPA project begun in 1971 and "an objective of 1,000 words spoken in a quiet room by a limited number of people, using a restricted subject vocabulary."[162] Like the MT projects in the 1960s, SUR lost its funding in 1976. But in the 1980s, SUR enjoyed a renaissance in defense dollars such that by the 1990s, "the results of this research have been incorporated into the products of established companies" from IBM to Microsoft. Analysts in the late 1990s were optimistic that "personal-use voice recognition technologies" could be a $4 billion market by 2001.[163]

It is important to note, however, that "speech understanding" and "voice recognition" are two very different things. Feasible speech understanding would demand that a computer interpret continuous, speaker-independent, large-vocabulary utterances; more limited voice-recognition could get by with discrete words, speaker-dependent training, and small-vocabulary domains.[164] Even in very restricted domains of speech, like commenting on a soccer match, automated systems have proven difficult to perfect. In June 2002 a BBC experiment in speech-recognition during one game "renamed Poland as Holland, changed a referee's name from Hugh to Huge and rendered the Portuguese player Jao Pinto's name as 'So Pointed.'"[165] Even as the A. G. Bell Association for the Deaf advocated to Congress in the late 1990s that "voice-to-text transcription would be the ultimate step toward independence for students who are deaf or hard of hearing, because it would end their reliance on notetakers and interpreters," court reporters stated confidently that "reliable, consistent, accurate, large-vocabulary, natural-speech recognition systems are still at least a decade away."[166]

What these critics had not counted on was the voice-to-print innovators stealing a technique from realtime CAT. Since about 1940, courts in many locations (especially the military) had long tolerated a particular practice of electronic recording called "stenomask" recording, where a person sat to the side of the proceedings wearing a microphone in a mask covering the bottom of the face. Every time a witness, lawyer, or judge uttered something in the court that needed to be recorded on the verbatim record, the stenomask reporter repeated it, quietly but clearly, into the mask's microphone.[167] Explained one stenomask reporter, "When training to use a stenomask, you learn to modulate your voice so that you are not heard by the other people in the room." You don't whisper, but "You dictate quietly, with controlled breathing, like you're in a room with a sleeping baby."[168] The mask served both to muffle the stenomask reporter's words from being heard by others and to keep extraneous noises in the courtroom from showing up on the taped recording.[169] Normally, a stenomasker's tape-recorded dictation would still have to be transcribed and indexed by hand later, keeping the practice cost- and labor-competitive with stenotype court reporting. But as stenomasker Anita Glover explained in 1995, the training demands of stenomasking were much shorter than with stenotyping, such that "many stenomask reporters are college graduates who have changed careers." And although "thousands" of stenomaskers were at work in the mid-1990s, because of certification differences and uneven labor markets in various states, these reporters had been concentrated "in the Southeast, the Northeast, states like Texas and Louisiana, and the Midwest."[170] Court reporters had long made rather facile arguments against stenomasking, conflating it with other forms of electronic recording. Wrote one reporter in 1974, three years after the founding of the National Stenomask Verbatim Reporters Association (NSVRA), "If Stenomask were a viable alternative to the shorthand reporter it would have been more widely recognized and utilized than it has been."[171] But now computer speech recognition—even relatively discrete, speaker-dependent, and limited-vocabulary recognition—could be combined with the stenomask input, just in the same way that computer language translation had been combined with stenotype input in CAT decades before. In this way, around the turn of the millennium, computer-aided realtime "voice writing"—combining stenomask input and computer-translated output—was born.[172]

There were some key differences between realtime voice writing and realtime CAT. First was the method of "dictionary-building" for the computer-translation algorithm. As explained in the *Journal of Court Reporting* in 2003, "The stenotypist focuses on building words with the use of prefixes, roots, and suffixes and entering those in a personal dictionary," but "realtime voice writers build vocabulary models by reading text modules that are provided by the software company into the system." In other words, stenotype reporters had to learn a stenographic theory and apply that theory to add new stroke patterns to their computer dictionaries; but voice writers *trained* their computers on new words, adding those new words to their dictionaries automatically. Different skills were required for the different input technologies as well: "Stenographic reporters learn accurate stroking. Voice writers learn to breathe properly."[173] But there were some important similarities between the two methods of turning speech into text. Both could reach a high rate of speed—voice writers were said to top out at 225 words per minute, much like steno writers—and both required additional scoping labor to produce an error-free transcript.[174] Transcripts could be turned around quickly with voice writing, with an estimated two hours of transcription for every hour of recording.[175] And equipment costs were (finally) comparable: "Machines, costing around $9,000 each, can usually transcribe continuous speech by a voice-writer who has spent several hundred hours training the systems to recognize his or her speech."[176]

Learning to voice-write would take only "several hundred hours" rather than several years of training, which especially worried the court reporters—as described above, they had staked the future of their profession on rebuilding their training infrastructure (with federal subsidy). Now voice writers, who were often self-taught, threatened to appear as a less-costly realtime transcription alternative before a cash-strapped Congress. By 2002 the voice writers had even organized professionally into the "800-member National Verbatim Reporters' Association" (NVRA).[177] Studying the threat in 2003, the NCRA concluded that "realtime voice writers cannot yet meet the realtime quality standards expected of stenographic realtime writers" because "the faster the voice writer talks, the more likely the words will run together, leading to more mistranslations by the speech recognition engine." Ironically, the NCRA applied the same standard to voice-writer "accuracy" as it had resisted in the past with respect to electronic recording: "To get a real picture of accuracy, the text must be compared to the digital audio recording of the event." But even the court reporters were forced to acknowledge "many well-qualified voice writers with the proper training, education, work ethic, and discipline can perform as well as stenographic reporters."[178]

In August 2001, the crucial issue had been put to a vote of NCRA members: should the 27,000 association court reporters admit voice writers into their group? The reporters answered no, "by an almost four-to-one vote." One reporter argued that if voice writers became part of the NCRA, "no one would opt for machine shorthand because of the shorter training time." But another pointed out, "Reporters that go into captioning and CART are leaving gaping holes in litigation reporting." Perhaps the most interesting point in the debate came from CAT pioneer and captioning entrepreneur Marty Block, who spoke up for the voice writers to be included: "Voice writing works. Trust me. I was right time and again when I was told CAT didn't work. It worked, and we made it work. And when I was told realtime was impossible, and a court reporter, a stenotypist could not sit down at the machine and write realtime at 99 and a half percent and put it out over the air, I made it happen, and others followed me to do a job better than I ever did." Even if the NCRA staved off the threat of voice writing today, Block predicted, "someday the vote will come up again, and we will be in a much weaker position at that time to deal with this issue and control it than we are today."[179]

Just as the NCRA was voting to exclude the voice writers from its court-reporting association—an attempt to preserve its jurisdiction over the production of the official transcript in the courtroom—another development occurred to illustrate how court reporters had extended their jurisdiction over the production of the realtime translation in television. Earlier that summer, Senator John Breaux (D-LA) had introduced resolution SR 95, resolving that August 3, 2001, would be known as National Court Reporting and Captioning Day. Breaux justified this proclamation by arguing that "court reporters and captioners translate the spoken word into text and preserve our history" and that "whether called the scribes of yesterday, court reporters of today, or real time captioners of tomorrow, the individuals that preserve our Nation's history are truly the guardians of the record."[180] By the turn of the millennium the NCRA had succeeded in conflating the professions of court reporting and closed captioning in the minds of many legislators. But this blurring of occupations was not only a rhetorical strategy—the economic landscapes of both court reporting and closed captioning had indeed converged during the 1990s. These two formerly distinct areas of technology and practice only threatened to fragment and intertwine further with the increasing expansion of both practices into global cyberspace.

The various actors working to reproduce the realtime speech-to-text industry across space and time—from professional associations and private firms to media activists and individual laborers—demonstrate that the geographies emerging at the intersection of new technologies, new regulations, and new careers are often contradictory and contested. While individual court-reporting graduates search both locally and globally for the right mix of subcontracted work to sustain their own reproduction, individual court-reporting agencies search both locally and globally for the right mix of client contracts and labor pools to sustain their own profitability. At the same time, deaf and hard-of-hearing clients seeking media justice, quality education, and communication freedom must negotiate the resulting uneven geography of offline and online captioning providers, finding a site within the space of flows where audio and visual information flow together with text. And the state, after asserting the legal power to mandate universal provision of these services across the nation, has found itself exerting the budgetary power to target the growth of these services only in selected areas.

Ever since the 1960s, many in the legal profession had assumed that court reporting was a labor practice destined to be replaced by automated-recording technology. In a similar way, the labor requirements of closed captioning were regularly underestimated in comparison to the technological challenges of building such a system; as one public-affairs programming expert argued in 1974, "personnel will be expensive but a comparatively minor problem."[181] The emerging potential of voice-writing technology in the late 1990s has kept visions of a technological fix to this "comparatively minor" labor problem—"realtime, speaker-independent, portable, continuous speech language recognition"—alive into the new century.[182] One expert recently predicted in the *Volta Review* that "in the near future, speech recognition will progress to where portable and accurate processors will be feasible," resulting in "personal captioning devices for people who are deaf or hard of hearing."[183] But even if voice writers succeed in joining—or usurping—their stenographic colleagues in the courtroom, broadcast booth, and classroom sites of speech-to-text labor, their efforts as human mediators between language and technology will remain necessary. If recent history is any guide, those efforts will also remain fragmented, contradictory, and, more often than not, hidden.

# Conclusion:

# The Value of Turning Speech into Text

Video tapes, captioning consoles, and computers all conjure up an image of captioning as a cold, mechanized process. However, the central element in the closed-captioning process is very warm and human—the caption editor.

*NCI* Caption, *1980*

In 1882, an article in the *American Journal of Otology* described a wondrous new invention by Amadeo Gentilli called the "glossograph," a machine that allegedly performed "the automatic transcription, in the form of an easily translatable record, of the human speech at its ordinary rate of utterance." Gentilli's device was anything but "easy" for the person using it, however. His first design involved inserting a "false palate of gutta percha having metallic discs set in its under surface" into the mouth, which, when a person spoke, made electrical contact with further "metal contacts on the surface of the tongue and lips." These sent a signal along wires coming out of the mouth to an external recording machine. But because "the fluids of the mouth soon impaired the perfection of contact," Gentilli tried another method. His next device used "a series of light arms, or levers, passing into the mouth, resting lightly in contact with the parts whose movements were to be recorded, and communicating these movements to a corresponding series of slide bars set in the frame-work of the machine, and in their turn moving pivoted arms, carrying at the farther end of each a pencil." Thus through six inscribed pencil lines a supposed exact record of the movements of the mouth during speech was made, the reading of which, Gentilli claimed, was "easily acquired."[1]

Whether or not such a contraption was actually built, the story of the glossograph illustrates several important points about the history of sociotechnical systems for speech-to-text translation in the cinema, the classroom, the courtroom, and the captioning booth. First, the goal of "automatic" recording of speech through technology has inevitably involved realtime human mental and physical labor, usually in three near-simultaneous steps: (1) careful attention to an original spoken utterance, either for phonetic structure or semantic meaning or both; (2) cognitive transformation of that utterance along a number of axes, whether condensing for speed, editing for literacy level, translating from one language to another, or simply rephrasing for clarity and convention; and (3) reproducing the modified utterance in a form that can be converted to text, either through manual writing, mechanical keyboarding, or controlled revoicing into a microphone. Even a worker "taking down" the record simply by monitoring a tape recorder in a courtroom would have to perform these steps to some degree: listening to testimony, recognizing changing speakers and nonverbal responses, and noting these moments of context on a written sheet to accompany the taped record.

Second, the process of turning the encoded record of a live event into an English transcript always involves further labor in time and space—even when a computer performs much of that labor. Longhand stenograph symbols had to be retranslated by a court reporter, dictated into a tape recorder, and typed up by a transcriber. Keyboard-printed stenotype characters could be interpreted by a competent notereader, saving the dictation step and hastening the transcribing step. Digitally encoded stenoforms within a computer-aided transcription (CAT) system needed to be proofread by a scopist—a computer-savvy notereader—before being sent to the laser printer. Cutting corners on any of these "writing up" steps in an effort to reach realtime speed would result in a less-than-verbatim English representation of the original speech event—adequate for rapid-fire live television captioning or classroom interpretation, perhaps, but unacceptable for official courtroom work. Even subtitle or caption sentences edited down from the original English were in a sense later written up by being carefully timed and placed with scene cuts and speaker changes. From the glossograph of legend to the computer-aided, speech-recognizing, voice-writing system of tomorrow (see chapter 6), systems to convert speech to text have never been able to shirk such labors.

At different moments in time and space, however, different systems for transforming speech to text, serving different social, political, and economic needs, have repositioned, reconstructed, and revalued such labors. Foreign-film subtitling in the United States remained essentially a single-person enterprise for over a quarter-century after the invention of the "talkies." The subtitling of films for deaf entertainment was carried on by self-taught educators even after such efforts were molded into a national-scale, federally funded program. The transcription of legal proceedings invited a series of new technologies that were sometimes resisted by court reporters (electronic audio and video recording) and sometimes embraced by reporters (stenotype and CAT systems) but which always helped subdivide the court-reporting labor market along lines of skill, wage, and gender. And the effort to caption television alternately demanded the aesthetic and language skills of the postproduction captionist and the speed and attention skills of the realtime stenographer. Digital convergence with computer and communication infrastructures helped drive all these stories but not in any sort of deterministic or predictable way. Film subtitling, television closed captioning, and court reporting had all blurred together by the start of the new millennium. But the work they demand in turning sound into text still remains largely hidden in cyberspace.

Just because such work is hidden, however, does not mean it is unimportant. One captioner I talked to—a person trained in both offline and online aspects of her profession—made it very clear to me that she valued the ability to produce the highest-quality captions possible, which necessarily meant putting in a lot of time. She resisted the idea that TV captions did not have to be produced to high quality standards of accuracy, readability, and art because the deaf and hard-of-hearing (D/HOH) audience was simply "grateful to have any kind of captioning." She could quickly tell, just by watching a bit of a captioned program, how much time and effort must have gone into creating the captions. This captioner described her own hidden labor as "tedious, tedious, tedious . . . but I love it!"[2]

As this story illustrates, the "construction" of closed captioning has been less about the technologies themselves (closed-captioning chips, stenographic keyboards, or computer translation programs) and more about the people involved in producing, reproducing, and utilizing those technologies. Over the last half century, the technology and labor system of closed captioning was "materially constructed"—created, tested, debated, and finally fixed over time and space—through processes of state-funded technological development, nonprofit educational innovation, market-based corporate broadcasting, and grassroots minority activism. But in the course of this history, closed captioning was also "socially constructed" in that the purpose and meaning behind these processes was similarly developed, contested, and (at least temporarily) settled. At various times, and by various groups, the fight for closed captioning was framed as a system to provide emergency information, a tool to teach English literacy, a vehicle for marketing to underserved consumer demographics, and a precondition for full cultural citizenship.

Through this history of material projects and cultural debates, a pattern of digital convergence unfolded, as the analog broadcast system of television converged with the digital data-processing capabilities of computers. Closed captioning represented a new way to hide digital text within an analog image frame. Caption decoders were digital consumer electronics that could render computer-generated characters directly on the screens of analog televisions. And live captioning was only possible once analog stenographic keyboards could be interfaced with digital computers. But captioning represented another type of convergence as well: the temporal and spatial convergence between the different modes of information of "print" and "video." Synchronizing text to both the auditory and visual elements of the television broadcast involves temporal and spatial simultaneity in its production aspects (realtime captioners must stenographically type words as an event unfolds), its distribution and storage aspects (both digital text signals and analog video signals need to travel through the airwaves or get encoded onto videotape together), and its consumption aspects (both the text and the image must be rendered on a single screen, at a comfortable reading and viewing speed, without obscuring each other). Finally, convergence of a third sort took place as the previously distinct labor processes of postproduction film subtitling and realtime courtroom reporting came together in live television captioning. In the early days of television in the mid-twentieth century, it was educators and activists, targeting the needs of the deaf and hard-of-hearing community, who both set the agenda and supplied the labor for nationwide captioned-media efforts. But as television moves to a new digital infrastructure at the start of the twenty-first century, it is professional court reporters, trained in the skills and technologies of stenography, who are now the main producers and advocates for broadcast closed captioning.

This concluding chapter considers how these speech-to-text products and services—and those who produce them—have been valued, devalued, and revalued over recent years, in three specific contexts: the construction of distributed multimedia databases, the academic study of translation, and the development of high-definition television. Such questions are increasingly important as film subtitling, classroom interpreting, court reporting, and especially television captioning complete their transitions from their original analog media of celluloid, print, and videotape to the digital realms of distributed data storage, DVD, and high-definition broadcast.

## From Cost-Sink Captioning to Value-Added Indexing

Encoded captioning text can have value beyond its initial use for broadcast, courtroom, or classroom display. As described in chapter 4, when court reporters who had tried their hand in the realtime captioning booth in the 1980s returned to their former profession in the 1990s, they did so with a new set of credentialing standards for certified realtime reporting and a new vision of a computer-integrated courtroom. As one reporter pointed out in 1995, "Nationwide, federal reporters generate in excess of six million pages a year, and most of the information is not usable as it exists now. So we need to develop ways that attorneys can have access to large amounts of information through electronic databases."[3] No longer as fearful of the threat of digital video recording as they were before, reporters rebranded themselves as managers and indexers of this digital sound and image record, realizing that their realtime textual transcription could be matched moment-for-moment to the realtime video display to create a new "emergent commodity" with greater value to the legal community than each of the previously separate services. The court reporter's transcript would now serve as "metadata"—data describing other data—providing a searchable text index to persons and utterances represented on video.[4]

With transcripts reconceptualized as time-coded textual metadata for the videotape record, and with video translated into a digital format accessible through an online database, suddenly a court-reporting firm could replace its locked back room full of untranscribed steno notes, printed transcripts, and stacked-up VHS tapes with a globally accessible, boolean-searchable, time-coded digital video archive.[5] New open and extensible standards for tagging and formatting transcripts such as "legal XML" may even allow carefully written computer algorithms to search testimony all by themselves.[6] Freelance agencies that can assemble or enhance such systems stand to reap a new revenue stream from the added value that the text transcript brings when combined with the video transcript. Already, the global legal firm WordWave—the inheritor of the original for-profit American Data Captioning (see chapter 6)—announced in 2000 that it would launch a new division "aimed at making it easier to search film and sound clips over the Internet for specific content," charging "a few hundred dollars per hour of tape being transcribed, plus fees for hosting it on a server and making it available for streaming to users' computers."[7]

This asset management industry grew rapidly in the 1990s. Before the digitization of videotaped content was possible, "media asset management" (MAM) systems for video could involve as little as an organized shelving system for VHS tapes and a database of program titles stored on a personal computer.[8] The goal was simply to avoid any turnaround costs that might arise from not being able to locate audiovisual "assets" in an orderly fashion.[9] But converting the archived video to a digital format encouraged more complicated "digital asset management" (DAM) systems, where large databases of computer-stored video preview clips could be searched over a remote computer before an actual physical tape was pulled. The latest step in this development is the "multimedia on demand" (MOD) system, where the full digital video is stored on the computer at such a high level of quality that it becomes a "live electronic library" to be used directly by in-house professionals or sold to external consumers. On the production side, MOD systems promise video professionals "an end-to-end production chain from story creation through editing and playout," automatically "leaving a digital trail for payment and security purposes."[10] On the consumption side, MOD systems promise end users the ability to customize their video news diet in realtime—a "personalized digital news recorder that is capable of recording news stories of interest to a particular user and filtering out others," as one researcher put it.[11] With a market size of over $500 million in 1998, MOD systems were estimated to be a $5 billion industry by 2004.[12]

But what is the basis for all this new value? First is the shift from a "monomedia" system—a database meant to "focus on a single media object such as a video clip, audio track, or textual document"—to a "multimedia" system—storing video, audio, and text together. Multimedia databases enable "cross-modal information retrieval," the ability to search textual information in order to retrieve video information—or, ideally, vice versa. However, this is only possible if the different media in the database are synchronized. Synchronization can happen in three ways: (1) "inherent" synchronization occurs when two different forms of media are "born" together, as when video and audio tracks are simultaneously recorded; (2) "computed" synchronization occurs when a processing algorithm detects change points between two sets of data, such as video scene changes or textual speaker changes; and (3) "manual" synchronization occurs through human intervention, as when written annotations are linked by hand to a particular point in the video. "Text-to-speech alignment" is one of the most important multimedia synchronizations to produce.[13] With enough textual media, one can cheaply, quickly, and accurately retrieve any desired portion of the audiovisual media: "The more thorough the cataloging information, the more potential value for internal production and external client research and footage sales."[14] However, producing that cataloging information beforehand (either manually or computationally) needs to be inexpensive and rapid as well.

Any preexisting closed-captioning text—physically stored with the broadcast signal on an analog videotape—can be synchronized during the "ingest and index" phase when the analog video is digitized.[15] Thus program producers and distributors often have a ready-made, easily accessed, and perhaps even state-subsidized storehouse of synchronized text available before they even decide to digitize and repurpose analog content. In 1997 the Library of Congress used a $265 piece of hardware called the "hubcap," which "sits on a TV and captures closed captioned information and makes it available to a PC file," in order to monitor and index daily C-SPAN broadcasts.[16] Even when automated techniques are involved, such as computer speech recognition applied to the soundtrack of a video segment or computer character recognition applied to the words superimposed over the visual image, such solutions are enhanced by having a preexisting closed-captioning track available for vocabulary reference—as a "clue" to indicate where the automated process should begin, as one researcher put it.[17] Geoff Bowker has argued that any technological "archive" inevitably incorporates political and social factors, down to its very data and metadata design.[18] The political and social provision of free captioning labor for use as rough-and-ready metadata has been a precondition for the expansion and revaluing of MOD systems.

As described in chapter 4, much of this free captioning labor was originally undertaken to translate the countless hours of television news produced and broadcast locally and nationally across the country each day. The value of this "morgue" of old footage has only recently been recognized.[19] In the days before videotape, local news would be shot on inexpensive 16mm film, with key sequences recut and re-edited so much that "the footage literally could be used up."[20] By the mid-1970s, videotape had replaced film, but although this inexpensive medium offered new archiving potential, the fact that it could be overwritten and reused meant that much footage was lost forever.[21] Vanderbilt University recognized the value of such material and began archiving national daily news footage, some of which is now downloadable over the Web.[22] But it was not until the mid-1980s that archivists recognized the local newscasts produced by 1,500 or so television stations across the country as "an invaluable record of the day-to-day events and activities that have shaped the social and cultural fabric of both urban and rural communities across the United States."[23] As it had done with closed captioning, PBS affiliate WGBH Boston led this effort with the establishment in 1990 of a Media Archives and Preservation Center, which over the next decade collected "an estimated 160,000 hours of television programming, 20,000 hours of film, and more than 40,000 hours of radio master programming" for preservation.[24] As recently as 2000, however, some still lamented that local TV news was, "for practical purposes, inaccessible to scholars."[25]

Today the television news morgue is gaining new life as a source of knowledge, value, and profit. CNN, for example, created an MOD system encompassing "the entire creative process, from script to air."[26] Their $20 million system was designed to keep up with an ongoing production schedule of "more than 25 video programs an hour, 24 hours a day," in order to allow "more than 300 CNN editors, writers and producers at three Atlanta-based networks to search and browse video from up to 40 simultaneous satellite feeds in real time at their desktops." But the system also had to ingest some 100,000 hours of legacy content on analog videotape—content which, since 1991, had been stenocaptioned.[27] On a smaller scale, PBS teamed up with the vendor Virage to create a MOD system that "allows viewers to search the closed caption text of the NewsHour and retrieve short segments that contain the entered search terms."[28] Their "ingest" software not only parsed the captions but noted "changes in visual content, such as pans or zooms," synchronizing these together with the transcript.[29] The end result, claimed one PBS representative, was that "every public television station is evolving into a digital library."[30] Public service combines with profit potential; at the WGBH archive, "footage and research assets are also marketed to non-WGBH producers and outside researchers to generate revenue to support the department's primary mission."[31] By 2005, even Web search firm Google was offering Google Video, which could return time-indexed screen captures and snippets of text transcripts from television shows on all the major networks—primarily through a search of the stored closed captioning from those shows.[32]

Live closed-captioning text is of course often paraphrased, incomplete, or just plain wrong. But addressing these shortcomings only invites a further division of labor. Today a subscriber to the LexisNexis information service can search and retrieve "summaries of segments of most local news broadcasts," but to obtain full transcripts, one must deal directly with the "media-monitoring service" behind this LexisNexis feature, Video Monitoring Services of America (VMS).[33] VMS claims to provide added value to its corporate customers by bringing humans back into the media-monitoring loop, not removing them: "Closed captioning often misspells words and captures phrases incorrectly, making sentences unintelligible," so "VMS' editors and media specialists personally review all broadcast media content in all 210 US markets."[34] Because some stations use live stenocaptioners and other stations use teleprompter-linked electronic-newsroom systems, closed-captioning quality can vary greatly from one news market to another, requiring varying levels of post-editing to produce useful transcripts and indexes from this text. But apart from for-profit vendors like VMS, few organizations seem willing or able to take on this burden in service of the public interest. As recently as 1997, a major volume on "television newsfilm and videotape collections" by video and film archivists made no mention of capturing closed-captioning data during acquisition or using it as metadata for indexing.[35] One observer has noted, "A typical commercial TV station generates approximately four hours of local news daily. Even at several hundred words per minute, this requires less than 200 megabytes of storage capacity per year." With cheap text-storage technology readily at hand, and the expense of local news captioning already justified, "the failure to archive closed captions of local news, despite the trivial cost of doing so and the great value of an easily searchable news database, vividly illustrates the need to update public policy toward news archives."[36]

## Subtitling, Translation, and Power

Captioning also is valued differently depending on whether it is conceptualized as "transcoding" or "translation." A European subtitler wrote in the early 1990s, "It is extraordinary that an activity involving such large volumes has attracted so little attention and is regarded with such disdain."[37] As described in chapter 1, around the globe, cinematic subtitling is often performed after the fact on a piece-wage basis by subcontracted firms or individuals not connected with the initial production process and who do not reap royalties from subsequent distribution. Subtitlers rarely share in the royalties from a program they have worked on, nor do they retain copyright over their work if they are employed by large production house.[38] One might say the same about the invisibility and devaluing of captioning and CART, intellectual and manual labor that constructs emergent, ephemeral texts but is rarely considered to be "authorship" itself. Only court reporters have so far retained ownership of the product of their speech-to-text labor, in the official transcript that they produce, though it remains to be seen what will happen when a video transcript and its textual representation are produced and "owned" by different individuals yet accessible (and marketable) through the same multimedia database.

The case of cinematic language subtitling, however, illuminates another labor relation besides that of authorship. Subtitling can be seen as existing somewhere between literary translation and language interpretation. Film subtitlers face similar problems of conveying meaning as do literary translators and live interpreters, but with different "restraints relating to time and space" due to the material nature of screens and the cultural expectations of audiences.[39] "Sometimes a subtitler may be faced with difficult, perhaps even insoluble, problems, which a translator of a book would solve by a long explanatory translation, a footnote or by simply ignoring the problem in the case of an untranslatable joke, pun or double entendre connected with pronunciation."[40] Yet for the sake of the viewing audience, subtitlers must write *something* to fit the pace of the spoken audio track and the appearance of the actors' moving mouths—twin imperatives quite different from the main demand put upon the literary translator "whose first loyalty is to the author."[41] Because of such considerations, cinematic subtitling—along with dubbing—has long been referred to as "adaptation" rather than "translation."[42] In fact, the only time subtitling is referred to as translation is when that subtitling is considered inadequate, such as in 1988 when "Federico Fellini sued his French film distributor and demanded that both the dubbed and the subtitled version of his film *Intervista* be withdrawn on the grounds that the 'brutally free' translation was outrageous and distorted his work."[43]

A quick look at the field of print translation, however, reveals substantial similarities with the speech-to-text occupations. Unlike in other parts of the world (especially Western Europe), translation as a profession in the United States has had a rather short history. Only in 1930 was the Society of Federal Translators (later, the Society of Federal Linguists) established in Washington, DC, as a professional association for government-employed translators. The profession remained small until the second half of the twentieth century, "under the triple impact of the second World War, the expansion of US business interests and government commitments abroad and the international cross-fertilization of scientific and political thought." Still, when the American Translators Association was established in 1959, catering to both government and freelance translators (similar to the institutional polarization in court reporting), it attracted barely two hundred members in its first few years.[44] One observer argued in 1968 that "the quaint old n[o]tion that translators are essentially typists whose parents were born in another country still lingers on in many circles."[45] (Such perceptions may have contributed to the fears of a "translation crisis" that were used to justify machine translation efforts in the 1950s and the 1960s, as described in chapter 3.) By 1970, echoing professionalization strategies in court reporting, one translator argued for "the professionalization of the craft; the training, education, and authoritative accreditation of the literary translator" at the university level.[46]

Translation was split not only in terms of government service versus private business but in the kind of content translators specialized in. The large pool of scientific and business translators (focused on technical accuracy) contrasted with the smaller group of literary translators (focused on aesthetic integrity).[47] This split echoed the polarizations emerging in captioning over content (news versus entertainment) and style (verbatim versus edited). In scientific and business translation, because of the dominance of particular nations and language communities in international commerce and research, a slate of languages similar to those common to film subtitling were in high demand for print translation in the United States—Spanish, French, German, Portuguese,

and Italian. Another parallel to subtitling was that the center of technical translation remained New York City, with "approximately 100 translation agencies, ranging in size from one to a half dozen or more
full-time resident translators and subcontracting for the services of perhaps altogether 400–600 . . . self-employed (full-time or part-time) translators" in the 1960s. But unlike the subtitlers, who in the United States
specialized in producing only English translations of foreign materials,
print translators were expected to work in both directions: "Translators
qualified to translate from English into a foreign language (usually for
publication) earn more than those who translate from a foreign language into English (primarily for purposes of information)."[48] The translator's employment relations matched those of court reporters more
closely than those of subtitlers. Observers in the 1960s mentioned incomes between $6,000 and $8,000 per year, but translators debated
whether they were selling a commodity (a finished translation, priced at
perhaps $16 per thousand words) or a service (their translation labor,
billed by the hour).[49] Either way, translators recommended a gendered
division of labor similar to that of court reporters so that they could
split the mental work of translation from the supposedly manual work
of production. One joked, "For a young, male translator, my advice
would be to marry a girl who is a competent typist."[50]

The terms used to talk about literary translation were reminiscent
of both subtitling and captioning, with debates focused on the degree
of editing, the needs of the audience in understanding the author's intent, and the production-line turnaround time pressures that miti-
gated against artistic craftsmanship. In the 1960s a freelance translator
might be granted three or four months to translate a full-length
novel.[51] By 1990, the fee for translating a three-hundred-page novel
ranged from $3,000 to $6,000, so translators needed to turn manu-
scripts around quickly in order to make a decent living.[52] This pressure
was a problem because literary translators claimed more of a creative
and artistic role than technical translators: "The [literary] translator is
a writer, or he is nothing; and if he has art enough, and a great text to
work with, he is a creative writer."[53] Yet like subtitlers and captioners,
they were employed under "work made for hire" arrangements, without any intellectual property rights of their own.[54] Thus, also like captioners and subtitlers, literary translators lamented their invisibility:
"The able and professional literary translator works harder, gets less
recognition from the critics, and is paid less than any writer in the

world of literature."[55] A successful translation, as with a successful speech-to-text transcription, was by definition one that "conceals the numerous conditions under which the translation is made, starting with the translator's crucial intervention in the foreign text."[56]

Today, those who study the labor of print translation are beginning to see some of these parallels with speech-to-text labor. Starting in the late 1970s, a new international and interdisciplinary field of "translation studies" began to emerge, incorporating insights from "linguistics, literary study, history, anthropology, psychology, and economics."[57] From this crucible a broader definition of translation developed, as creating a "text that makes an 'image' of another text" or claims to represent another text.[58] Scholars in translation studies grant the translation "its status as a text in its own right, derivative but nonetheless independent as a work of signification."[59] Rather than teaching students to improve the scientific "equivalence" of their translations with the originals, modern translation-studies scholars emphasize the "function" that a translation plays in linking disparate worlds—"how the translated text is connected to the receiving language and culture."[60] The new focus recalls James Carey's division between the "transmission" model of communication and the "ritual" model of communication: the meaning of a text is not fixed but must be considered all through the different and related moments of production, circulation, and consumption of that text.[61]

With such a definition, both language subtitling and vocal dubbing in cinematic film qualify as translation.[62] As one translation studies scholar recently commented, "for many cultures in Europe . . . a very substantial part, if not most of the mass media messages in circulation have undergone some process of translation" and so "translation processes in mass communication play a very effective part in both the shaping of cultures and the relations between them."[63] Recent work has examined how attitudes such as humor and politeness are translated across cultures through film and television.[64] But even same-language television captioning, as NCI itself pointed out in 1991, "is more than just putting words on the TV screen. It's preserving, through captions, a program's excitement, drama, suspense or comedy for everyone to enjoy."[65] Under the new cultural research agenda of translation studies, the captioning and CARTing of speech for deaf audiences both qualify as "translation" as well—if one takes a cultural view of deafness. After all, both involve moving a text from hearing culture to Deaf culture for the specific purposes of equity and understanding, rather than for achieving some sort of mathematical "equivalence" between the two forms of communication.

Writers in translation studies point out that such cultural translations are "never innocent" but always involve power relations: "There is always a context in which the translation takes place, always a history from which a text emerges and into which a text is transposed."[66] The long struggle for the standardization, institutionalization, and regulation of closed captioning in the television-production process provides a clear historical example of the ways that state, corporate, and activist groups jockey for power in defining the social relations of translation. As "rewritings" of truths originally produced for one audience for the benefit of another, translation projects involve what scholar André Lefevere calls "authority" in at least four ways: (1) "the authority of the person or institution commissioning or, later, publishing the translation: the patron"; (2) "the authority of the text to be translated"; (3) "the authority of the writer of the original"; and (4) "the authority of the culture that receives the translation."[67] Consider, for example, the relations of authority in captioning for D/HOH populations: (1) patron authority of government, corporate, and charitable sponsors; (2) the authority of original soundtracks which much be edited dramatically for the space and time limitations of television print; (3) the authority of program authors who retain copyright, and royalties, to their work; and (4) the very limited authority of D/HOH viewers who "receive" captioned translations, augmented by the increased authority of young readers, ESL learners, and hearing people in public places who are increasingly cast as the mainstream captioning audience.

But authority, whether in translation, transcription, or adaptation, can also be challenged, undermined, and appropriated. In cinematic subtitling, the same technologies that have enabled the digital convergence, global extension, and economic concentration of the film industry have also enabled a diverse fan culture to emerge with the goal of creating its own subtitles—a "do-it-yourself" movement that recalls the early pioneering efforts of deaf educators in developing their own subtitles after 1930. The tool of choice for today's home subtitler is not the 16mm film projector, however, but the laptop-computer DVD player. With the ability of PCs to process the audio, video, and text from both analog and digital cultural commodities (whether legally or illegally), an underground movement for retitling these products has emerged. Enthusiasts "rip" (extract) the subtitle tracks from commercial DVDs, alter them for new purposes (understanding, analysis, or parody) and then freely trade the new subtitle tracks online to be reinserted into copied DVDs by others.[68] Sometimes the new text contains commentary on the original film. Film critic Roger Ebert mused in 2002, "I'd love to hear a commentary track by someone who hates a movie, ripping it to shreds. Or a track by an expert who disagrees with the facts in a film. Or a track by someone with a moral or philosophical argument to make. Or even a Wayne's World–style track from dudes down in the basement who think *The Mummy Returns* is way cool."[69] Other enthusiasts use their handmade subtitle tracks to translate films that have not yet officially made it to the US market, such as the vast catalog of Japanese anime. As one observer explained, such subtitles break with professional conventions in many ways: "In scenes with overlapping dialogue, they use different colored subtitles. Confronted with untranslatable words, they introduce the foreign word into the English language with a definition that sometimes fills the screen . . . They use different fonts, sizes, and colors to correspond to material aspects of language, from voice to dialect to written text within the frame. And they freely insert their 'subtitles' all over the screen."[70]

Unlike the amateur retitlers, more casual "tinkerers" with television closed captions can find add-on consumer appliances that enable them to significantly alter their audiovisual viewing experience. In 1999, "Arkansas father" Rick Bray developed a $150 aftermarket television "foul-language filter" called TVGuardian, which consumers could use to "look at all the words and block out the dirty ones before they can be spoken."[71] Hooked up to a television or VCR, the device relied on the closed-captioning text to mute what Bray thought of as objectionable material. He described that besides removing "exclamatory uses of God, Jesus and Christ" on the "strict" setting, his device "filters out ethnic slurs and it—really it filters out references to sex, period . . . It'll take the word 'sex' and change it to 'hug' so it still shows a sign of affection, but making it a little less explicit."[72] The text created by this automated-translation process is radically different from the original. Testing the device on the R-rated science-fiction film *Men in Black*, one journalist found that "to our surprise, TVGuardian was muting every sentence that contained a curse. (Will Smith was practically silenced!)"[73] Such do-it-yourself censorship has moved into the public realm as well. The Georgia-based seller of the $80 ProtecTV device, another captioning-based television sanitizer, donated 2,800 units to a Chattanooga, Tennessee, school district in 2002.[74] Echoing the caption-decoder distribution strategies used by D/HOH activists in the 1980s, ProtecTV was "marketed through churches and nonprofit organizations . . . with the organization receiving a profit for each device sold."[75] Consumer efforts like these, rooted in particular cultural communities and targeted to texts with particular cultural meanings, reveal that the "authority" to link text to speech on film and television no longer resides solely in the hands of the corporate patron or the professional translator.

## Boundary Objects and Brokers in Court Reporting and Captioning

Finally, a shift in value takes place when a court reporter's labor product, the printed transcript, is replaced by the captioner's labor service, the ephemeral translation. The threat of electronic audio recording (ER), made manifest in the courtrooms of Alaska in the 1960s, threw the court-reporting community into a turmoil that led directly to the adoption of computer-aided transcription (CAT) techniques (see chapter 3). But another technological threat shared the pages of court-reporting journals with the dreaded tape recorders during that decade: the electrostatic copier, commonly known as the "Xerox machine."[76] As one court reporter put it in 1966, "Have your favorite clients suddenly started ordering one copy of a deposition whereas in the past they always ordered two? Have your board or commission hearing transcript sales dwindled from nine to two? Have your plaintiff's attorneys stopped ordering the one copy they normally do? Electrostatic reproduction (Xerox, etc.) is the bugaboo!"[77] Similarly, a 1967 advertisement in the *National Shorthand Reporter* by the Pengad legal-supply corporation proclaimed, "Ever since the Model 914 Xerox Electrostatic Copier was made available to the general public in 1958, court reporters all over the nation have been suffering from shrinking profits."[78] "We need your help," implored another reporter to the readers of the *NSR*, "in trying to resolve this red dragon which has reared its ugly head."[79]

But this time the court-reporting community could come up with no technological fix to the problem. Rumors circulated about a "sensitized paper which when exposed to photocopying and Xeroxing would not reproduce," or "a colored paper that doesn't come up well when copied."[80] But even the court-reporting supply houses were skeptical.

Admitted Pengad, "We have talked to most of the paper mills in the country about this, but they point out that there are so many methods of copying now that one paper could not defeat all of them."[81] Legal recourse was (ironically) not possible either, since reporters could not copyright transcripts.[82] Court managers who, as described in chapter 3, had long argued against the court reporter's dual payment structure (wage or salary for "taking down," per-page piece rate for "writing up") gleefully observed that "it might make more sense to have a single charge per page—which would be higher than the present charge for an original—and have the court provide Xerox copies as required."[83] In the end, Pengad advised reporters to purchase Xerox machines themselves in order to reduce copying costs at the source, passing savings along to their clients.[84] The debate would resurface every few years. As late as 1994, one reporter admitted, "When we think it's necessary, we use Superglue [on the transcript] to melt the Velobind. Jerks who think they can pull apart the transcripts and Xerox them get really irritated, but they come around and finally order copies."[85] But today most seem to agree that "court reporters must do a better job of educating their clients about all the elbow grease that goes into making a transcript. Maybe then they can charge a fair price for it."[86]

The debate over the place of inexpensive electronic duplicating in court reporting highlighted one of the most important tensions in this speech-to-text industry: the question of whether reporters were really selling a product (the transcript itself) or a service (their labor as producer of the transcript). This question was fundamental to federal legislation affecting court reporters after 1993 when the Department of Labor, "in response to a request for a formal opinion from the Indiana Judicial Center," declared that any hours official court reporters spent outside of the courtroom writing up the transcript fell under the overtime provisions of the Fair Labor Standards Act. While other workers might have been happy to have their overtime labor compensated, the court reporters reacted by creating an FLSA task force to strike down the regulation. By November 1994, Senators Larry Pressler (R-SD) and Nancy Kassebaum (R-KS) had introduced "Court Reporter Fair Labor Amendments," which would exempt the official reporters from overtime pay.[87] Signed by President Clinton in September 1995, the act was a victory for the NCRA because it allowed reporters to continue to earn transcript fees for their own outside work. If courts had been forced to pay reporters overtime wages for this labor, then the transcript fee structure that reporters had grown to love might have changed radically.

How should this unusual lobbying effort be understood? Susan Leigh Star's concept of a "boundary object" is useful in exploring this contradiction between labor time and labor product.[88] As Etienne Wenger explains it, "Because artifacts can appear as self-contained objects, it is easy to overlook that they are in fact a nexus of perspectives, and that it is often in the meeting of these perspectives that artifacts obtain their meanings."[89] The speech-to-text transcript is an example of an object that is interpreted quite differently depending on one's point of view in the transcript-production process. To the plaintiffs, defendants, or lawyers who purchase a transcript using their own resources, it is an expensive but necessary component in the legal process, a precondition to any further appeal and the raw material for new arguments. To the public court, which must often subsidize the production of the transcript for poor citizens, it is an unjustifiably external procedure out of the direct time, space, and cost control of the bureaucratic structure that manages all other artifacts in the legal process—evidence, personnel, and electronic recordings. Finally, to the court reporter (and his or her own subcontracted labor force) the transcript represents the tangible embodiment of ephemeral physical and mental labors over the stenotype keyboard, a formal and official product that should command a premium above and beyond the price of hourly work and salaried benefits, precisely because of the contingent nature of the transcript's demand and usefulness. Only some transcripts will be ordered throughout the working life of the court reporter, but when those transcripts *are* ordered, their temporal demands threaten to throw the entire court reporting labor process into disarray. As with any boundary object, it the transcript which makes court reporting work *visible* in both positive and negative senses as Star describes: "On the one hand, visibility can mean legitimacy, rescue from obscurity or other aspects of exploitation. On the other, visibility can create reification of work, opportunities for surveillance, or come to increase group communication and process burdens."[90]

The moment of visibility in court reporting was altered by that profession's encounter with the computer. Realtime CAT had the contradictory effects of both hiding the reporter's knowledge within a computer database (the ever-growing CAT dictionary) and displaying the reporter's skill on the computer screen (the "dirty copy" of the unscoped CAT translation). But realtime CAT also made the new markets of closed captioning and CARTing visible to court reporters. Like both electronic recording and Xerox duplication, realtime CAT was what Andrew Abbott referred to as a "disturbance" within the court-reporting profession: "a new technology requiring professional judgment or a new technique for old professional work," leading to "readjustments" and "jurisdictional contests."[91] As a direct result of realtime CAT development, two sets of speech-to-text workers—offline captioners and online captioners—came into open competition over which could best handle the work of translating television to text. For now it appears that the realtime court reporters have won.

But this was not the only "professional jurisdiction" that court reporters and closed captioners competed over. At stake as well was the question of which profession would be visible enough to best represent the interests of the D/HOH community at large—and to profit from those interests by serving them in new ways, with new technologies, in new social contexts. Here online court reporters, rather than offline captioners, acted as successful "brokers" with regard to the different but related speech-to-text demands in the institutional contexts of the courtroom, the classroom, and the television studio. As Wenger put it, "Brokers are able to make new connections across communities of practice, enable coordination, and—if they are good brokers—open new possibilities for meaning." But brokering "requires enough legitimacy to influence the development of practice, mobilize attention, and address conflicting interests. It also requires the ability to link practices by facilitating transactions between them, and to cause learning by introducing into a practice elements of another."[92] One way that realtime reporters built such legitimacy was by engaging in what Abbott refers to as "abstraction." Put simply, "professions sometimes use their abstract knowledge to reduce the work of competitors to a version of their own."[93] In the case of court reporters, the longstanding abstract knowledge of steno theory acted as a starting point for new jurisdictional claims over captioning work. But further abstract knowledge—how to lobby effectively for federal aid, how to efficiently subcontract an information-manipulation labor process, how to legitimize profit-making ventures as public service, and how to deal collectively with the challenge of potentially labor-replacing technology—cemented the court reporters' claim to the role of informational captioning, whether in the lecture hall, through the cyber-meeting, or on the nightly news.

This jurisdiction may come to an end if the new profession of voice writing—and the new technological "disturbance" of personal speech-recognition computer software—can effectively change the meaning of the boundary object that is, simultaneously, the archived transcription of the record and the ephemeral translation of the proceeding (see chapter 6). But one further change in the spatial and temporal environment of captioning may serve as a disturbance to the jurisdiction of the court reporters as well: digital television (DTV). The transition to computer-mediated production, distribution, and presentation of television programming has been lauded as a boon to both speech-captioning and language-subtitling efforts, with the idea that consumers will be able to choose between dozens of different text tracks for each show. Digital television captions, for example, differ greatly from analog Line 21 captions. Through the late 1990s the Electronics Industry Association and the FCC developed a new digital standard, EIA-708, which would both extend the old EIA-608 analog captioning capabilities and remain backward-compatible with the now considerable library of caption-encoded video in use.[94] Digital captions can occupy nearly ten times the data bandwidth available in the old analog Line 21 format, with character transmission times increased from the current 960 bits per second (bps) to 8640 bps.[95] This increased capacity would allow multiple language and speech tracks to be transmitted simultaneously.[96] Digital captions also offer more display options for consumers, beyond the single-font, uppercase, white-on-black text that make up most analog captions: "altering the size, color and type of the font used for display, necessary for certain age groups, are all possible with digital systems."[97] (Such enhancements mirror the claims of CBS for its failed teletext captions two decades earlier, described in chapter 5.) Enthusiasts even predicted that "picture-in-picture" (PIP) viewing, where a person monitors two channels at once, could be granted its own captioning display as well, since a viewer cannot listen to two audio tracks simultaneously.[98]

Digital captions were not actually broadcast over the air until a November 1999 test in Boston (in a joint venture between WGBH and ABC), and the FCC only issued its final DTV captioning standards in July 2000.[99] Two years later the commission approved a requirement to include digital television tuners in nearly all televisions sold in the United States by 2007.[100] Yet by 2003, "fewer than 1 million of the nation's 106 million television households" had "sets capable of receiving HDTV."[101] Congress may let the analog cutoff date slide "if fewer than

85 percent of households have digital television sets."[102] The jury is still out on how the bulk of the television-viewing public will react to digital television and digital captioning. But clearly the new parameters of DTV captioning ignore the substantial question of labor in producing enough captions, in enough languages, and at enough reading speeds, to fill the promoted bandwidth. Caption length and placement concerns, long the province of offline postproduction captioners rather than online realtime captioners, assume a new urgency when viewers can change the size and shape of their captions at will. Further, during the period of public transition to digital television, many households will likely downgrade their digitally received broadcast into their existing analog television. In this case, new digital captions must be stripped down to fit back into the analog stream.[103] Some stations, especially news and sports properties, may use their increased picture definition to display more on-screen text anyway, which will inevitably conflict with closed-caption placement. And even though digital captions operate at a higher speed, resulting in an overall higher bandwidth, when encoding and decoding *realtime* captions, testing stations have found that an additional digital-conversion delay of nearly one second may actually be introduced.[104] All of these new space and time parameters of caption presentation and consumption in digital television may help swing the center of gravity in caption production work away from the much-publicized realtime writers, and back to the currently invisible offline caption editors and authors.

## Technological Systems, Labor Processes, and Media Justice

The new digital space-time terrain for speech-to-text work—in MOD indexing, DVD subtitling, and DTV captioning—has once again cast the process of television captioning as an object of communication research. For fiscal year 2003 the Department of Education listed as a priority "innovative research on the use of various approaches to captioning," especially in "enhancing the reading or literacy skills of deaf and hard of hearing children in kindergarten through grade 12." Both realtime captioning and realtime voice writing were named as technologies of interest in this funding call.[105] But in many cases it is the offline captioners—given the proper time and training—who have the best chance of applying such research findings to their captioning craft. For example, in a 1997 content analysis of the actual text of captioned television revealed that "there are more than 500,000 words in the English language, but a person who masters [the most frequently used] 250 words will recognize more than two-thirds of all words shown in television captions."[106] Does this indicate that the level of discourse on television is simplistic, or that caption writers are editing a diverse and complicated spoken dialogue down to a simplified text for ease of reading? As the caption-editing debates in the 1970s between PBS and WGBH illustrated (chapter 2), one would have to know the type of captioning (offline versus online), the audience for the captioning (children or adults), and the purpose of the captioning (information or entertainment) to make any guesses.

Other recent research has focused on the proper speed and placement of captioned text. One study in 1998 in which "video segments captioned at different speeds were shown to a group of 578 people that included deaf, hard of hearing, and hearing viewers" found that, for these adults, the most comfortable viewing speed for captions was about 145 words per minute. Surprisingly, neither age nor sex nor education mattered much to one's caption-reading speed, and, in fact, "hearing people wanted slightly slower captions."[107] A follow-up study in 2000 tracked the eye movements of deaf caption viewers in particular, revealing that "subjects gazed at the captions 84% of the time, at the video picture 14% of the time, and off the video 2% of the time," regardless of age, sex, and educational level.[108] Such findings could dramatically affect the work of offline captioners, leading to new standards for caption speed and placement and constraining the translation creativity of these artists in new ways. But for online captioners—simply trying to keep up in realtime, with their captions scrolling across the bottom of the screen as fast as they are typed—the ability to slow caption speed or adjust caption placement is beyond their reach.

A third thread of recent captioning research has dealt with the question of whether captions—offline and online—actually help people understand audiovisual media. One study attempted to measure "quality of perception"—"not only a user's satisfaction with multimedia clips, but also his/her ability to perceive, synthesize and analyse the informational content of such presentations"—finding that "when captions were used, neither hearing nor deaf participants experienced a significant difference in the level of information assimilation."[109] This startling result calls into question many long-standing claims about the value of captioning made by D/HOH educators. But the findings are seriously undermined by the study's lack of attention to the decades of caption research, development, and pilot testing that occurred in the 1970s and 1980s. The study reported no details regarding the size, placement, timing, or editing of its captions—which were produced by a custom Java applet and not through standard video techniques. The participants viewed well under an hour's worth of disconnected video clips in a 352 x 288 pixel window on a computer screen, an experimental setting much different from watching, say, a two-hour-long documentary or feature film on full-size TV. And the authors made no effort to differentiate the various subcultures within the D/HOH community, or to address the idea that one's view of deafness as either "medical" or "social" may affect one's use of captions.

This lack of attention to the diversity of the D/HOH community reflects the most important shift in recent captioning research: a fundamental change in the target audience for captioning. When captioning was reconceived as a "curb cut" in the 1990s, it went from being a specialized accommodation for a minority viewing audience to a multipurpose amenity for all to use. An Annenberg Public Policy Center study of closed-captioning consumption in 2003 surveyed not just the D/HOH community but the general public. Although 60 percent of the respondents used TV captions "sometimes," these users were still structured by individual difference. About 90 percent of the D/HOH respondents used captioning, as did 66 percent of the respondents who were learning to read English. But only 11–15 percent of the general (hearing) viewing audience used them.[110] Those who did report using captions were far from satisfied with the quality of the service, with "the captioning on local news . . . identified as having the poorest quality."[111] Because the study explored the conditions of caption consumption apart from the conditions of caption production, it could not answer whether such poor local news captioning was due to the labors of humans or to the use of teleprompters.

The key limitation to all of these studies is that a broad historical and geographical understanding of the production and reproduction of speech-to-text technological and labor systems in society is a necessary prerequisite to interpreting experimental captioning research, extrapolating future captioning trends, and making political-economic decisions over the regulation and subsidy of captioning labor. Projects to link text to sound and image within the time and space of the viewing

screen date back a century to the earliest days of film, and they continue today with the latest handheld iPod. At different moments, such
projects are pushed by different interests for different reasons—
sometimes to widen profitable markets through language translation,
sometimes to build state legitimacy through public service, and sometimes to obtain access and justice for an excluded minority. But the
outcomes of such projects are neither technologically nor socially determined; rather, they are subject to continual reinterpretation, renego-
tiation, and reproduction as information technologies, labor practices,
social beliefs, and political-economic structures all change over time.

Although defense contractors, broadcast executives, and court
reporters have all played their part in the construction of closed captioning, which began nearly a century ago, the deaf and hard-of-
hearing community has been the most consistent advocate for this
technological vision. Yet when this vision was finally instantiated in
both technological and legal infrastructure in the 1990s, it inspired in
this community a bit of nostalgia for the communication technologies
and practices of the past. As one deaf adult recalled in 1994, "When I
was little, before captioning came in, we used to watch TV as a family.
It was great. We would compare notes on what we saw and propose
our own versions of the plot. Later, if we saw other Deaf friends who
had seen the same show, we would all discuss what we thought had
happened. We would construct personalities and events from our
guesswork and imaginations; it was practice for the guesswork and humor we need to interpret the world. We laughed so much, and it
brought us close together."[112] Others began to speak fondly of the old
screenings of Captioned Films for the Deaf, something that only happened at deaf clubs and schools: "Before the days of captioned television, deaf people congregated at centers or clubs to find out the latest
news and events and to share their thoughts and opinions. This reliance
on each other for information promoted social contacts. Now, they
feel their peers are isolating themselves and relying more on the media
to get their information than from each other."[113] Some suggested that
trends of suburbanization and "the rise of the Deaf middle class"
might be to blame for such fading community spirit.[114] Cultural historian Cecelia Tichi has analyzed the "evolution of television from the
exotic to the commonplace" in mainstream American society and argued that it was the children of television's first adopters, those who
"never knew life without the small screen," who first "experienced

television as integral and natural."[115] But in the D/HOH community, what Tichi calls the "sociocultural naturalization" of television is still taking place.

The naturalization of closed captioning has largely been a household-based consumption experience. But projects to move captioning outside of the space and time of the household continue. As recently as 2000, in Washington, DC, three deaf men filed a "national class-action lawsuit" against the Loews Cineplex and AMC movie theater chains; in Portland, eight deaf and hard-of-hearing people filed a similar action against Regal Cinemas, Century Theatres, and Carmike Cinemas; and in Texas, a father of a deaf teen filed a class-action suit against twelve film-production companies and theater chains. Their demands? Seventy-five years after the "talkies" first stole the intertitles from the cinema, these deaf activists were still demanding the open-captioned screening of first-run films in theaters. The reply of the studios? "It will annoy hearing customers."[116]

This continuing debate can be understood as a clash of incompatible visions of what a just and equitable information society should mean. Writer and activist Harlan Lane has described an interesting folk tale in Deaf culture involving an alternate-world utopia called "Eyeth" (replacing the "Ear" in "Earth" with "Eye") where "everyone assumes you are Deaf, movies are all open-captioned, plays are all in ASL, school classes are in ASL, and kids are not sure who is Deaf and who hearing, nor do they care."[117] Similarly, a key moment in the 1986 Deaf utopian novel *Islay* comes when the residents of this planned city for the deaf amass the political and economic power to purchase their own television station and achieve "captioned news, movies, sports, cartoons, plays, everything captioned!"[118] In 1990, Gallaudet vice president for academic affairs and former NAD president Roslyn Rosen described what such a "Utopia" for Deaf people might be like: "There would be no difference in education, employment, communications and community life," with "total access, around the clock, on television, in movie theatres, over the phone, and in any human interactions . . . In the absence of attitudinal barriers, paternalism would fly out the window. There would be total acceptance of a multicultural society and valuing of natural differences in people. There would be true partnerships between deaf and non-deaf people in all walks of life."[119] The turn of the millennium in the United States saw deaf and nondeaf people finally partnering to make their most ubiquitous and profitable medium of communication, the audiovisual realm of television, textually accessible to all. Although this achievement in digital convergence certainly did not bring about an "eye"-centered Deaf utopia in all media, perhaps careful attention to the construction of closed captioning—and to the way that human labor, human skill, and human empathy become embedded in political, economic, and technological information systems—can help us through thorny questions of equality, equity, and justice in whatever shared information landscape we produce next.

# Abbreviations

| | |
|---|---|
| AAD | *American Annals of the Deaf* |
| AO | *Auditory Outlook* |
| AP | *Associated Press* |
| ARIST | *Annual Review of Information Science and Technology* |
| B&C | *Broadcasting & Cable* |
| BE | *Broadcast Engineering* |
| CFD | Captioned Films for the Deaf |
| COED | Commission on Education of the Deaf |
| DA | *Deaf American* |
| D/HOH | deaf/hard-of-hearing |
| E&PA | *Environment & Planning Annual* |
| ER | electronic (audio) recording |
| FCC | Federal Communications Commission |
| FJC | Federal Judicial Center |
| FR | *Federal Register* |
| GUA | Gallaudet University Archives |
| ICTs | information and communication technologies |
| JCR | *Journal of Court Reporting* |
| JOC | *Journal of Communication* |
| MSCF | Media Services and Captioned Films |
| NAD | National Association of the Deaf |
| NCI | National Captioning Institute |
| NCRA | National Court Reporters Association |
| NCSC | National Center for State Courts |
| NPBA | National Public Broadcasting Archives |
| NSR | *National Shorthand Reporter* |
| NSRA | National Shorthand Reporters Association |
| NTID | National Technical Institute for the Deaf |
| NYT | *New York Times* |
| PBR | *Public Broadcasting Report* |
| PBS | Public Broadcasting Service |
| PHG | *Progress in Human Geography* |

| | |
|---|---|
| *PTR* | *Public Telecommunications Review* |
| RMCD | Regional Media Center for the Deaf |
| SHHH | Self Help for Hard of Hearing People |
| *SW* | *Silent Worker* |
| *T&C* | *Technology and Culture* |
| *UPI* | *United Press International* |
| *VR* | *Volta Review* |
| *WSJ* | *Wall Street Journal* |

# Notes

## Introduction

*Epigraphs:* John D. Rhodes, cited in Horace A. Edgecomb, ed., "The changed conditions in the professional shorthand field," in NSRA, *Proceedings of the 22nd annual convention of the NSRA* (1921), 102–129, 128; Emil S. Ladner Jr., "Silent talkies," *AAD* 76 (1931), 323–325, 324.

1. Nielsen Media Research, as reported by the Television Bureau of Advertising, www.tvb.org (visited 31 Oct. 2005).

2. Thomas H. Kean and Lee H. Hamilton, *The 9/11 Commission report: Final report of the National Commission on Terrorist Attacks upon the United States* (Washington, DC: GPO, 2004).

3. On closed captioning regulations, see the FCC Consumer & Government Affairs Bureau, Disability Rights Office, www.fcc.gov/cgb/dro/caption.html (visited 31 Oct. 2005). On closed-captioning careers, see Gary D. Robson, *Alternative realtime careers: A guide to closed captioning and CART for court reporters* (Vienna, VA: NCRA, 2000).

4. Alessandra Stanley, "A struggle to keep up for TV caption writers," *NYT* (22 Nov. 2001); "United we stand," *JCR* (Jan. 2002), 52–56; "Closed captioning company, VITAC, captions September 11 anniversary coverage for millions of deaf and hard of hearing Americans," *PR Newswire* (28 Aug. 2002).

5. Karen Rivedal, "MATC adds two degree programs," *Wisconsin State Journal* (5 Sep. 2002).

6. "United we stand," 56.

7. On the "digital divide," see the US Department of Commerce, National Telecommunications and Information Administration, 2002–2004 reports on "A nation online," www.ntia.doc.gov/reports/anol/ (visited 31 Oct. 2005), and 1995–2000 reports on "Falling through the net," www.ntia.doc.gov/opad home/digitalnation/index_2002.html (visited 31 Oct. 2005).

8. See Greg Downey, "Virtual webs, physical technologies, and hidden workers: The spaces of labor in information internetworks," *T&C* 42 (2001), 209–35.

9. There is no published secondary history of television captioning, although there is an unpublished dissertation covering the institutional and technical innovations in closed captioning from 1970 to 1985: Mark G. Borchert, "The development of closed-captioned television: Technology, policy, and

cultural form" (Ph.D. thesis, University of Colorado, Boulder, 1998). There is also no published secondary history of either film subtitling or court reporting.

10. On interdisciplinary research claims, see Erica Schoenberger, "Interdisciplinarity and social power," *PHG* 25:3 (2001), 365–382.

11. On the history of technology, see John M. Staudenmaier, S.J., *Technology's storytellers: Reweaving the human fabric* (Cambridge, MA: MIT Press, 1985); Stephen H. Cutcliffe and Robert C. Post, eds., *In context: History and the history of technology* (Bethlehem, PA: Lehigh University Press, 1989).

12. Thomas P. Hughes, "Machines, megamachines, and systems," in Cutcliffe and Post (1989), 106–119; Gabrielle Hecht and Michael Thad Allen, "Introduction: Authority, political machines, and technology's history," in Michael Thad Allen and Gabrielle Hecht, eds., *Technologies of power: Essays in honor of Thomas Parke Hughes and Agatha Chipley Hughes* (Cambridge, MA: MIT Press, 2001), 1–23, 3; Paul Edwards, "Y2K: Millennial reflections on computers as infrastructure," *History & Technology* 15 (1998), 7–29.

13. Wiebe E. Bijker and John Law, eds., *Shaping technology/building society: Studies in sociotechnical change* (Cambridge, MA: MIT Press, 1992); Ronald R. Kline, *Consumers in the country: Technology and social change in rural America* (Baltimore: Johns Hopkins University Press, 2000).

14. Wiebe Bijker, Thomas Hughes, and Trevor Pinch, eds., *The social construction of technological systems: New directions in the sociology and history of technology* (Cambridge, MA: MIT Press, 1987); Merritt Roe Smith and Leo Marx, eds., *Does technology drive history? The dilemma of technological determinism* (Cambridge, MA: MIT Press, 1994).

15. Brian Woods and Nick Watson, "In pursuit of standardization: The British Ministry of Health's Model 8F Wheelchair, 1948–1962," *T&C* 45 (2004), 540–568.

16. Alfred D. Chandler Jr. and James W. Cortada, eds., *A nation transformed by information: How information has shaped the United States from colonial times to the present* (Oxford: Oxford University Press, 2000); Janet Abbate, *Inventing the Internet* (Cambridge, MA: MIT Press, 1999).

17. Frank Webster and Ensio Puoskari, eds., *The information society reader* (New York: Routledge, 2004); Noah Wardrip-Fruin and Nick Montfort, eds., *The new media reader* (Cambridge, MA: MIT Press, 2003).

18. Amy Slaton and Janet Abbate, "The hidden lives of standards: Technical prescriptions and the transformation of work in America," in Allen & Hecht (2001), 95–144, 95.

19. Ruth Schwartz Cowan, "The consumption junction: A proposal for research strategies in the sociology of technology," in Bijker, Hughes & Pinch (1987); Aad Blok and Greg Downey, eds., *Uncovering labour in information revolutions, 1750–2000* (Cambridge: Cambridge University Press, 2004).

20. Amy Sue Bix, *Inventing ourselves out of jobs? America's debate over techno-logical unemployment, 1929–1981* (Baltimore: Johns Hopkins University Press, 2000); Robert Kanigel, *The one best way: Frederick Winslow Taylor and the enigma of efficiency* (New York: Viking, 1997).

21. Harry Braverman, *Labor and monopoly capital: The degradation of work in the twentieth century* (New York: Monthly Review Press, 1974); Alvin Toffler, *The third wave* (New York: Bantam Doubleday, 1981); Shoshana Zuboff, *In the age of the smart machine: The future of work and power* (New York: Basic Books, 1988); Paul S. Adler, ed., *Technology and the future of work* (New York: Oxford University Press, 1992).

22. Greg Downey, "The place of labor in the history of information tech-nology revolutions," in Blok & Downey (2004), 225–261.

23. Greg Downey, *Telegraph messenger boys: Labor, technology, and geography, 1850–1950* (New York: Routledge, 2002).

24. On human geography, see Derek Gregory and John Urry, eds., *Social re-lations and spatial structures* (London: Macmillan, 1985); Richard Peet and Nigel Thrift, eds., *New models in geography: The political-economy perspective* (London: Unwin Hyman, 1989); John Agnew, David N. Livingstone and Alisdair Rogers, eds., *Human geography: An essential anthology* (Oxford: Blackwell, 1996); Erica Schoenberger, "Discourse and practice in human geography," *PHG* 22:1 (1998), 1–14; Michael Dear and Steven Flusty, eds., *The spaces of postmodernity: Readings in human geography* (Oxford: Blackwell, 2002); David Harvey, "On the history and present conditions of geography: An historical materialist manifesto," *Pro-fessional Geographer* 3 (1984), 1–11.

25. David Harvey, *The urban experience* (Baltimore: Johns Hopkins Univer-sity Press, 1989); David M. Smith, *Moral geographies: Ethics in a world of difference* (Edinburgh: Edinburgh University Press, 2000).

26. Neil Smith, *Uneven development: Nature, capital, and the production of space* (New York: Blackwell, 1984); Yuko Aoyama and Eric Sheppard, "The di-alectics of geographic and virtual space," *E&PA* 35 (2003), 1151–1156, 1151.

27. Stephen Kern, *The culture of time and space, 1880–1918* (Cambridge, MA: Harvard University Press, 1983).

28. David Harvey, *The condition of postmodernity: An enquiry into the origins of cultural change* (Cambridge, MA: Blackwell, 1989), 258.

29. Smith (1984).

30. Deborah C. Park, John P. Radford, and Michael H. Vickers, "Disability studies in human geography," *PHG* 22:2 (1998), 208–233.

31. Matthew Zook, *The geography of the Internet industry* (New York: Blackwell, 2005); Stephen Graham, "Information technologies and reconfigurations of ur-ban space," *International Journal of Urban and Regional Research* 25:2 (2001), 405–410; Rob Kitchin and James Kneale, "Science fiction or future fact? Exploring

imaginative geographies of the new millennium," *PHG* 25:1 (2001), 19–35; Greg Downey, "Human geography and information studies," *ARIST* 41 (2007).

32. Michael Benedikt, ed., *Cyberspace: First steps* (Cambridge, MA: MIT Press, 1991); Stephen Graham and Simon Marvin, *Telecommunications and the city: Electronic spaces, urban places* (London: Routledge, 1996); William H. Dutton, Jay G. Blumler, and Kenneth L. Kraemer, "Continuity and change in conceptions of the wired city," in *Wired cities: Shaping the future of communications* (Boston: G.K. Hall & Co., 1987), 3–26; William J. Mitchell, *City of bits: Space, place, and the infobahn* (Cambridge, MA: MIT Press, 1995).

33. David Ellis, Rachel Oldridge, and Ana Vasconcelos, "Community and virtual community," *ARIST* 38 (2004); Susan Hanson, "Reconceptualizing accessibility," in Donald G. Janelle and David C. Hodge, eds., *Information, place, and cyberspace: Issues in accessibility* (New York: Springer, 2000), 267–278.

34. Stephen Graham, "The end of geography or the explosion of place? Conceptualising space, place and information technology," *PHG* 22 (1998), 165–185, 167, 174.

35. Manuel Castells, *The informational city: Information technology, economic restructuring, and the urban-regional process* (New York: Blackwell, 1989); Manuel Castells, "Grassrooting the space of flows," in James O. Wheeler, Yuko Aoyama and Barney Warf, eds., *Cities in the telecommunications age: The fracturing of geographies* (New York: Routledge, 2000), 18–27.

36. Bijker, Hughes & Pinch (1987); Smith & Marx (1994).

37. David Harvey, *The limits to capital* (Oxford: Oxford University Press, 1982); David Harvey, "Retrospect on *The limits to capital,*" *Antipode* (2004), 544–549, 544.

## Chapter 1. Subtitling Film for the Cinema Audience

*Epigraph:* Abe Mark Nornes, "For an abusive subtitling," *Film quarterly* (spring 1999), 17–34.

1. Herman G. Weinberg [interviewed by Colin Edwards], *Film subtitling art & business* (Berkeley, CA: University of California Extension Media Center, 1969); Herman G. Weinberg, *Manhattan odyssey: A memoir* (New York: Anthology Film Archives, 1982), 107–108. On *The Jazz Singer,* see Karel Dibbets, "The introduction of sound," in Geoffrey Nowell-Smith, ed., *The Oxford history of world cinema* (New York: Oxford University Press, 1996), 211–219.

2. Weinberg (1969); Weinberg (1982), 107, 108.

3. Nornes (1999).

4. Harry M. Geduld, *The birth of the talkies: From Edison to Jolson* (Bloomington: Indiana University Press, 1975), 30–31, 42; Norman King, "The sound of silents," *Screen* 25:3 (1984), 2–15, 2.

5. Michael P. du Monceau, "A descriptive study of television utilization in communication and instruction for the deaf and hearing impaired, 1947–1976" (Ph.D. thesis, University of Maryland, 1978), 31.

6. Fotios Karamitroglou, *Towards a methodology for the investigation of norms in audiovisual translation: The choice between subtitling and revoicing in Greece* (Amsterdam: Rodopi, 2000), 7.

7. Michael North, "Words in motion: The movies, the readies, and 'the revolution of the word'" *Modernism/Modernity* 9:2 (2002), 205–223, 207.

8. Alice T. Terry, "Moving pictures and the deaf," *SW* 30:9 (Jun. 1918), 154.

9. George Gelzer [as told to Frank L. Smith], "The deaf and the talkies: The advent of sound destroyed an absorbing amusement for the deaf," *Films in Review* 8:7 (1957), 315–318, 329, 315.

10. "A night in Arcady," *SW* 23:2 (Nov. 1910).

11. "The moving picture," *SW* 23:10 (Jul. 1911).

12. "Moving pictures," *AAD* 65 (1920), 137–141, 138.

13. Lucile M. Moore, "The deaf child and the motion picture," *AAD* 63 (1918), 467–475, 472.

14. "Moving pictures" (1920), 137.

15. John S. Schuchman, *Hollywood speaks: Deafness and the film entertainment industry* (Urbana: University of Illinois Press, 1988), 22.

16. "Moving pictures versus reading," *AAD* 64 (1919), 451.

17. T. Hayward, "Cinemas and 'signs,'" *SW* (Oct. 1916).

18. Terry (1918), 154.

19. Robert E. Maynard, "The Owl," *SW* (Feb. 1910).

20. "Preserving a famous film," *SW* (Apr. 1912).

21. George W. Veditz, "The preservation of the sign language" (1913); reprinted in Lois Bragg, ed., *Deaf world: A historical reader and primary sourcebook* (New York: New York University Press, 2001), 83–85, 85.

22. Schuchman (1988), 21.

23. Jan Ivarsson, *Subtitling for the media: A handbook of an art* (Stockholm: Transedit, 1992), 15.

24. Richard Maltby and Ruth Vasey, "The international language problem: European reactions to Hollywood's conversion to sound," in David W. Ellwood and Rob Kroes, eds., *Hollywood in Europe: Experiences of a cultural hegemony* (Amsterdam: VU University Press, 1994), 68–79, 78.

25. Kristin Thompson, "National or international films?: The European debate during the 1920s," *Film History* 8:3 (1996), 281–96, 282.

26. Maltby & Vasey (1994), 78.

27. Thomas Guback, *The international film industry: Western Europe and America since 1945* (Bloomington: Indiana University Press, 1969), 8.

28. Geduld (1975), 9, citing *Scientific American* (22 Dec. 1877).

29. Edward W. Kellogg, "History of sound motion pictures," *Journal of the Society of Motion Picture and Television Engineers* 64 (Jun.–Aug. 1955); reprinted in Raymond Fielding, ed., *A technological history of motion pictures and television* (Berkeley: University of California Press, 1983), 174–220, 186.

30. Geduld (1975), 195.

31. Schuchman (1988), 42.

32. Henry Jenkins III, " 'Shall we make it for New York or for distribution?':
Eddie Cantor, Whoopee, and regional resistance to the talkies," *Cinema Journal*
29:3 (1990), 32–52, 38.

33. Alexander Walker, *The shattered silents: How the talkies came to stay* (New
York: Morrow, 1979), ix.

34. Jenkins (1990), 38.

35. Schuchman (1988), 42. See also Emily Martin, *The soundscape of moder-
nity: Architectural acoustics and the culture of listening in America, 1900–1933*
(Cambridge, MA: MIT Press, 2002).

36. Harriet Andrews Montague, "My Dear Grievances," *VR* 31 (1929), 8–11, 8.

37. Gelzer (1957), 317.

38. Jane K. Bigelow, " 'Talkies?' Not for us!" *VR* 31 (1929), 221–222, 221.

39. Albert Ballin, "Coming to California," from *The deaf mute howls* (Los
Angeles: Grafton, 1930); reprinted in Bragg (2001), 27–32, 30.

40. Ladner (1931), 323–324.

41. George Veditz, "Motion pictures and the deaf," *AAD* 76 (1931),
299–300, 300.

42. Bigelow (1929), 221.

43. Montague (1929), 9.

44. "Notice!" *VR* 31 (1929), 280.

45. Bigelow (1929), 222.

46. Laura Stovel, "You may still go to the movies," *VR* 31 (1929), 283–284,
283.

47. "Silent films," *VR* 31 (1929), 342; Stovel (1929), 284.

48. Gelzer (1957), 318.

49. John A. Ferrall, "Motion pictures for meetings," *AO* 1:6 (Sep. 1930),
277.

50. Jerome D. Schein, "The demography of deafness," in Paul C. Higgins
and Jeffrey E. Nash, eds., *Understanding deafness socially: Continuities in research
and theory,* 2nd ed. (Springfield, IL: Charles C. Thomas, 1996), 21–43, 21–22,
40. On the social construction of disability, see P. Abberley, "The concept of op-
pression and the development of a social theory of disability," *Disability, Handicap & Society* 2 (1987), 5–19; M. Oliver, *The politics of disablement* (New York: St.
Martin's, 1990); Gary L. Albrecht, *The disability business: Rehabilitation in America* (Newbury Park, CA: SAGE, 1992); J. P. Shapiro, *No pity: People with disabilities
forging a new civil rights movement* (New York: Times Books, 1993).

51. Schein (1996), 36.

52. Jerome D. Schein and Marcus T. Delk Jr., *The deaf population of the United
States* (Silver Spring, MD: NAD, 1974), 4.

53. The terms "oralist" and "manualist" represent broad characterizations of historically specific philosophies toward deaf education and social integration which are often more complicated and contradictory than these polarized concepts might suggest. On the history of deafness in the United States, see Harlan Lane, *When the mind hears: A history of the deaf* (New York: Random House, 1984); John V. Van Cleve, ed., *The Gallaudet encyclopedia of deaf people and deafness* (New York: McGraw-Hill, 1987); Beryl Lieff Benderly, *Dancing without music: Deafness in America* (Washington, DC: Gallaudet University Press, 1990); and Douglas S. Baynton, *Forbidden signs: American culture and the campaign against sign language* (Chicago: University of Chicago Press, 1996).

54. "Social Features," *VR* (Oct. 1929), 657.

55. Harriet Andrews Montague, "I enjoy the theatre," *AO* 1:6 (Sep. 1930), 253–256, 253.

56. Advertisement for E.A. Myers & Sons "Radioear," *AO* (Dec. 1930), 520.

57. "The Editor's Page," *VR* 31 (1929), 344.

58. Terry (1918), 154.

59. Ladner (1931), 325.

60. "Again the talkies!" *VR* 31 (1929), 450.

61. Myrtle Rea, "Movies—A blessing or a curse?" *VR* 38 (1936), 327–328.

62. Wilma I. Nelson, "Advancing the use of visual aids," *VR* 44 (1942), 11–12.

63. Gilbert Hunsinger, "The deaf family in Indianapolis—II," *AAD* 88 (1943), 183–219, 192–193.

64. Frederick Rukdeshal, "J. Pierre Rakow: Pioneer in captioned films," *SW* (May 1962), 3.

65. Maltby & Vasey (1994), 69; Thompson (1996), 295.

66. Antje Ascheid, "Speaking tongues: Voice dubbing in the cinema as cultural ventriloquism," *Velvet Light Trap* (fall 1997), 32–41, 32.

67. Kristin Thompson, *Exporting entertainment: America in the world film market, 1907–1934* (London: BFI Publishing, 1985), 158; Guback (1969), 9; Maltby & Vasey (1994), 83.

68. Thompson (1985), 158–159.

69. Georg-Michael Luyken, *Overcoming language barriers in television: Dubbing and subtitling for the European audience* (Manchester: European Institute for the Media, 1991), 30. Douglas Gomery, *Shared pleasures: A history of movie presentation in the United States* (Madison: University of Wisconsin Press, 1992), 176–177.

70. Thompson (1985), 160–161.

71. Ginette Vincendeau, "Hollywood Babel: The coming of sound and the multiple language version," *Screen* 29:2 (1988): 24–39, 29.

72. Thompson (1985), 159.

73. Ivarsson (1992), 24–25.

74. Maltby & Vasey (1994), 89; Josephine Dries, *Dubbing and subtitling: Guidelines for production and distribution* (Düsseldorf: European Institute for the Media, 1995), 36; Ivarsson (1992), 10.

75. Weinberg (1969); John Minchinton, "Fitting titles: The subtitler's art and the threat of Euro titles," *Sight and Sound* 56:4 (1987), 279–282, 281; Luyken (1991), 55.

76. Ivarsson (1992); Luyken (1991); Dries (1995); Helene J. B. Reid, "Subtitling, the intelligent solution," in P. A. Horguelin, ed., *La traduction, une profession* (Ottawa: Canadian Translators and Interpreters Council, 1978), 420–428; Minchinton (1987); "In brief," *New Scientist* (21 Sep. 2002); Hillel Italie, "Nuances and subtleties: The art of film subtitling," *AP* (19 Feb. 1991).

77. Weinberg (1969); Mike A. Jones, *The development of television services for deaf people* (n.p., 1984), 47.

78. Thompson (1985), 160; Luyken (1991), 96.

79. Basil Hatim and Ian Mason, "Politeness in screen translating," in Lawrence Venuti, ed., *The translation studies reader* (London: Routledge, 2000), 430.

80. Robert Baker, *Subtitling television for deaf children* (Southampton: University of Southampton, Dept. of Teaching Media, 1985), 1, Baker's words; Jan Ivarsson and Mary Carroll, *Subtitling* (Simrishamn: TransEdit HB, 1998), v. Descriptions drawn from Ivarsson (1992), 39–54; Ivarsson & Carroll (1998), 100; and Luyken (1991), 44–45.

81. Baker (1985), 1.

82. Ella Shochat and Robert Stam, "The cinema after Babel: Language, difference, power," *Screen* 26:3–4 (1985), 35–58, 48.

83. Ascheid (1997), 34.

84. Weinberg (1969), paraphrase of original.

85. Ivarsson (1992), 8.

86. Luyken (1991), 31.

87. Dries (1995), 9.

88. Thompson (1985), 158.

89. Maltby & Vasey (1994), 89.

90. Luyken (1991), 31.

91. Maltby & Vasey (1994), 89.

92. Dries (1995), 13–15.

93. Ivarsson (1992), 18.

94. Maltby & Vasey (1994), 89.

95. Thomas L. Rowe, "The English dubbing text," *Babel* 6:3 (1960), 116–120, 116.

96. Dries (1995), 9.

97. Rowe (1960), 116.

98. Thompson (1985), 160.

99. Ascheid (1997), 34.

100. Richard Kilborn, " 'Speak my language': Current attitudes to television subtitling and dubbing," *Media, Culture & Society* 15:4 (1993), 641–661, 643.

101. Dries (1995), 10.

102. Kilborn (1993), 644.

103. Ascheid (1997), 39; Luyken (1991), 31; Dries (1995), 26.

104. Ivarsson (1992), 35.

105. Hans Vöge, "The translation of films: Sub-titling versus dubbing," *Babel* 23:3 (1977), 120–25, 122.

106. Kilborn (1993), 644.

107. Dries (1995), 27–28.

108. Dries (1995), 14.

109. Luyken (1991), 94.

110. Thomas H. Guback, "Film as international business," *JOC* 24:1 (1974), 90–101, 91.

111. Luyken (1991), 12.

112. Weinberg (1969); Italie (1991).

113. Dries (1995), 27, 30; Luyken (1991), 171.

114. Dries (1995), 11, 27.

115. Ivarsson (1992), 75.

116. Ivarsson (1992), 128; Luyken (1991), 94; Dries (1995), 30.

117. Minchinton (1987), 281.

118. Guback (1969), 70–75; Gomery (1992), 173.

119. Gomery (1992), 177.

120. Gomery (1992), 180–181.

121. Guback (1969), 70–71.

122. Rowe (1960), 117.

123. Mark Betz, "The name above the (sub)title: Internationalism, coproduction, and polyglot European art cinema," *Camera Obscura* 16:1 (2001), 1–44; Fausto F. Pauluzzi, "Subtitles vs. dubbing: The *New York Times* polemic, 1960–66," in Douglas Radcliff-Umstead, ed., *Holding the vision: Essays on film* (Kent, OH: International Film Society, Kent State University, 1983).

124. Bosley Crowther, "Subtitles must go!" *NYT* (7 Aug. 1960).

125. Bosley Crowther, "Dubbing (continued)" *NYT* (21 Aug. 1960).

126. Betz (2001), 5.

127. Bosley Crowther, "The Tower of Babel again" *NYT* (14 Aug. 1966).

128. Pauluzzi (1983), 135.

129. Weinberg (1969).

130. Rowe (1960), 120.

131. Weinberg (1982), 109.

132. Crowther (21 Aug. 1960).

133. Robert L. Swain, "Foreign film theater in New York City attracts world's largest deaf audience," *DA* (Jul.–Aug. 1969) 7–9.

134. "Film catalog available," *VR* 60 (1958), 181.

135. Edmund B. Boatner, "Captioned films for the deaf," *AAD* 96 (1951), 346–352, 347.

136. Grace Heider and Fritz Heider, "Motion pictures in class room work," *VR* 37 (1935), 71–76, 71; Nannie S. Davison, "Visual aids in schools for the deaf" (M.A. thesis, Hampton Institute, 1945), 25.

137. Stanley D. Roth, "Motion pictures as a contributory means toward visual education in schools of the deaf" (M.A. thesis, Gallaudet College, 1934), 1.

138. Nettie McDaniel, "Visual education: Introduction," *VR* 34 (1932), 60–62, 61; William J. McClure, "Visual education and the deaf," *AAD* 86 (1941), 166–180, 175; Sylvia Wudel Sanders, "The use of visual aids in teaching the deaf" (M.A. thesis, Gallaudet College, 1946), 18.

139. "Use of motion pictures in teaching lip-reading," *AAD* 86 (1941), 386–387; Margaret Yoder, "Use of the Harvard Reading Films with preparatory class students at Gallaudet College," *AAD* 87 (1942), 395–407, 397.

140. Heider & Heider (1935), 72–76.

141. Sanders (1946), 21.

142. Malcolm Norwood, "Captioning for deaf people: An historical overview," in Judith E. Harkins and Barbara M. Virvan, eds., *Speech to text: Today and tomorrow* (Washington, DC: Gallaudet Research Institute, 1989), 133–138, 133.

143. Marilyn Lundell Majeska, *Talking books: Pioneering and beyond* (Washington, DC: Library of Congress, 1988), 1–4.

144. Majeska (1988), 9–18.

145. Patricia Cory, *Report of [the 4th] Conference on the Utilization of Captioned Films for the Deaf* (Washington, DC: US Dept. of Health, Education, and Welfare, 1960), 1–2; George Propp, "The history of learning technology in the education of deaf children," in C. R. Vest, ed., *Learning technology for the deaf* (Warrenton, VA: Society for Applied Learning Technology, 1978), 4–9, 5.

146. "Movies with sub-titles," *VR* 49 (1947), 524; Emerson Romero, "Sound films for the deaf," *VR* 50 (1948), 259–260, 259; Schuchman (1988), 23, 26–27; Edmund B. Boatner, "Captioned films for the deaf," *AAD* 126 (1981), 520–525, 520; Dorene Romero, "Emerson Romero: That's my pop," *DA* (Jul.–Aug. 1965), 9–10.

147. Romero (1948), 259.

148. Andrew Solomon, "Away from language," *NYT* (28 Aug. 1994).

149. Boatner (1951), 349.

150. Propp (1978), 5; "Destry rides again (for the deaf)," *SW* (Dec. 1961), 3–5, 3.

151. Rukdeshal (1962), 3.

152. Rukdeshal (1962), 3; Boatner (1981), 521; "Motion picture subtitling corporation violated 'no solicitation' provision of final judgment," *New York Law Journal* (14 Nov. 2001).

153. Boatner (1981), 522; "Destry rides again" (1961), 3; Rukdeshal (1962), 3; *The Noose Hangs High,* entry in Internet Movie Database, www.imdb.com/title/tt0040652/ (visited 26 Mar 2004).

154. Boatner (1981), 521; Cory (1960), 1–2.

155. Boatner (1981), 522; NAD, "Captioning and accessibility information: Captioned Media Program," www.nad.org (visited 9 Aug. 2002); Rukdeshal (1962), 3.

156. Cory (1960), 2; Boatner (1981), 522.

157. Boatner (1951), 346–350.

158. Boatner (1981), 523.

159. Cory (1960), 2; Boatner (1981), 523.

160. "Minutes of the 65th annual meeting, Alexander Graham Bell Association for the Deaf," *VR* 57 (1955), 313.

161. John Gough, "Captioned films for the deaf," *SW* (Aug. 1960), 18–20, 18.

162. Cory (1960), 2.

163. Boatner (1981), 524.

164. Du Monceau (1978), 40; Borchert (1998), 151; "The editor's page," *SW* (Mar. 1959), 2; "The editor's page," *SW* (Oct. 1959), 2.

165. "Captioned films for the deaf," *AAD* 104 (1959), 398–399; Robert Panara and John Panara, "Malcolm Norwood: Captioned media specialist," *Great deaf Americans* (Silver Spring, MD: TJ Publishers, 1989), 99–103, 101.

166. Gough (1960), 19; "Film fare," *SW* (Jan. 1961), 13.

167. Panara & Panara (1989), 100–101.

168. "The editor's page," *SW* (Oct. 1959), 2.

169. "Trainees start films captioning at Gallaudet," *SW* (Jan. 1960), 32.

170. "Film fare," *SW* (Apr. 1961), 30.

171. Rukdeshal (1962), 3; Gough (1960), 19; "Film fare," *SW* (Nov. 1960), 24.

172. "Destry rides again" (1961), 4.

173. "Film fare," *SW* (Jan. 1961), 13; "Destry rides again" (1961), 4.

174. "Destry rides again" (1961), 4; Cory (1960), 11.

175. "Film fare," *SW* (May 1961), 24.

176. "Film fare" (Apr. 1961), 30.

177. Cory (1960), 11; "Film fare" (Jan. 1961), 13.

178. "Film fare" (Jan. 1961), 13.

179. Cory (1960), 9; "The editor's page," *SW* (Mar. 1960), 2.

180. "Film fare," *SW* (Oct. 1960), 22.

181. "Film fare," *SW* (Aug. 1961), 8.

182. "Film fare" (Oct. 1960), 22; "Film fare," *SW* (Dec. 1960), 28.

183. Boatner (1981), 523.

184. "Film fare," *SW* (Nov. 1960), 24.

185. "Film fare" (Nov. 1960), 24.

186. Cory (1960), 3.

187. "Film fare" (Jan. 1961), 13.

188. "Film Fare," *SW* (Oct.–Nov. 1962), 10.

189. Gough (1960), 19; John A. Gough, "Captioned films for the deaf: The new program," *VR* 65 (1963), 24–25, 24.

190. "Production and distribution of captioned films for the deaf," *FR* (1 Dec. 1964), 15955–15956.

191. Gough (1963), 24.

192. "Film fare," *SW* (May 1962), 5; "Film fare," *SW* (Apr. 1963), 12.

193. "Film fare," *SW* (May 1963), 9; Cory (1960), 9.

194. "Indiana School selected as national depository for captioned films," *AAD* 106 (1961), 487.

195. "Film fare" (Aug. 1961), 8; "Film fare," *SW* (Oct. 1961), 12; "Indiana School selected" (1961), 487; "Film fare" (May 1962), 5.

196. "Film fare," *SW* (Feb. 1963), 6.

197. "Captioned films, an up-to-date report," *DA* (Jun. 1965), 13–14.

198. "Film fare," *DA* (Apr. 1966), 11.

199. Gough (1966), 407–410.

200. "Film fare," *DA* (May 1966), 9; "Captioned films" (Jun. 1965), 13–14.

201. "Film fare," *SW* (Feb. 1964), 7; "Captioned films for the deaf," *AAD* 110 (1965), 323.

202. "Film fare," *DA* (May 1966), 9.

203. Propp (1978), 5.

204. Robert E. Stepp Jr., "A technological metamorphosis in the education of deaf students," *AAD* 139 (1994), 14–17, 15.

205. "Filmed lessons to teach keypunch operation to the deaf," *VR* 66 (1964), 341.

206. Jerome D. Schein and John J. Kubis, *A survey of visual aids in schools and classes for the deaf in the United States* (Washington DC: Gallaudet College, 1962), 3, 10.

207. Schein & Kubis (1962), 16.

208. US Congress, House Committee on Education and Labor, *Handicapped individuals services and training act,* H. Rep. 97-950 (9 Dec. 1982), 8.

209. James J. Kundert, "Media services and captioned films," *AAD* 114 (1969), 680–692, 680; Robert Root, "A survey of the reaction of hearing individuals to captioned television to benefit the hearing impaired," *AAD* 115 (1970), 566–68, 566.

210. "Film fare," *DA* (Mar. 1965), 12–13; "Production and distribution of captioned films for the deaf," *FR* (1 Dec. 1964), 15955–15956.

211. Gough (1960), 20.

212. "John A. Gough—Eduactor extraordinary," *DA* (Feb. 1969), 3–5, 3.

213. "Production and distribution of captioned films for the deaf" (1964), 15955; "Film fare," *DA* (Mar. 1965), 13.

214. William Jackson and Roger Perkins, "Television for deaf learners: A utilization quandary," *AAD* 119 (1974), 537–548, 538.

215. Robert E. Stepp Jr., "Educational media and technology for the hearing-impaired learner: An historical overview," *VR* 83 (1981), 265–274, 267; Kundert (1969), 692.

216. John A. Gough, "Report from Captioned films for the deaf," *AAD* 112 (1967), 642–649, 644.

217. "Attention television fans," *DA* (Nov. 1965), 2.

218. Jackson & Perkins (1974), 539.

219. Panara & Panara (1989), 102; Root (1970), 566.

220. Nornes (1999).

221. Donald Richie, "Subtitling Japanese films," *Mangajin* (1991), 16–17.

222. Minchinton (1987), 279.

223. Kilborn (1993), 647.

224. Ivarsson (1992), 132.

225. Betz (2001), 6.

226. Dirk Delabastita, "Translation and the mass media," in Susan Bassnett and André Lefevere, *Translation, history and culture* (London: Pinter, 1990); Ascheid (1997); Ivarsson & Carroll (1998), v, 30.

227. Dries (1995), 4.

228. Shochat & Stam (1985), 48.

229. Maltby & Vasey (1994), 90.

230. NTID, "Captioning of motion picture films," ts. (Rochester, NY: NTID, 1976).

231. "Movie Captioned for the Deaf Closes," *NYT* (7 Jun. 1979).

232. Schuchman (1988), 7.

233. Roger Ebert, *Chicago Sun-Times* (3 Oct. 1986).

234. NAD, "Captioning and accessibility information: Captioned Media Program," www.nad.org (visited 9 Aug. 2002).

## Chapter 2. Captioning Television for the Deaf Population

*Epigraph:* Doris C. Caldwell, "The caption-production process at PBS," in Barbara Braverman and Barry Jay Cronin, eds., *Captioning: Shared perspectives* (Rochester, NY: NTID, 1979), 54–59, 54.

1. Thomas Freebairn, *Television for deaf people: Selected projects* (New York: NYU Deafness Research and Training Center, 1974), 1.

2. Helen S. Lane, "Television for the deaf," *VR* 53 (Aug. 1951), 345–347, 345.

3. Donald Torr, "Gallaudet update," in Robert E. Stepp, ed., *Update 74: A decade of progress* (Lincoln: University of Nebraska, 1974).

4. Donald V. Torr, "Captioning philosophy: Verbatim captions," in Braverman & Cronin (1979), 15–22, 17; Sandra L. White and Bill Pugin, "The captioning process: Production techniques," in Braverman & Cronin (1979), 65–71, 70.

5. "Captioning evaluation forms 1971," GUA, Donald V. Torr papers (MSS 28), b.1, f.2.

6. Sue Kelley, "Television for the deaf," *VR* 51 (1949), 445–446.

7. Alvin Weiner, "Television is a blessing," *VR* 54 (1952), 170.

8. Lucile Cypreansen and Jack McBride, "Lipreading lessons on television," *VR* 58 (1956), 346–348, 346; "Television for deaf children in Chicago," *VR* 64 (1962), 30; B.G. Cross, "At work and play: Television in the lives of the deaf and the hard of hearing," *VR* 69 (March 1967), 203; Gallaudet College, *Now see this: A list of TV stations presenting broadcasts captioned or interpreted for the deaf television audience* (Washington, DC: Gallaudet College, 1972).

9. "The editor's page," *SW* (May 1958), 2.

10. Jackson & Perkins (1974), 538.

11. B. G. Cross, "Television news for the hearing impaired," *VR* 71 (1969), 542b–c.

12. "White House reacts favorably to TV proposals for the deaf," *VR* 71 (1969), 224a; "Television for the hearing impaired," *VR* 71 (1969), 400–401.

13. William Jackson, "1970 survey of instructional television in programs for the deaf," *AAD* 115 (1970), 615–618, 618.

14. Southern RMCD, *1968 summary report: Implications for the use of television in schools for the deaf* (Knoxville, TN: Southern RMCD, 1968), 14, 28; Jackson & Perkins (1974), 539.

15. Southern RMCD, *1969 summary report: Video technology in schools for the deaf* (Knoxville, TN: Southern RMCD, 1969), iii.

16. George Propp, "The history of learning technology in the education of deaf children," in Vest (1978), 4–9, 6.

17. "A campus wide closed circuit television system," *VR* 72 (1970), 60; Michael Z. Waugh, "Captioned television for hearing impaired viewers at The Kansas School for the Deaf: Method and applications," in Braverman & Cronin (1979), 46–53, 46–67.

18. "Symposium on Research and Utilization of Educational Media for Teaching the Deaf: Communicative Television and the Deaf Student," *AAD* 115 (1970), 543–545; George Propp, "Introduction," *AAD* 115 (1970), 547–551, 549; Jackson, "1970 survey" (1970), 615.

19. E. Jack Goforth, *Suggestions and guidelines for development of television facilities in schools for the deaf* (Knoxville, TN: Southern RMCD, 1968), 5, 22.

20. J. A. Gough, "Educational media and the handicapped child," *Exceptional children* 34 (1968), 561–564, 562.

21. William Jackson, "Designing a prototype television studio-laboratory," *AAD* 115 (1970), 558–561, 560; Joel Ziev, "*Sesame Street* captioned for the deaf," *AAD* 115 (1970), 630–31.

22. G. Ofiesh, "The potential of television for teaching the deaf," *AAD* 115 (1970), 577–586, 581.

23. Gary L. Albrecht, *The disability business: Rehabilitation in America* (Newbury Park, CA: SAGE, 1992), 19.

24. Andrew Solomon, "Away from language," *NYT* (28 Aug. 1994); William C. Stokoe, *Sign language structure: The first linguistic analysis of American sign language,* rev. ed. (Silver Spring, MD: Linstok Press, 1978).

25. Harlan Lane, Robert Hoffmeister, and Ben Bahan, *A journey into the Deaf-World* (San Diego: DawnSign Press, 1996), 63.

26. Baynton (1996), 11–12.

27. Albrecht (1992), 21.

28. Baynton (1996), 2.

29. Albrecht (1992), 21.

30. Susan Foster, "Doing research in deafness: Some considerations and strategies," in Higgins & Nash (1996), 3–20, 4–5.

31. Jerome D. Schein, "The demography of deafness," in Higgins & Nash (1996), 21–43, 23.

32. Baynton (1996), 2.

33. Schein (1996), 23.

34. Harlan Lane, "Constructions of deafness," *Disability & Society* 10 (1995), 171–189, 173.

35. Lane, Hoffmeister & Bahan (1996), 6.

36. Sherman Wilcox and Phyllis Wilcox, *Learning to see: American Sign Language as a second language* (Englewood Cliffs, NJ: Prentice-Hall, 1991), 66–67.

37. Katherine A. Jankowski, *Deaf empowerment: Emergence, struggle, and rhetoric* (Washington, DC: Gallaudet University Press, 1997), 80.

38. Carl J. Jensema, "Telecommunications for the deaf: Echoes of the past—a glimpse of the future," *AAD* 139 (1994), 22–27, 22.

39. Barry Strassler, "Telecommunications: Telephone services," in Van Cleve (1987), v. 3, 262–264, 262; Benderly (1990), 237.

40. Jensema (1994), 22.

41. Strassler (1987), 263.

42. Jensema (1994), 22–23. See also Harry G. Lang, *A phone of our own: The deaf insurrection against Ma Bell* (Washington, DC: Gallaudet University Press, 2000).

43. Jerome D. Schein, Thomas Freebairn, Bertram Sund, and Susan Hooker, *Television for deaf audiences: A summary of the current status* (New York: Deafness Research & Training Center, 1972), 1.

44. Jerome D. Schein, "The deaf television audience," *DA* (Feb. 1972), 13.

45. Schein quoted in Roger Perkins, ed., *Proceedings of the first National Conference on Television for the Hearing-Impaired* (Knoxville: Southern RMCD, University of Tennessee, 1972), 34.

46. J. D. Schein and S. Hooker, "Television for the deaf," *Hearing & Speech News* (Jul.–Aug. 1972), 13.

47. Schein et al. (1972), 13.

48. Robert Jackson, "Television and the deaf," *DA* (Apr. 1974), 10–11.

49. Schuchman (1988), 9; Lane, Hoffmeister & Bahan (1996), 270.

50. Betsy Montandon, "The whys and wherefores of captioned media," in Nan Decker and Betsy Montandon, *Captioned media in the classroom* (Silver Spring, MD: NAD, 1984), 1–12, 2.

51. Baynton (1996), 12.

52. Thomas Forrester and Stuart Fletcher, *Real time visual interpretation of televised network news for the deaf: A feasibility study* (Rochester, NY: NTID Division of Research & Training, 1971), 11.

53. Nancy Rigg and Earl Higgins, "The sound of silence: Television is at last beginning to serve the needs of the deaf," *Videography* (May 1977), 16–20, 17.

54. Jerome D. Schein and Ronald Hamilton, *IMPACT 1980: Telecommunications and deafness* (Silver Spring, MD: NAD, 1980), 23.

55. Schein & Hamilton (1980), 23.

56. George Rosen, "Garroway 'Today' off to boff start as revolutionary news concept," *Variety* (16 Jan. 1952), 29.

57. "ABC-TV News captioning news bulletins," *DA* (Sep. 1965), 2.

58. "Television for the hearing impaired," *VR* 71 (1969), 400–401.

59. "Federal Communications Commission; Captioning of Emergency Messages on Television," *FR* (31 Dec. 1975), 60080–60081.

60. "Federal Communications Commission issues public notice regarding use of telecasts to inform and alert the deaf," *DA* (Feb. 1971), 4–5; "Implementing the FCC recommendation on TV for hearing impaired," *VR* 73 (1971), 288g.

61. Gallaudet College, *Now see this* (1972).

62. James L. Baughman, *The republic of mass culture: Journalism, filmmaking, and broadcasting in America since 1941* (Baltimore: Johns Hopkins University Press, 1992), 59.

63. William Hoynes, *Public television for sale: Media, the market, and the public sphere* (Boulder, CO: Westview Press, 1994), 1.

64. Carnegie Commission on Educational Television, *Public television: A program for action* (New York: Bantam, 1967).

65. Robert J. Blakely, *To serve the public interest: Educational broadcasting in the United States* (Syracuse, NY: Syracuse University Press, 1979), 200.

66. Hoynes (1994), 14.

67. "The Caption Center turns 20," *Deaf Life* (Mar. 1992), 26; Mardi Loeterman, "The Caption Center at nineteen," *DA* (Spring 1989), 9–15, 9.

68. Sharon Earley, "Captioning at WGBH-TV," *AAD* 123 (1978), 655–662, 657; Malcolm J. Norwood, "Media services and captioned films," *AAD* 117 (1972), 553–555, 554.

69. Norwood (1972), 554.

70. Malcolm Norwood, "Captioning for deaf people: An historical overview," in Harkins & Virvan (1989), 133–138, 134; Robert Root, "A survey of the reaction of hearing individuals to captioned television to benefit the hearing impaired," *AAD* 115 (1970), 566–68, 566.

71. Root (1970), 567.

72. Robert L. Shayon, "Hearing and listening," *Saturday Review* (6 Feb. 1971), 47.

73. Root (1970), 568.

74. Schein & Delk (1974), 1.

75. US Census Bureau, *Decennial census of population, vol. 1: Characteristics of the population, part 1: United States summary* (Washington, DC: GPO, 1970).

76. Schein & Delk (1974), 1.

77. Schein quoted in Perkins (1972), 6.

78. Schein & Delk (1974), 4.

79. Augustine Gentile, *Persons with impaired hearing: United States 1971* (Rockville: US Dept. of Health, Education, and Welfare, 1975), 1.

80. "The Caption Center turns 20" (1992), 19.

81. Earley (1978), 657.

82. Loeterman (1989), 9.

83. Earley (1978), 657.

84. Virginia Murphy-Berman and Jill Shulman, "The evolution of caption writing techniques for children at the WGBH Caption Center: An integration of research and practical application," in George Propp, ed., *1980s schools . . . Portals to century 21* (Silver Spring, MD: Convention of American Instructors of the Deaf, 1980), 149–155, 150.

85. Earley (1978), 657–658.

86. Earley (1978), 659.

87. Sandra L. Danielson and David A. Howe, "Use of the television vertical interval to broadcast time for everyone and program captions for the deaf," *Communications Society* 11:5 (1973), 3–6, 4.

88. Maureen Goss, "Breaking the sound barrier," *PTR* (Jan.–Feb. 1979), 53–55, 53.

89. Goforth (1968), 26–29.

90. Danielson & Howe (1973), 4.

91. Danielson & Howe (1973), 3.

92. Sandra Howe, *NBS time and frequency dissemination services* (Washington, DC: US Dept. of Commerce, National Bureau of Standards, 1976), 1.

93. Normal Felsenthal, "Closed captioning," in Horace Newcomb, ed., *Museum of Broadcast Communication Encyclopedia of Television* (Chicago: Fitzroy Dearborn, 1997), 1:384–85.

94. Danielson & Howe (1973), 3.

95. Borchert (1998), 131, petition text.

96. Borchert (1998), 125.

97. Danielson & Howe (1973), 3.

98. "ABC Inc. cited by HEW secretary Califano for eight-year effort to develop industry-wide closed-captioning program to serve deaf," *DA* (Apr. 1979), 5–7, 5.

99. Perkins (1972), ii–iii.

100. Malcolm J. Norwood, "Developments in captioned television," *DA* (Sep. 1973), 7–8.

101. Perkins (1972), 9–11; Danielson & Howe (1973), 4; Borchert (1998), 139–40.

102. "ABC Inc. cited by HEW secretary Califano" (1979), 5–7, 6.

103. Letter from Julius Barnathan to Donald V. Torr (10 Apr. 1979), GUA, Donald V. Torr papers (MSS 28), b.2, f.4.

104. Norwood (1972), 554.

105. Perkins (1972), 18.

106. "ABC Inc. cited by HEW secretary Califano" (1979), 5–7, 6.

107. Borchert (1998), 98, Borchert's words.

108. Borchert (1998), 140–141.

109. "Department of Commerce; National Bureau of Standards; TvTime Signal Format; Broadcast authorization," *FR* (4 Jan. 1973), 813–814.

110. Blakely (1979), 205.

111. Hoynes (1994), 3.

112. Norwood (1989), 135; "Federal Communications Commission; Television broadcast signal for captioning for the deaf," *FR* (10 Feb. 1976), 5834–5836, 5834.

113. Michael P. du Monceau, "A descriptive study of television utilization in communication and instruction for the deaf and hearing impaired, 1947–1976" (Ph.D. thesis, University of Maryland, 1978), 59.

114. "12. Captioning for the deaf: Report" from PBS Boards of Governors/Managers meeting (9 Feb. 1977), NPBA, Arthur A. Paul papers, s.1, b.3, f.5.

115. Donald V. Torr, *Captioning project evaluation: Final report* (Washington, DC: Gallaudet College, 1974), 5.

116. "PBS spearheading captioning for hearing impaired," *Broadcasting* (28 Feb. 1977), 40–43, 43.

117. Jackson & Perkins (1974), 541; Torr, *Captioning project evaluation* (1974), 8–9.

118. "PBS captioning staff" (Jan. 1976), NPBA, Donald R. Quayle papers, s.1, b.1, f.21.

119. David Sillman, "Line 21, closed captioning of television programs—a progress report," *AAD* 123 (1978), 726–729, 727; Les Brown, "Captions for deaf to be shown on TV," *NYT* (28 Feb. 1974).

120. Brown (1974).

121. Torr, *Captioning project evaluation* (1974), 6, 12–13.

122. "PBS spearheading captioning" (1977), 43.

123. Petition for rulemaking from PBS before the FCC in the matter of "Amendment of Subpart E of Part 73 of the Commission's Rules and Regulations to Reserve Line 21 of the Vertical Blanking Interval of the Television Broadcast Signal for Captioning for the Deaf," RM-2616 (Nov. 1975), GUA, Donald V. Torr papers (MSS 28), b.2, f.5.

124. "Federal Communications Commission; Television broadcast signal for captioning for the deaf," *FR* (10 Feb. 1976), 5834–5836, 5835.

125. "PBS's caption plan inflexible, costly, contend opponents: EIA, broadcasters urge FCC to do more research," *Broadcasting* (17 May 1976), 56.

126. "Federal Communications Commission; Radio broadcast services; Vertical blanking interval of the television broadcast signal for captioning for the deaf," *FR* (28 Dec. 1976), 56321–56324.

127. Danielson & Howe (1973), 4.

128. "FCC gives its approval to hearing-aid captions," *NYT* (13 Dec. 1976).

129. Patrick J. Leahy and Charles H. Percy, "The fate of line 21: Will the deaf 'hear' TV?" *Washington Post* (7 Sep. 1976); reprinted in *DA* (Oct. 1976), 7.

130. "Federal Communications Commission; Radio broadcast services" (1976), 56323.

131. "PBS president calls for unified broadcaster action in developing TV service for the hearing impaired," *DA* (Apr. 1977), 7–8.

132. Goforth (1968), 27.

133. Doris C. Caldwell, Research Associate, PBS, "Captioning production costs," Appendix F (Oct. 1975), GUA, Donald V. Torr papers (MSS 28), b.2, f.5.

134. "Federal Communications Commission; Radio broadcast services" (1976), 56323.

135. "PBS's caption plan inflexible" (1976), 56.

136. "PBS president calls for unified broadcaster action" (1977), 7.

137. Leahy & Percy (1976).

138. "Federal Communications Commission; Television broadcast signal for captioning for the deaf" (1976), 5834.

139. "PBS keeps up heat under petition for TV captioning," *Broadcasting* (18 Oct. 1976), 47; "Federal Communications Commission; Radio broadcast services" (1976), 56324.

140. Torr, *Captioning project evaluation* (1974), 12–15.

141. "Federal Communications Commission; Radio broadcast services" (1976), 56324.

142. Barbara B. Braverman and Barry Jay Cronin, "Television and the deaf," *Journal of Educational Technology Systems* 7:1 (1978–1979), 9; Barbara B. Braverman and Barry Jay Cronin, "Television and the deaf," in Vest (1978), 61–66, 61.

143. Schein (1972), 13.

144. "Federal Communications Commission; Radio broadcast services" (1976), 56323.

145. "PBS keeps up heat" (1976), 47; "FCC gives its approval" (1976).

146. "Federal Communications Commission; Radio broadcast services" (1976), 56324–56325.

147. Gallaudet College, Public Service Programs, *Now see this: A survey of television stations programming for deaf viewers* (Washington, DC: Gallaudet College, 1977), viii.

148. "PBS president calls for unified broadcaster action" (1977), 8.

149. "Captioning for the Deaf—Background" (1977), NPBA, Donald R. Quayle papers, s.1, b.1, f.21.

150. Freebairn (1974), 28.

151. R. Evans Wetmore, Asst. Manager of Transmission Engineering, PBS, "Computer assisted captioning of television programs" (Sep. 1975), GUA, Donald V. Torr papers (MSS 28), b.2, f.5.

152. Caldwell (1979), 55.

153. Linda Carson, "The captioning process: Production techniques at the National Technical Institute for the Deaf," in Braverman & Cronin (1979), 25–36, 26; Waugh (1979), 48.

154. Leonard Bickman, Thomas Roth, Ron Szoc, Janice Normoyle, H. B. Shutterly, and W. D. Wallace, *An evaluation of captioned television for the deaf: Final report* (Evanston: Westinghouse Evaluation Institute, 1979), 5-1-5-4.

155. "Captioned TV demonstrated at conference," *VR* 74 (1972), 96a–b.

156. "Federal Communications Commission; Television broadcast signal for captioning for the deaf" (1976), 5836.

157. Edmund A. Williams, Transmission engineer, PBS, "Cost estimates for a captioning decoder" (Oct. 1975), GUA, Donald V. Torr papers (MSS 28), b.2, f.5; Bickman et al. (1979), 3–5.

158. Danielson & Howe (1973), 4.

159. "PBS spearheading captioning" (1977), 40; Norwood (1989), 136.

160. "PBS spearheading captioning" (1977), 43.

161. "PBS keeps up heat" (1976), 47.

162. Gentile (1975), 16.

163. J. R. Lucyk, *Television and the hearing-impaired* (Ottawa: Broadcasting and Social Policy Branch, Dept. of Communications, 1979), 10.

164. Caldwell (1979), 54.

165. "Captioning for the Deaf—Background" (1977).

166. Bickman et al. (1979), 3–5.

167. "PBS Interim Captioning Service" (7 Jan. 1976), NPBA, Donald R. Quayle papers, s.1, b.1, f.21.

168. "PBS Interim Captioning Service" (1976).

169. Letter from Michael S. Rice to John Montgomery (1 Apr. 1976), NPBA, Donald R. Quayle papers, s. 1, b.1, f.21.

170. Letter from Michael S. Rice to John Montgomery (30 Apr. 1976), NPBA, Donald R. Quayle papers, s.1, b.1, f.21.

171. Edward Dolnick, "Deafness as culture," *Atlantic Monthly* (Sep. 1993), 37–53, 40.

172. Torr, "Gallaudet update" (1974).

173. Torr (1979), 16–17.

174. Letter from Donald V. Torr to Patria Forsythe (11 Jul. 1975), GUA, Donald V. Torr papers (MSS 28), b.2, f.4.

175. Letter from Donald V. Torr to "Superintendents, Principals, Media directors, Librarians" (Dec. 1976), GUA, Donald V. Torr papers (MSS 28), b.2, f.4.

176. White & Pugin (1979), 66.

177. Earley (1978), 655; Sharon Earley, "The philosophy of edited captions," in Braverman & Cronin (1979), 10–14.

178. George D. Grant, "Open and closed television captioning for the deaf" (M.S. thesis, University of Wisconsin–Stout, 1978), 31.

179. Earley (1978), 659.

180. Schein & Hamilton (1980), 34–35.

181. Earley (1978), 659.

182. Loeterman (1989), 9.

183. "Television news with subtitles planned in Boston for the deaf," *NYT* (1 Dec. 1973); "Harry and Howard for the hearing-impaired: PBS stations set to start carrying captioned 'ABC Evening News,'" *Broadcasting* (3 Dec. 1973), 28.

184. Jeff Hutchins and Carole Osterer, "Captioning process at The Caption Center, WGBH," in Braverman & Cronin (1979), 37–45, 38–42.

185. Earley (1978), 660.

186. Carl Jensema, Ralph McCann, and Scott Ramsey, "Closed captioned television presentation speed and vocabulary," *AAD* 141 (1996), 284–292, 285.

187. Earley (1978), 656.

188. "The Caption Center turns 20," *Deaf Life* (Mar. 1992), 27.

189. "News notes," *VR* 77 (1975), 383; Gallaudet College (1977), vii.

190. "The Captioned ABC Evening News: Common questions and accurate answers," *DA* (Mar. 1975), 3–4.

191. "ABC Evening News with captions for deaf on PBS," *NYT* (1 Oct. 1974), 83.

192. "Deafness Research & Training Center," *DA* (May 1975), 3–10, 5.

193. Hutchins & Osterer (1979), 44.

194. Bickman et al. (1979), 1–5; "The Caption Center completes 5 years of programming for hearing impaired," *DA* 31:4 (1978), 5–6.

195. S. Earley, "Captions bring ZOOM to hearing-impaired children," *DA* (May 1976), 15; "PBS spearheading captioning" (1977), 43.

196. Janice H. VanGorden, "Audience consideration and its impact on captioning at the National Technical Institute for the Deaf," in Braverman & Cronin (1979), 72–75, 73.

197. Carson (1979), 27–29.

198. Doris Cooper Caldwell, "Use of graded captions with instructional television for deaf learners," *AAD* 118 (1973), 500–507, 501.

199. Caldwell (1979), 55.

200. Letter from Michael S. Rice to John Montgomery (1 Apr. 1976).

201. Letter from Philip W. Collyer to Doris Caldwell (23 Apr. 1976), NPBA, Donald R. Quayle papers, s.1, b.1, f.21.

202. Letter from Doris C. Caldwell to Philip W. Collyer (4 May 1976), NPBA, Donald R. Quayle papers, s.1, b.1, f.21.

203. Letter from Philip W. Collyer to Doris Caldwell (23 Apr. 1976).

204. Letter from Doris C. Caldwell to Philip W. Collyer (4 May 1976).

205. Sharon Earley and Hillary Kimmel Lyons, "Broadcast captioning: 1979 update," *AAD* 124 (1979), 624–626, 624.

206. Bickman et al. (1979), 2-2-2-5.

207. Memo from Dan Wells to Don Quayle on "Captioning for the Deaf" (4 Jan. 1977), NPBA, Donald R. Quayle papers, s.1, b.1, f.21.

208. "PBS president calls for unified broadcaster action" (1977), 8; "12. Captioning for the Deaf: Report" (1977).

209. Letter from Frederick C. Schreiber to Don Torr (26 Jan. 1977), GUA, Donald V. Torr papers (MSS 28), b.1, f.3.

210. "PBS spearheading captioning" (1977), 40.

211. Letter from Frederick C. Schreiber to "Participants of the Feb. Line 21 Meeting" (2 Mar. 1977), GUA, Donald V. Torr papers (MSS 28), b.1, f.3.

212. "Carter gets mixed response on TV for deaf," *Broadcasting* (14 Mar. 1977), 61.

213. Montandon (1984), 2.

214. Letter from D. V. Torr to E. C. Merrill Jr. (26 May 1977), GUA, Donald V. Torr papers (MSS 28), b.1, f.3.

215. "Califano calls summit meeting to push closed captioning," *Broadcasting* (26 Sep. 1977), 36.

216. "Stern words from Lee on captioning," *Broadcasting* (20 Nov. 1978), 54.

217. John E. D. Ball, "National Captioning Institute," in Van Cleve (1987), 2:221–223, 222.

218. Norwood (1989), 136.

219. "Captioned TV for deaf soon to be announced," *Broadcasting* (12 Mar. 1979), 42.

220. Press release from Joseph A. Califano Jr. on new closed captioning project (23 Mar. 1979), GUA, Donald V. Torr papers (MSS 28), b.1, f.4; Ball (1987), 221; "The caption race: Who's ahead and what's ahead," *Deaf Life* (Feb. 1990), 9–14, 10.

221. Carl Jensema and Molly Fitzgerald, "Background and initial audience characteristics of the closed-captioned television system," *AAD* 126 (1981), 32–36, 32; "John E. D. Ball named president of National Captioning Institute, Inc." *DA* (Dec. 1979), 21; "Door is opened on closed captioning," *Broadcasting* (26 Mar. 1979), 32–33; "National Captioning Institute names hearing impaired advisory boards," *DA* (Apr. 1980), 4; "John Ball: TV's wordsmith," *Broadcasting* (27 Aug. 1990), 87.

222. Press release from Joseph A. Califano Jr. (1979).

223. Joseph A. Califano Jr., "Statement [ . . . ] March 23, 1979," *DA* (Apr. 1979), 3–4; "Captioned TV for deaf soon to be announced" (1979), 42; "Door is opened on closed captioning" (1979), 32; "Closed-captioning set to start up in the spring," *Broadcasting* (28 Jan. 1980), 99.

224. "NBC and its TV affiliates to participate in industry effort to provide captioned programs for hearing-impaired viewers," *DA* (Apr. 1979), 8–10, 10; "Door is opened on closed captioning" (1979), 32.

225. Lucyk (1979), 3; "Door is opened on closed captioning" (1979), 32; Ball (1987), 222.

226. Jones (1984), 30–31; John E. D. Ball, "Reaching millions of viewers through closed captioning," *PTR* (Nov.–Dec. 1980), 30–32, 38, 31; "NCI's nongrowing pains with captioning," *Broadcasting* (16 Mar. 1981), 202–211, 202.

227. "Door is opened on closed captioning" (1979), 33.

228. "Closed-captioning for the hearing impaired," *DA* (May 1979), 18–19; "Closed-captioning set to start up in the spring" (1980), 99.

229. "NBC and its TV affiliates to participate" (1979), 9.

230. Loeterman (1989), 10.

231. Carole Osterer, letter to the editor, *AAD* (1980), 453.

232. Loeterman (1989), 11; "The Caption Center turns 20," *Deaf Life* (Mar. 1992), 28.

233. Ball (1987), 222; "Captioning now network reality," *Broadcasting* (17 Mar. 1980), 86.

234. Norman Black, "TV networks begin deaf captioning service," *AP* (15 Mar. 1980).

235. Jensema & Fitzgerald (1981), 32–33.

236. Memo from Donald V. Torr to Edward C. Merrill, "Estimate of number of persons who would be interested in purchasing a line 21 decoder" (10 Sep. 1980), GUA, Donald V. Torr papers (MSS 28), b.2, f.3.

237. "Door is opened on closed captioning" (1979), 33; Norman Black, "Deaf will soon be able to follow prime time television," *AP* (6 Jan. 1980).

238. "NCI's non-growing pains with captioning" (1981), 202; Norman Black, "One year later, some successes, some setbacks," *AP* (16 Mar. 1981).

239. NCI, *Closed-caption decoder sales and the "core" of the hearing-impaired community,* report 81-5 (Falls Church, VA: NCI, 1981).

240. NCI, *A comparison of three groups interested in closed captioning,* report 81-7 (Falls Church, VA: NCI, 1981).

241. "Closed caption television: What can be done to expand it?" *SHHH Journal* (Sep.–Oct. 1981), 3.

242. NCI, *A demographic profile of households with closed captioned TV,* report 81-12 (Falls Church, VA: NCI, 1981).

243. Black (1981).

244. "Closed caption television: What can be done to expand it?" (1981).

245. Marsha B. Liss and Debora Price, "What, when and why deaf children watch television," *AAD* 126 (1981), 493–98, 494–495.

246. Renee Z. Sherman and Joel D. Sherman, *Analysis of demand for decoders of television captioning for deaf and hearing-impaired children and adults* (Washington, DC: Educational Resources Information Center, 1989), iii; Ball (1980), 30.

247. House Committee on Appropriations, "Departments of Labor, Health and Human Services, Education, and Related Agencies Appropriations for 1984," part 8, CIS-NO: 83-H181-75 (5, 9–12, 16 May 1983), 98th Cong., 1st sess., 368; "NCI's non-growing pains with captioning" (1981), 202.

248. Sherman & Sherman (1989), 42.

249. Jeff Hutchins, "Real time closed captioning at NCI," *BE* (May 1983), 19, 22–30, 30.

250. Gary D. Robson, "Closed captioning FAQ," www.robson.org/capfaq/ (visited 14 Jul. 2002).

251. Linda Carson, "More questions and answers on real-time captioning," in Harkins & Virvan (1989), 157–160, 158.

252. Robson (2002).

253. Black (1981).

254. Erik Barnouw, "Television: Another aid materializing," *AO* (Jan. 1932), 751.

255. Kenneth Lipartito, "Picturephone and the information age: The social meaning of failure," *T&C* 44:1 (2003).

256. Borchert (1998), 4.

## Chapter 3. Stenographic Reporting for the Court System

*Epigraph:* Sandra W. McFate, "Will women eventually dominate reporting?" *NSR* (Dec. 1973), 13–14.

1. Morris L. Fried, "The shorthand reporter: A study in occupational sociology" (Ph.D. thesis, New School for Social Research, 1964), 24.

2. Charles G. Foster, "Sound recording—1962," *Transcript* (Mar. 1962), 2–4, 3; US General Accounting Office, *Federal court reporting system: Outdated and loosely supervised* (Washington, DC: GAO, 1982), 4.

3. Lynda Batchelor, "ER versus shorthand in Alaska," *NSR* (Jun. 1989), 32–33.

4. Harry L. Libby, "Report of survey of the Alaskan Courts System," *NSR* (Jan. 1964), 137–143, 137–138; Batchelor (1989), 33.

5. Lynne Reaves, "Tape or type? Court reporters fear for jobs," *ABA Journal* (Jan. 1984), 29; Libby, "Report of survey" (1964), 139, 142; Samuel M. Blumberg Jr., "Latest survey of electrical recording in Alaska," *NSR* (Mar. 1970), 12–21, 13.

6. Michael F. Crowley, "Employment opportunities for shorthand reporters," *Occupational Outlook Quarterly* (Dec. 1966), 28–30, 30.

7. Foster (1962), 2.

8. Harry L. Libby, "Conclusions from study of electrical recording," *NSR* (Jun. 1964), 348–349.

9. Merle P. Martin and David Johnson, *Electronic court reporting in Alaska* (Anchorage: Alaska Court System, 1979), 4.

10. George Kraft, "Displacement by sound?" *NSR* (Mar. 1967), 13–16, 13.

11. Anthony N. Morphy, *How to be a court reporter* (Bayonne, NJ: Pengad, 1980), 4.

12. Fried (1964), 6; NSRA, *Celebrating our heritage* (Arlington, VA: NSRA, 1976), 18.

13. Institute for Career Research [ICR], *Career as a court reporter* (Chicago: ICR, 2001), 4–5; NSRA (1976), 18.

14. "From cave painting to realtime: 100 years of the NCRA," *JCR* (Jul.–Aug. 1999), 19; Samuel A. Fitz-Henley, "Pitman shorthand: An historical perspective," in NSRA, *Celebrating our heritage* (Arlington, VA: NSRA, 1976), 19–22, 22; Charles Dickens, *David Copperfield* (1850).

15. David J. Saari, *The court and free-lance reporter profession: Improved management strategies* (New York: Quorum Books, 1988), 26.

16. Fried (1964), 20.

17. ICR (2001), 5; Fitz-Henley (1976), 19–21; "From cave painting to realtime" (1999), 19; Fried (1964), 16.

18. NSRA (1976), 23–24; "From cave painting to realtime" (1999), 20; Fried (1964), 17.

19. NSRA (1976), 24.

20. Max F. Meyer, "Shorthand as a typically social phenomenon—with special reference to the education of the deaf during the pre-school period," *AAD* 78 (1933), 248–257, 255.

21. NSRA (1976), 24.

22. Andre Millard, *Edison and the business of innovation* (Baltimore: Johns Hopkins University Press, 1990), 260.

23. Lisa M. Fine, *The souls of the skyscraper: Female clerical workers in Chicago, 1870–1930* (Philadelphia: Temple University Press, 1990), 7, citing John Allen Rider, "A history of the male stenographer in the United States" (Ph.D. thesis, University of Nebraska, 1966).

24. Kate Bolick, "The stenographer as archetype, from Dostoyevsky to Rob Reiner," *NYT* (22 Jun. 2003); Helen M. McCabe and Estelle L. Popham, *Word processing: A systems approach to the office* (New York: Harcourt Brace Jovanovich, 1977), 6.

25. Angel Kwolek-Folland, *Engendering business: Men and women in the corporate office, 1870–1930* (Baltimore: Johns Hopkins University Press, 1994), 43.

26. Margery Davies, *Woman's place is at the typewriter: Office work and office workers, 1870–1930* (Philadelphia: Temple University Press, 1982), 5.

27. Fine (1990), 26.

28. Fine (1990), 48.

29. McCabe & Popham (1977), 6.

30. David Morton, *Off the record: The technology and culture of sound recording in America* (New Brunswick, NJ: Rutgers University Press, 2000), 75.

31. Davies (1982), 30.

32. Fine (1990), 88.

33. Davies (1982), 130, citing Ellen Lane Spencer, *The efficient secretary* (New York: Frederick A. Stokes, 1916).

34. Morphy (1980), 5.

35. Mary H. Knapp and Robert W. McCormick, *The complete court reporter's handbook,* 3rd ed. (Upper Saddle River, NJ: Prentice-Hall, 1999), 151.

36. Bolick (2003); Roy E. Fuller, *Requirements for court reporting* (Chicago, 1944), 5.

37. Fried (1964), 23; Morton (2000).

38. Saari (1988), 43, emphasis Saari's; Cynthia Moore Callahan, "The verbatim principle: A linguistic analysis of court reporter practices" (M.A. thesis, North Carolina State University, 1995), 34.

39. Jesse T. Cantrill and William H. Hoffman, *Analysis of the work of the official court reporter: Prepared for the National Court Reporters Foundation* (Washington, DC: Hay Management Consultants, 1992), 3.

40. Saari (1988), 47.

41. Callahan (1995), abstract.

42. John Collins, "A view from the stenographer's table," *The Green Bag* 15 (1903), 311–312.

43. Fried (1964), 40.

44. Callahan (1995), 10.

45. Anne G. Walker, "The verbatim record: The myth and the reality," in Susan Fisher and Alexandra Todd, eds., *Discourse and institutional authority: Medicine, education and law* (Norwood, NJ: Ablex, 1986), 205–222, 206.

46. Walker (1986), 221.

47. Charles Parker Jr. and Norman R. Tharp, *The court reporting system in United States District Courts, 1960* (Washington, DC: Administrative Office of the US Courts, 1960), 6.

48. J. Michael Greenwood and Douglas C. Dodge, *Management of court reporting services* (Denver, CO: NCSC, 1976), 18.

49. John R. Reily, *"Read that back, please!": Memoirs of a court reporter* (Santa Barbara, CA: Fithian Press, 1999), 27, 235.

50. Dana Chipkin, *Successful freelance court reporting* (Albany, NY: West/Thomson Learning, 2001), 56.

51. Robert A. Pearson, "Federation of Shorthand Reporters: Forty years in the free-lance field," *Transcript* (Summer 1987), 28–29; Pat Barkley, "Court reporting done right," *Trial* (Jun. 2001), 67–69; Knapp & McCormick (1999), 52.

52. Chipkin (2001), 81.

53. Knapp & McCormick (1999), 63; Chipkin (2001), 187; Denise L. Doucette, "Feast or famine: What to do?" *JCR* (Jun. 2000), 54–57, 55.

54. Chipkin (2001), 82.

55. Chipkin (2001), 81, 186, 259; Knapp & McCormick (1999), 63; Doucette (2000), 55.

56. Doucette (2000), 55.

57. Morphy (1980), 59.

58. Carl Adrian Riffe, *Verbatim court reporting—there ain't such a thing: A common sense approach to becoming and being a court reporter* (Anniston, AL: Higginbotham, 1986), 32.

59. Pearson (1987), 29; Fried (1964), 141; Chipkin (2001), 93, 204; Knapp & McCormick (1999), 180.

60. Fried (1964), 46; Saari (1988), 35, 76; Sacha Pfeiffer, "Court reporting firms' deals draw fire" *Boston Globe* (29 Jul. 1999); Knapp & McCormick (1999), 53; Chipkin (2001), 95, 259–260.

61. Chipkin (2001), 2.

62. Parker & Tharp (1960), 8; Kacen (1976), 8.

63. Albert Glotzer, "The GAP in the free-lance field," *Transcript* (Jun. 1962), 49–50; Irwin R. Stone, "Shorthand reporters in New York City," *NSR* (May 1971), 15–16; Knapp & McCormick (1999), 62.

64. Parker & Tharp (1960), 12, 47.

65. Parker & Tharp (1960), 15–18.

66. Fried (1964), 51–54.

67. Edward J. Van Allen, *Your future as a shorthand reporter* (New York: Richards Rosen, 1969), 126.

68. Fried (1964), 51–54.

69. Alex Kacen, "The latest word about shorthand reporting," *Occupational Outlook Quarterly* (summer 1976), 8–9; Saari (1988), 39.

70. Van Allen (1969), 46.

71. Saari (1988), 110.

72. Crowley (1966), 29.

73. Fried (1964), 47.

74. Parker & Tharp (1960), 56.

75. William Boniface, "The state takeover in New York City," *Transcript* (Jun. 1977), 3–8, 4; Mary F. O'Leary, "The twentieth anniversary of the Association of Surrogates and Supreme Court Reporters within the City of New York," *Transcript* (summer 1987), 24; "New contract ends free lance strike," *Transcript* (Dec. 1968), 21.

76. Crowley (1966), 29; Kacen (1976), 8–9.

77. "NCRA: The national organization for reporters," *JCR* (Jul.–Aug. 1999), 22–34, 23; NSRA (1976), 47–48.

78. Edgecomb (1921), 112.

79. Chas. E. Weller, "Report of committee on statistics," in NSRA (1921), 52–61, 61.

80. "NCRA: The national organization" (1999), 24–26.

81. Fitz-Henley (1976), 19.

82. NSRA (1976), 30.

83. NSRA (1976), 40.

84. Vernon W. Stone, "Optical character recognition applied to phonotypy," *Data Processing* (Aug. 1961), reprinted in *Transcript* (Oct. 1962), 62–65, 62.

85. NSRA (1976), 26, 31, 42, 57; Joseph A. Miller, "American shorthand machines and their inventors," *Transcript* (Jun. 1966), 36–42, 37, 41–42; "From cave painting to realtime" (1999), 20.

86. Cantrill & Hoffman (1992), 5.

87. Parker & Tharp (1960), 23.

88. Gerard Salton, "The automatic transcription of machine shorthand," *Proceedings of the Eastern Joint Computer Conference* (Dec. 1959), 148–159, 148.

89. Reily (1999), 65.

90. Crowley (1966), 29.

91. Fried (1964), 119–121; NSRA (1976), 41.

92. Fried (1964), 122.

93. Glotzer (1962), 50.

94. Fried (1964), 33 n.51.

95. Norman Tulin, "The challenge to reporting in California," *NSR* (Jul. 1963), 374–376, 376.

96. "NCRA: The national organization" (1999), 27.

97. "Attention, wives," *Transcript* (Jul. 1961), 397–398.

98. Fried (1964), 53.

99. Fried (1964), 53; Crowley (1966), 28; Van Allen (1969), 26.

100. Van Allen (1969), 25.

101. Sandra W. McFate, "Men wanted [follow up]," *NSR* (May 1976), 27.

102. Kraft (1967).

103. Edward J. Van Allen, *Why not a $15,000-a-year court reporting career?* (Mineola, NY: Reportorial Press, 1965).

104. Block's words, in Marshall Jorpeland, "Captain Marty Block sets a courts for reporting excellence," *NSR* (Nov. 1989), 78–80, 79.

105. J. Michael Greenwood and Jerry R. Tollar, *Evaluation guidebook to computer-aided transcription* (Denver: NCSC, 1975), 7.

106. Fried (1964), 121.

107. Joseph L. Ebersole, *Improving court reporting services* (Washington, DC: FJC, 1972), 7.

108. Fried (1964), 124 n.173.

109. Alfred Stern, "The reporter and the stenotype notereader," *NSR* (May 1967), 20–21.

110. Morphy (1980), 51.

111. Stern (1967), 20.

112. John J. Murtha, "Dictating . . . then and now," *NSR* (Jun. 1965), 374–377, 374.

113. Morphy (1980), 2, citing *NSR* (Jan. 1951).

114. Parker & Tharp (1960), 29.

115. Ernest H. Short, ed., *A study of court reporting systems* (Gaithersburg, MD: National Bureau of Standards, 1971), 6.

116. Edgecomb (1921), 129, G. Russell Leonard quote.

117. Edgecomb (1921), 128, John D. Rhodes quote.

118. Mary Louise Gilman, "Transcribing through the years," *NSR* (Mar. 1977), 16–19, 19; Lynne Davenport, *Home transcribing: Could you? Should you?* (Sacramento, CA: Dalyn Press, 1984), 68.

119. Murtha (1965), 374.

120. J. Michael Greenwood, *Computer-aided transcription: A survey of federal court reporters' perceptions* (Washington, DC: FJC, 1981), 19.

121. Fuller (1944), 24.

122. Murtha (1965), 374.

123. Morphy (1980), 50; Davenport (1984), xi; Reily (1999), 68.

124. Parker & Tharp (1960), 58.

125. Morphy (1980), 13; Reily (1999), 28; Knapp & McCormick (1999), 181.

126. Parker & Tharp (1960), 30.

127. NSRA, Committee on Computer Aided Transcription, *CAT needs assessment: Buying the right CAT for you* (Vienna, VA: NSRA, 1985), 45; Greenwood & Dodge (1976), 33.

128. David W. Louisell and Maynard E. Pirsig, "The significance of verbatim recording of proceedings in American adjudication," *Minnesota Law Review* 38 (1953), 29–45, 45.

129. Parker & Tharp (1960), 77.

130. Parker & Tharp (1960), 80–83.

131. Vincent De Ciucis, "Is there a future to shorthand reporting?" *NSR* (Nov. 1965), 11–12.

132. William W. Tremaine, "Handbook on electrical recording," *NSR* (May 1965), 337–338.

133. Foster (1962), 3–4; Alan Roberts, "Sound recording in the New Jersey courts," *NSR* (May 1965), 326–330, 327.

134. Charles Lee Swem, "The future of reporting," *Transcript* (Jan. 1962), 145–148, 146.

135. Libby, "Conclusions from study" (1964), 348.

136. Libby, "Conclusions from study" (1964), 348.

137. C. G. Neese, "How technology aids court functions," *NSR* (Apr. 1969), 17–18.

138. Tremaine (1965), 337.

139. Ray Geauthreaux, "Task force recommends machines replace hearing reporters," *NSR* (Jun. 1968), 43–44.

140. Blumberg (1970), 13.

141. De Ciucis (1965), 11.

142. Libby, "Conclusions from study" (1964), 348.

143. Van Allen (1969), 9.

144. Parker & Tharp (1960), 70, 104.

145. Blumberg (1970), 16.

146. Blumberg (1970), 17.

147. Roberts (1965), 328.

148. Fried (1964), 55 n.79.

149. Blumberg (1970), 17.

150. Parker & Tharp (1960), 24.

151. Swem (1962), 147.

152. "Committee to Evaluate Electronic Recording Techniques (report)," *NSR* (Feb. 1972), 34–37, 34.

153. Philip H. Burt, "Eight decades of court reporting," parts 1–2, *NSR* (Feb. 1964, 173–175; Mar. 1964, 213–214), 214.

154. Morphy (1980), 69.

155. William Cohen, "Shorthand, yes! Mechanical aids, no!" *Transcript* (May 1958), 29–31, 31.

156. "NCRA: The national organization" (1999), 31.

157. Joseph J. Sweeney, "Sound recording in the federal courts," *NSR* (Nov. 1965), 18.

158. Crowley (1966), 30.

159. Van Allen (1965), 82.

160. Saari (1988), 131.

161. "Committee to Evaluate Electronic Recording Techniques" (1972), 36.

162. Margaret Bredeman, "Court reporters—Who needs 'em?" *NSR* (May 1973), 21–25, 23; Glenn J. Leathersich, "Alleviating the court reporting shortage," *Junior College Journal* (May 1972), 32–34, 32.

163. De Ciucis (1965), 11.

164. Greenwood & Dodge (1976), 33.

165. Robert A. Lowe and David C. Steelman, *A literature search and analysis of evaluations of alternative court reporting technologies* (North Andover, MA: NCSC, 1988), 2.

166. Short (1971), 1.

167. NCSC, State Judicial Information Systems Project, *Taking the court record: A review of the issues* (Williamsburg, VA: NCSC, 1981), 1, citing National Advisory Commission on Criminal Justice Standards and Goals, *Courts* (Washington, DC: GPO, 1973), recommendation 6.1, 140–141.

168. J. Michael Greenwood and Jerry R. Tollar, *User's guidebook to computer-aided transcription* (Williamsburg, VA: NCSC, 1977), 4.

169. Ebersole (1972), 2.

170. Greenwood & Tollar (1975), 1.

171. Alexander B. Aikman, "Measuring court reporter income and productivity," *State Court Journal* (winter 1979), 16–23, 23.

172. See Mark Bowles, "Liquifying information: Controlling the flood in the cold war and beyond," in Miriam R. Levin, ed., *Cultures of control* (Amsterdam: Harwood, 2000) for more on the general "information crisis" of the Cold War.

173. Warren Weaver, memorandum on machine translation (16 Jul. 1949), reprinted as "Translation," in William N. Locke and A. Donald Booth, eds., *Machine translation of languages* (New York: Wiley, 1955).

174. A. D. Booth, "Mechanical translation," *Computers and automation* 2:4 (1955), reprinted in Sergei Nirenburg, Harold Somers, and Yorick Wilks, eds., *Readings in machine translation* (Cambridge, MA: MIT Press, 2003), 19–20, 19.

175. Weaver (1949), 13.

176. B. Buchmann, "Early history of machine translation," in Margaret King, ed., *Machine translation today: The state of the art* (Edinburgh: Edinburgh University Press, 1987), 3–21, 7, 14.

177. Erwin Reifler, "The mechanical determination of meaning," in Locke & Booth (1955), reprinted in Nirenburg, Somers & Wilks (2003), 21–36, 21.

178. Yehoshua Bar-Hillel, "The present status of automatic translation of languages," in Franz L. Alt, ed., *Advances in computers*, vol. 1 (New York: Academic Press, 1960), reprinted in Nirenburg, Somers & Wilks (2003), 45–76.

179. Charles J. Bashe, Lyle R. Johnson, John H. Palmer, and Emerson W. Pugh, *IBM's early computers* (Cambridge, MA: MIT Press, 1986), 565.

180. J. L. Craft, E. H. Goldman, and W. B Strohm, "A table look-up machine for processing of natural languages," *IBM Journal of Research and Development* (Jul. 1961), 192–203, 193.

181. Erwin Reifler, "Current research at the University of Washington," in H. P. Edmundson, ed., *Proceedings of the National Symposium on Machine Translation* (Englewood Cliffs, NJ: Prentice-Hall, 1961), 155–159, 155–156.

182. Bashe et al. (1986), 559.

183. W. J. Hutchins, *Machine translation: Past, present, future* (Chichester, UK: Ellis Horwood, 1986), 67.

184. Bashe et al. (1986), 565, citing *IBM Research News* (May 1960).

185. Bashe et al. (1986), 565, citing *IBM Business Machines* (Jul. 1960).

186. D. M. Bowers and M. B. Fisk, "The World's Fair machine translator," *Computer Design* (Apr. 1965), 16–29, 17–18.

187. National Research Council and National Academy of Sciences, Automatic Language Processing Advisory Committee (ALPAC), *Language and machines: Computers in translation and linguistics* (Washington, DC: NRC, 1966), 43, 111.

188. Reifler (1955), 23.

189. Bar-Hillel (1960), 46.

190. ALPAC (1966), 64.

191. Buchmann (1987), 14; Gilbert King, "Functions required of a translation system," in Edmundson (1961), 53–62, 54.

192. L. A. Kamentsky and C.-N. Liu, "Computer-automated design of multifont print recognition logic," *IBM Journal of Research and Development* 7:1 (1963), 2–13, 12; King (1961), 58; G. O. Tarnawsky and S. A. Walker, *The IBM Russian-English automatic translation program* (Yorktown Heights, NY: IBM Watson Research Center, 1964), 1.

193. Salton (1959), 148, 152.

194. E. J. Galli, "The Stenowriter—a system for the lexical processing of stenotypy," *IRE Transactions on Electronic Computers* (Apr. 1962), 187–199, 188.

195. Stenocomp Inc., *Philadelphia court-operated computer-aided transcription service center* (Washington, DC: Stenocomp, 1974), 44; Cassie Lee Bichy and Monette Benoit, "Sharpening the cutting edge: From CAT to the Internet," *JCR* (Aug.–Sep. 1998), 56–61, 56.

196. Galli (1962), 196.

197. J. W. Newitt and A. Odarchenko, "A structure for real-time Steno-type transcription," *IBM Systems Journal* 9:1 (1970), 25, citing Bowers & Fisk (1965).

198. Galli (1962), 189; Short (1971), 15; Greenwood & Tollar (1977), 64; William E. Hewitt and Jill Berman Levy, *Computer aided transcription: Current technology and court applications* (Williamsburg, VA: NCSC, 1994), 17–20.

199. Galli (1962), 189.

200. Sandra W. McFate, "Computer transcription is on the brink of revolutionizing court reporting [interview with Frank Nelson]," *NSR* (Mar. 1975), 16–19, 16.

201. Ed Phillipi, "Stenotype tape to finished transcript via computer," part 1, *NSR* (May 1966), 4–5.

202. [W.] John Hutchins, "ALPAC: The (in)famous report," in *MT News International* (Jun. 1996), reprinted in Nirenburg, Somers & Wilks (2003), 131–136, 131; W. John Hutchins, "Machine translation: A brief history," in E. F. K. Koerner and R. E. Asher, eds., *Concise history of the language sciences: From the Sumerians to the cognitivists* (Oxford: Pergamon Press, 1995), 431–445.

203. ALPAC (1966), 28, 44, 12.

204. Hutchins (1986), 68; [W.] John Hutchins, "Gilbert W. King and the USAF Translator," in W. J. Hutchins, ed., *Early years in machine translation: Memoirs and biographies of pioneers* (Amsterdam: John Benjamins, 2000), 171–176, 176; Jonathan E. Lewis, *Spy capitalism: Itek and the CIA* (New Haven: Yale University Press, 2002), 244.

205. Stenocomp (1974), 7–8.

206. Stenocomp (1974), 8; Newitt & Odarchenko (1970), 25.

207. Newitt & Odarchenko (1970), 25; Ed Phillipi, "Stenotype tape to finished transcript via computer," part 3, *NSR* (Jul. 1966), 4–6, 4.

208. Joseph Hanlon, "CIA helps develop steno tape reader," *Computerworld* (13 Nov. 1968), 9.

209. Phillipi (Jul. 1966), 5.

210. Hanlon (1968), 9.

211. Doris O. Wong and Robert T. Wright, "Computer transcripts," *NSR* (May 1967), 11–12; Phillipi (May 1966), 4.

212. Wong & Wright (1967), 11; Stern (1967), 21.

213. Stenocomp (1974), 3; McFate, "Computer transcription is on the brink" (1975), 16; Bichy & Benoit (1998), 56.

214. Irving Kosky, "Report of the Committee on Computer Transcription," *Transcript* (Oct. 1971), 2–4, 2; Michael Brentano, "A history of CAT: From the CIA to the CIC," *NSR* (Jul. 1990), 24–26, 25.

215. Kosky (1971), 4; Irving Kosky, "Report of the Committee on Computer Transcription," *NSR* (Nov. 1973), 18–20, 18.

216. Short (1971), iii, 18–21.

217. Ebersole (1972), 9.

218. Bichy & Benoit (1998), 58.

219. Irving Kosky, "Report of the Committee on Computer Transcription," *Transcript* (Dec. 1972), 2–5, 3.

220. NSRA (1976), 67.

221. Martin Greenberger, "The computers of tomorrow," *Atlantic Monthly* (May 1964), 63–66, 65.

222. Kosky (1973), 20.

223. McFate, "Computer transcription is on the brink" (1975), 17–18.

224. McFate, "Computer transcription is on the brink" (1975), 18.

225. Doris O. Wong, "The CAT corner," *NSR* (Jan. 1975), 21.

226. McFate, "Computer transcription is on the brink" (1975), 18–19.

227. Greenwood & Tollar (1977), 1.

228. Kosky (1973), 19.

229. Greenwood & Tollar (1977), 27–30.

230. Greenwood & Tollar (1975), 33, 43.

231. Greenwood & Tollar (1977), 10.

232. Doris O. Wong, "The CAT corner," *NSR* (Jun. 1975), 24; Greenwood & Tollar (1977), 6.

233. Stenocomp (1974), 27–28; Greenwood & Tollar (1977), xi.

234. Dan W. Morrison, "Writing for the computer," *Transcript* (Dec. 1977), 22–23.

235. Greenwood & Tollar (1975), 4–5, 43.

236. Greenwood & Tollar (1977), xi, 32–34, 55, 63.

237. Greenwood & Tollar (1975), 10.

238. Martin Winkler, "Old man running scared," *NSR* (Jun. 1975), 25.

239. Bob Hopkins, cartoon, *NSR* (Nov. 1977), 28.

240. Frank O. Nelson, "In defense of fair criticism of CAT," *NSR* (Nov. 1976), 38.

241. Jill Berman Wilson, *Introduction to computer-aided transcription* (Vienna, VA: NSRA, 1985), 1:5.

242. Wong (Jan. 1975), 21; Salvatore Rao, "The future of CAT reporters—a dissenting opinion," *Transcript* (Jun. 1977), 10–11.

243. Alan I. Penn, "The daily copy team: Computer, telephone and notereaders," *NSR* (Dec. 1974), 20–21; Alexander Rose, "A shady affair," *NSR* (Apr. 1975), 39.

244. Charles W. Bartlett, "Judicial comfort in re: Xerox, ER, and Itek," *NSR* (May 1967), 21–22.

245. Alfred A. Betz, "Life with CAT," *NSR* (Apr. 1977), 42–43.

246. Winkler (1975), 25.

247. Baron Data Systems, "CAT tales [advertisement]," *NSR* (Nov. 1977), 43–46, 43.

248. Betz (1977), 42.

249. Gilbert Frank Halasz, "CAT at work," part 1, *NSR* (May 1975), 26; Mc-Fate, "Computer transcription is on the brink" (1975), 19; Rao (1977), 10.

250. Grant E. Perry, "Computer-aided transcription: Genie in a bottle or pig in a poke?" *NSR* (Jan. 1976), 20–21.

251. Edmund J. Lenahan, "Manual court reporting here to stay," *Transcript* (Mar. 1979), 9–11.

252. Rao (1977), 11.

253. Kosky (1973), 20.

254. Sandra McFate, "Reporting in the year 2000," *NSR* (Nov. 1975), 17.

255. Baron Data Systems, "CAT tales [advertisement]," *NSR* (Mar. 1978), 45–48, 48, emphasis theirs.

256. Perry (1976), 21.

257. Kosky (1972), 2.

258. Lester P. Kane, "Computer-aided transcription: Is it the answer?" *NSR* (Jun. 1976), 13–15, 14.

259. Halasz (May 1975), 27; Rao (1977), 11.

260. Baron Data Systems, "CAT tales [advertisement]," *NSR* (Jan. 1977), 43–46, 43.

261. McCabe & Popham (1977), vi.

262. NCSC, *Computer-aided transcription in the courts* (Williamsburg, VA: NCSC, 1981), 58, 66.

263. Fried (1964), 141.

264. Baron Data (Jan. 1977), 43.

265. Daniel C. Heath, "Scoping—the indispensable CAT connections," *NSR* (Jun. 1990), 34; William Y. Sober and Linda A. Knipes-Sober, "Secondary services and sensible scoping," *NSR* (Jun. 1990), 40–41.

266. Marcia Goldman and Leigh Harrod, "Starting a scoping business," *NSR* (Jul. 1990), 42; Maura Baldocchi, "Realtime: Another president's message," *Transcript* (spring 1992), 13–15, 13.

267. Kevin R. Hunt, *Management of computer aided transcription in a freelance office* (Vienna, VA: NSRA, 1985), 13, 15–16.

268. Hunt (1985), 17.

269. Shari L. Buie, "Scopist spouse," *NSR* (Jun. 1990), 42; Angela Frewer Payne, "Scopist? What is a scopist?" *NSR* (Jun. 1990), 39.

270. C. Ray Beebe, "Should you use a scopist? Some questions to answer first," *NSR* (Jul. 1990), 37; Goldman & Harrod (1990), 43.

271. William Braun, "CAT in the future: Ownership of the record," *NSR* (Mar. 1976), 52.

272. Goldman & Harrod (1990), 42.

273. Reily (1999), 71.

274. FJC, "Computer transcription comes of age," *NSR* (Nov. 1974), 42–43.

275. Morphy (1980), 51.

276. Stenographic Machines Inc., "Writing for computer transcription," *Transcript* (Jun. 1975), 2–7, 6.

277. Ben Hyatt, "Computer-aided transcription: Questions and answers," in Jill Berman Wilson, *Computer-aided transcription: An introduction and systems analysis* (Vienna, VA: NSRA, 1981), 31–36, 36.

278. Rao (1977), 11.

279. Kane (1976), 15.

280. Robert B. Kurnitz, "The CAT that lurks," *Transcript* (Oct. 1978), 20.

281. McFate, "Men wanted [follow up]" (1976), 26.

282. Kacen (1976), 8; Sandra W. McFate, "Men wanted," *NSR* (Feb. 1976), 13.

283. Quoted in McFate (1973), 13.

284. Quoted in McFate, "Men wanted" (1976), 15.

285. Quoted in McFate, "Men wanted [follow up]" (1976), 27.

286. Eugene A. Sattler, "The reporting systems division," *NSR* (Nov. 1974), 45; Bichy & Benoit (1998), 59.

287. McFate (1973), 13, McFate's words.

288. Kerry L. Gillett, "Our feminine majority—The positive side," *NSR* (Nov. 1977), 32–33, Gillett's words; Richard Tuttle, "CAT goes to the convention," *NSR* (Oct. 1975), 47.

289. McFate, "Men wanted" (1976), 15.

290. McFate, "Men wanted [follow up]" (1976), 26.

291. Leonard M. Apcar, "A special news report on people and their jobs in offices, fields and factories," *WSJ* (10 Sep. 1985), 1.

292. Joseph O. Inquagiato, "Feeding the CAT," *Transcript* (Dec. 1978), 18–19.

293. Halasz (May 1975), 27; Jill Berman Wilson, "Can CAT help court reporters speed transcript production and reduce appellate delay?" *Judicature* 69 (1985), 185, 245, 245; Baron Data Systems, "CAT tales [advertisement]," *NSR* (May 1978), 41–44, 41.

294. NCSC, *Computer-aided transcription in the courts* (1981), 33.

295. Wilson (1981), 114.

296. Wilson (1981), 110.

297. Wilson (1981), 118.

298. Wilson (1981), 110, 1.

299. NCSC, *Computer-aided transcription in the courts* (1981), 30, 242–243.

300. NCSC, *Computer-aided transcription in the courts* (1981), 109.

301. Greenwood (1981), 20.

302. Douglas W. Du Brul, "CAT can be as personal as a pocket calculator," *NSR* (May 1976), 62–64, 62

303. US GAO (1982), abstract, 7.

304. Randall A. Czerenda, "Man vs. machine," *Transcript* (Jan. 1984), 16–20, 16.

305. J. Michael Greenwood, Julie Horney, M. Daniel Jacoubovitch, Frances B. Lowenstein, and Russell R. Wheeler, *A comparative evaluation of stenographic and audiotape methods for United States District Court reporting* (Washington, DC: FJC, 1983), xi–xii.

306. Czerenda (1984), 16; Wilson, "Can CAT help court reporters" (1985), 185.

307. Czerenda (1984), 17–18.

308. Czerenda (1984), 18; Reaves (1984), 29.

309. Bruce A. Kotzan, *Court reporting practices among the state court systems* (Williamsburg, VA: NCSC, 1984).

310. Kotzan (1984), 1, 25–26, 35.

311. Gilman (1977), 17, citing *Stenographer* (Dec. 1910); *Shorthand Writer* (Jul. 1911).

312. George C. Trovato, "Realtime—A paintbrush that colors the future," *Transcript* (summer 1995), 13.

313. Gary D. Robson, "Paperless writers come of age," *JRCP* (Oct. 2003), 60–62, 60.

314. Sober & Knipes-Sober (1990), 40.

315. Joseph R. Garber, "Disabilities Act bonanza," *Forbes* (2 Aug. 1993), 124.

316. Henry S. Sanders, "The stenographer's vision—A look into the future," in NSRA (1921), 135–139, 136.

317. Fuller (1944), 31.

318. Wilson, "Can CAT help court reporters" (1985), 245; Elmer deWitt, "The courtroom of the future," *Time* (4 Aug. 1986), 60.

319. Saari (1988), 34, 118.

320. Joyce P. Jacobsen, "Results from a national survey of court reporters," *JCR* (Jun. 1993), 42–47, 44.

321. "Court reporters of today," *JCR* (Feb. 1998), 34–35.

322. Thom Weidlich, "How should reporters be compensated?" *JCR* (May 2003), 55–57, 56.

323. Jacobsen (1993), 42–43.

324. "Trends in freelance reporting," *JCR* (Nov. 1997), 54–56, 55.

325. Jacobsen (1993), 44–46.

326. "Roundtable discussion: Issues in reporter education," *JCR* (Jan. 1998), 48–51, 51.

327. Langdon Winner, *The whale and the reactor: A search for limits in an age of high technology* (Chicago: University of Chicago, 1986), 10.

## Chapter 4. Realtime Captioning for News, Education, and the Court

*Epigraph:* Marshall S. Jorpeland, "Court reporter helps make the silent screen speak," *NSR* 43:6 (1982), 20–22, 20.

1. Ben Rogner, "The O. J. Simpson trial reporters tell their story," *JCR* (Jan. 1996), 34–37, 34.

2. Susan Wollenweber, "Real-time reporting: Making instant transcription work for you," *Trial* (Jul. 1995), 68–73, 68.

3. Rita Jensen, "Dollars and a sense of transcript ownership," *JCR* (Jul. 1995), 32–35, 35.

4. Rogner (1996), 35–36.

5. Rogner (1996), 35–36; Patrick Rogers and Lyndon Stambler, "Paper tigers," *People* (12 Jun. 1995), 85–86.

6. Peter Wacht, "The O.J. Simpson trial: Bringing out the best in court reporting technology," *JCR* (Dec. 1994), 50–53, 51.

7. Rogers & Stambler (1995).

8. "Sponsors rush to caption TV commercials," *Business week* (2 Jun. 1980), 104.

9. Donald Torr, "Gallaudet Update," in Robert E. Stepp, ed., *Update 74: A decade of progress* (Lincoln: University of Nebraska, 1974).

10. G. Canon, "Television captioning at the Clarke School for the Deaf," *AAD* 125:6 (1980), 643–653, 648.

11. Leonard Bickman, Thomas Roth, Ron Szoc, Janice Normoyle, H. B. Shutterly, and W. D. Wallace, *An evaluation of captioned television for the deaf: Final report* (Evanston, IL: Westinghouse Evaluation Institute, 1979), 2–2.

12. NCI, *A survey of the opinions of closed caption decoder owners,* report 81-6 (Falls Church, VA: NCI, 1981); NCI, *A survey of captioned television news program preference,* report 81-9 (Falls Church, VA: NCI, 1981).

13. NCI, *Audience reactions to the closed captioned* ABC World News Tonight, report 82-4 (Falls Church, VA: NCI, 1982).

14. Schein et al. (1972), 3.

15. "'Telegraphic language' experiment supplements support services," *NTID Focus* (Sep.–Oct. 1972), 6–7.

16. E. Ross Stuckless and T. Alan Hurwitz, "Reading speech in real-time print: Dream or reality?" *DA* (Sep. 1982), 10–15, 12.

17. Edward C. Merrill Jr., travel report re: London trip (14 Feb. 1977), GUA, Donald V. Torr papers (MSS 28), b.2, f.1.

18. Malcolm J. Norwood, "Just don't scramble the wrong egg," in Braverman & Cronin (1979), 1–9, 2; Emily McCoy, "Real-time TV captioning: A progress report," in Braverman & Cronin (1979), 116–123, 116; Malcolm Nor-

wood, "Captioning for deaf people: An historical overview," in Harkins & Virvan (1989), 133–138, 137.

19. Emily McCoy and Robert Shumway, "Real-time captioning: Promise for the future," *AAD* 124:5 (1979), 681–690, 682–683; Jeff Hutchins, "Real-time captioning: The current technology," in Harkins & Virvan (1989), 139–142, 140.

20. Norwood (1989), 137; Martin H. Block and Marc Okrand, "Real-time closed captioned television as an educational tool," *AAD* 128:5 (Sep. 1983), 636–641, 638; Linda D. Miller, "From the courtroom to the classroom," *NSR* 46:9 (1985), 32; Brentano (1990), 24.

21. Translation Systems, Inc. (Rockville MD), "TSI TomCAT! [advertisement]," *NSR* (May 1980), 33; Cassie Lee Bichy and Monette Benoit, "Sharpening the cutting edge: From CAT to the Internet," *JCR* (Aug.–Sep. 1998), 56–61, 56.

22. Block & Okrand (1983), 638.

23. Jeff Hutchins, "Real time closed captioning at NCI," *BE* (May 1983), 19, 22–30, 26; Hutchins (1989), 140.

24. Jorpeland (1982), 20; Block & Okrand (1983), 636.

25. Marshall Jorpeland, "Captain Marty Block sets a course for reporting excellence," *NSR* (Nov. 1989), 78–80, 78–79.

26. Matt Dellinger, "Meet Oscar's transcriber," *New Yorker* (27 Mar. 2000), 39.

27. Jorpeland (1982), 22.

28. William Oliver, "Real-time captioning: Training and employment," in Harkins & Virvan (1989), 143–148, 143.

29. Block & Okrand (1983), 641.

30. Hutchins (1983), 24; Jennifer Cooke Elcano, "Training manuals for part-time and full-time text editors, National Captioning Institute" (M.A. thesis, George Mason University, 1990), 73.

31. Marty Block and Jeff Hutchins, "The politics of captioning," *NSR* (May 1989), 20–21.

32. Dan Heath, "The thrill of it all [interview with Peggy Belflower]," *JCR* (Jul. 1991), 28–29; Judith H. Brentano, "Who or what is 'real-time ready'?" *JCR* (Feb. 1992), 24–25.

33. "ABC goes live: World News for hearing impaired," *Broadcasting* (23 Nov. 1981); Norman Black, "Hearing-impaired to get access to captioned evening news," *AP* (17 Nov. 1981); NCI, report 82-4 (1982).

34. Hutchins (1983), 27.

35. "ABC goes live" (1981), 55.

36. NCI, report 82-4 (1982).

37. Dena Kleiman, "First deaf lawyer goes before Supreme Court," *NYT* (24 Mar. 1982).

38. Jeanne Clare Feron, "Court ruling seen as benefit for deaf," *NYT* (28 Feb. 1982); Joseph R. Karlovits, "Equal justice—the deaf and court-reporting,"

*NSR* 42:9 (1982), 23–24; Jane Seaberry, "Deaf attorney gets electronic help," *Washington Post* (23 Mar. 1982).

39. Karlovits (1982), 24; Seaberry (1982).

40. Seaberry (1982).

41. Block & Okrand (1983), 637.

42. Edward Carney and Ruth Verlinde, "Caption decoders: Expanding options for hearing impaired children and adults," *AAD* 132:2 (1987), 73–77, 75; NCI, *Reactions to captioned news services,* report 83-2 (Falls Church, VA: NCI, 1983); Hutchins (1983), 24.

43. Elcano (1990), 63.

44. Norman Black, "Instant captions for deaf on nightly news," *AP* (10 Oct. 1982).

45. NCI, report 83-2 (1983).

46. Marjorie Boone, "The coming of age of TV captioning," *SHHH journal* (Jan.–Feb. 1985), 22–23.

47. Black (1982).

48. Quoted in Jorpeland (1982), 22.

49. Patricia Brennan, "National Captioning Institute," *Washington Post* (29 Sep. 1985).

50. Salvatore J. Parlato, *Watch your language: Captioned media for literacy* (Silver Spring, MD: T.J. Publishers, 1986), 1.

51. Linda Carson, "More questions and answers on real-time captioning," in Harkins & Virvan (1989), 157–160, 159.

52. NCI, report 83-2 (1983).

53. "The caption race: Who's ahead and what's ahead," *Deaf Life* (Feb. 1990), 9–14, 13.

54. Patricia S. Koskinen and Robert M. Wilson, *Have you read any good TV lately? A guide for using captioned television in the teaching of reading* (Falls Church, VA: NCI, 1987), 31; Deborah Mesce, "Rather-Bush exchange challenged first night's captioning efforts," *AP* (26 Jan. 1988).

55. House Committee on Appropriations, "Departments of Labor, Health and Human Services, Education, and Related Agencies Appropriations for 1985," part 9, CIS-NO: 84-H181-87 (1–3, 7 May 1984), 98th Cong., 2d sess., 799; Norman Black, "ABC morning news programs to have 'real-time' captioning," *AP* (23 Oct. 23 1984); Norman Black, "ABC election coverage will be closed captioned for the deaf," *AP* (5 Nov. 1984).

56. *Caption* (fall 1987).

57. *Caption* (fall 1987); *Caption* (fall 1988).

58. AP, "Closed-captioning for Boston news," *NYT* (28 Dec. 1985), 46; Mardi Loeterman, "The Caption Center at nineteen," *DA* (spring 1989), 9–15, 11.

59. Carney & Verlinde (1987), 74; US Congress, House Committee on Appropriations, "Departments of Labor, Health and Human Services, Education,

and Related Agencies Appropriations for 1989," part 9, CIS-NO: 88-H181-76 (28 Apr., 2, 3 May 1988), 100th Cong., 2d sess., 105.

60. WGBH Caption Center, *Closed-captioned local news: Getting started in your town* (Boston: The Center, 1990).

61. Vincent Dollard, "Tuned in to the news," *NTID Focus* (summer 1988), 3–5, 5.

62. Patricia Brennan, "News (CC) words from 7," *Washington Post* (13 Sep. 1987).

63. Gary D. Robson and Karen A. George, "Captioning municipal governments," *JCR* (Feb. 1993), 38–39; Gus Morrison, "Fremont cable channel servers hearing impaired," *Nation's Cities Weekly* (7 Aug. 1995), 6.

64. Susan Eastman, "Government opens up with captioning," *St. Petersburg Times* (28 Oct. 1993); Tom Scherberger, "Glitch in TV service for deaf," *St. Petersburg Times* (7 Oct. 1992); Jennifer L. Stevenson, "Council delays a decision on TV captions," *St. Petersburg Times* (13 Dec. 1991).

65. WGBH Caption Center, [advertisement], *NSR* (May 1989), 77.

66. Peter L. Jepsen, "The world is tuning in to realtime," *JCR* (Mar. 1994), 37–38.

67. Darlene Leasure, "Local captioning," *JCR* (Jul. 1991), 37.

68. "Radio and television notes," *NYT* (24 Jan. 1952).

69. Joe Fedele, "Local stations should consider more captioning," *Electronic Media* (21 Jun. 1993).

70. Martin Block, "Association gears up to train and test real-time writers," *JCR* (Feb. 1992), 27; Paulette Dininny, "Closed-captioned news spreading," *BPI Entertainment News Wire* (16 Jan. 1991); "The Caption Center turns 20," *Deaf Life* (Mar. 1992), 21.

71. Brennan (1985).

72. Brentano (1992), 25.

73. Karen A. Finkelstein, "Training real-time captioners," *NSR* (Mar. 1990), 31.

74. Brentano (1992), 25.

75. Finkelstein (1990), 31.

76. Gracia L. Roemer, "The making of a 'writer,'" *NSR* (May 1989), 28–29.

77. Peter Wacht, "Captioning: Opportunities, challenges and the future," *JCR* (Jul.–Aug. 2000), 52–59, 58.

78. Jorpeland (1982), 20.

79. "Telegraphic language' experiment supplements support services" (1972), 6–7.

80. Schuchman (1988), 11; Janet L. Bailey, "Sign and oral interpreters: The who, what, when, and how," in Mark Ross, ed., *Communication access for persons with hearing loss: Compliance with the Americans with Disabilities Act* (Baltimore: York Press, 1994), 183–196, 184–187.

81. Author interview (23 Jan. 2004), Madison, WI.

82. Stuckless & Hurwitz (1982), 10–11; M. Stinson, E. R. Stuckless, J. Henderson, and L. Miller, "Perceptions of hearing-impaired college students toward real-time speech to print," *VR* 90:7 (1988), 339–348, 340.

83. Miller (1985), 32.

84. Miller (1985), 32; Ann Kanter, "From courtroom to classroom," *NTID Focus* (summer 1986), 6–7.

85. Stinson et al. (1988), 339.

86. Kanter (1986), 7.

87. Jerome E. Miller, "Some thank-yous," *NSR* (May 1989), 7.

88. B. J. Shorak, "NSRA takes part in Deaf Way conference," *NSR* (Nov. 1989), 69–70; Jeff Hutchins and Joe Karlovits, "The view from up front," *NSR* (Nov. 1989), 71.

89. Judith H. Brentano, "Access through real-time reporting," *SHHH Journal* (Sep.–Oct. 1993), 34–36, 36.

90. Melanie Humphrey-Sonntag, Sue Deer Hall, and B. J. Quinn, "Realtime tips for judicial reporters, CART providers, and captioners," *JRCP* (Oct. 2003), 66–69, 68; author interview (23 Jan. 2004), Madison, WI.

91. Deborah Dasch, "Realtiming for a large audience," *JCR* (Jan. 2000), 52–56, 53.

92. Nancy L. Eaton, "Cross training to become a successful CART provider," *JCR* (Apr. 2002), 68–71, 69.

93. Public Law 101-336.

94. Sy Dubow, "The Television Decoder Circuitry Act—TV for All," *Temple Law Review* (summer 1991) 609–618, 609 n.3.

95. Alliance for Technology Access, *Computer and web resources for people with disabilities: A guide to exploring today's assistive technology,* 3rd ed. (Alameda, CA: Hunter House, 2000), 66.

96. Carlos Suarez, "On-line captioning systems," *BE* (1994) 36:9, 70–74, 70.

97. Blair Boardman, "Court reporting is aiding deaf in classrooms," *Columbus* (OH) *Dispatch* (18 Jun. 1992), 12.

98. Deanna Baker, "The real-time market: Wide open, and not that scary," *JCR* (Aug. 1992), 36–37.

99. Oliver, "Real-time captioning" (1989), 144; William Cutler, "Captioning as an interpretive medium," in Harkins & Virvan (1989), 149–152, 151.

100. E. Ross Stuckless, "Developments in real-time speech-to-text communication for people with impaired hearing," in Ross (1994), 197–226, 218.

101. Robson (2000), 161.

102. Bailey (1994), 188.

103. Denise L. Doucette, "Feast or famine: What to do?" *JCR* (Jun. 2000), 54–57, 56.

104. Catherine Bauer, website, www.mindspring.com/~cathryn/catmain .htm (visited 22 Jul. 2003).

105. Julie Erickson, "Steno interpreting opens new doors," *NSR* (Feb. 1991), 29.

106. Humphrey-Sonntag, Hall & Quinn (2003), 67.

107. Peter Wacht, "A CART on the move: The growing market for qualified CART providers," *JCR* (Mar. 2001), 52–60, 58; Marybeth Everhart, "The ethics of realtime, part II," *JCR* (Jan. 2001), 66–67.

108. Bauer (2003).

109. Author interview (23 Jan. 2004), Madison, WI.

110. Marcia Simmons, "Dream real-time with me," *JCR* (May 1992), 42–43.

111. William Graham, "What real-time captioning means to late-deafened adults," *NSR* (May 1989), 26–27.

112. William Graham, "Personal captioning," *NSR* (Nov. 1989), 76.

113. Jonathan Young, "Communicating with realtime," *JCR* (Apr. 1997), 30–35, 34.

114. Karen Pope, "A new order in the courts: Reporters call on technology," *Crain's Detroit Business* (6 Dec. 1993), 1.

115. Nancy L. Catuogno, "An introduction to CART," *JCR* (Apr. 1999), 55–59, 56.

116. Humphrey-Sonntag, Hall & Quinn (2003), 67.

117. Baker (1992), 37.

118. Wacht, "A CART on the move" (2001), 52–59.

119. Peter Wacht, "Standing up for their rights," *JCR* (Jan. 2001), 58–61, 58.

120. Susan Foster, "Doing research in deafness: Some considerations and strategies," in Higgins & Nash (1996), 3–20, 18 n.2.

121. Harry G. Lang, "Higher education for deaf students: Research priorities in the new millennium," *Journal of Deaf Studies and Deaf Education* (Oct. 2002), 267.

122. Judith H. Brentano, "Hearing—and seeing—testimony," *National Law Journal* (21 Jun. 1993).

123. Michael Arnone, "U. of California settles federal lawsuit," *Chronicle of Higher Education* (22 Nov. 2002), 28.

124. Lang (2002), 267.

125. Aaron Steinfeld, "The benefit of real-time captioning in a mainstream classroom as measured by working memory," *VR* 100:1 (1998), 29–44, 30.

126. Michael S. Stinson and E. Ross Stuckless, "Recent developments in speech-to-print transcription systems for deaf students," in Amatzia Weisel, ed., *Issues unresolved: New perspectives on language and deaf education* (Washington, DC: Gallaudet University Press, 1998), 126–132, 128.

127. Barbara Virvan, "You don't have to hate meetings—try computer-assisted notetaking" *SHHH Journal* (Jan.–Feb. 1991), 25–27, 25–26.

128. Karen Youdelman and Carol Messerly, "Computer-assisted notetaking for mainstreamed hearing-impaired students," *VR* 98:4 (1996).

129. Stinson & Stuckless (1998), 129.

130. Susan Hahaj, "CART in the public school system," *JCR* (Sep. 2002), 72–74, 72.

131. William C. Oliver, "Dream the possible dream" *NSR* (May 1989), 25.

132. Baynton (1996), 1.

133. Andrew Solomon, "Away from language," *NYT* (28 Aug. 1994); Donald F. Moores, "An historical perspective on school placement," in Thomas N. Kluwin, Donald F. Moores, and Martha Gonter Gaustad, eds., *Toward effective public school programs for deaf students: Context, process, and outcomes* (New York: Teachers College Press, 1992), 7–29, 16.

134. Baynton (1996), 5.

135. Moores (1992), 17.

136. S. Richard Silverman, "Introduction," in National Advisory Committee on Education of the Deaf [NACOED], *Education of the deaf: The challenge and the charge* (Washington, DC: GPO, 1968), 1–4, 1.

137. Homer Babbidge Jr., "A view from the outside," in NACOED (1968), 5–12, 8.

138. Harlan Lane, "Constructions of deafness," *Disability & Society* 10:2 (1995), 171–189, 173.

139. Jerome D. Schein, "The demography of deafness," in Higgins & Nash (1996), 21–43, 33; Lane (1995), 173.

140. Katherine A. Jankowski, *Deaf empowerment: Emergence, struggle, and rhetoric* (Washington, DC: Gallaudet University Press, 1997), 19.

141. Moores (1992), 23. On the broader history of "special education" in the United States, see Margret A. Winzer, *The history of special education: From isolation to integration* (Washington, DC: Gallaudet University Press, 1993).

142. Public Law 94-142; Moores (1992), 23, Moores's words.

143. Joe D. Stedt, "Issues of educational interpreting," in Kluwin, Moores & Gaustad (1992), 83–99, 83.

144. Lane, Hoffmeister & Bahan (1996), 244; Frank B. Withrow, "Jericho: The walls come tumbling down!" *AAD* 139 (1994), 18–21, 18.

145. Michael L. Deninger, "Public Law 94-142: Issues and trends," in Propp (1980), 58–61, 59.

146. Donald F. Moores, "The future of education of deaf children: Implications of population projections," *AAD* 149:1 (2004), 3–4.

147. Public Law 101-476.

148. Benderly (1990), 242.

149. Commission on Education of the Deaf (COED), *Toward equality: Education of the deaf* (Washington, DC: GPO, 1988), xvi.

150. Solomon (1994).

151. Thomas N. Kluwin, "Considering the efficacy of mainstreaming from the classroom perspective," in Kluwin, Moores & Gaustad (1992), 175–193, 188.

152. Susan Foster, "Reflections of a group of deaf adults on their experiences in mainstream and residential school programs in the United States," *Disability, Handicap & Society* 4:1 (1989), 37–56, 38; Edward Dolnick, "Deafness as culture," *Atlantic Monthly* (Sep. 1993), 37–53, 46.

153. Foster (1989), 39.

154. Baynton (1996), 154.

155. Thomas N. Kluwin, "Conclusion: Some reflections on defining an effective program," in Kluwin, Moores & Gaustad (1992), 243–252, 244.

156. Claire L. Ramsey, *Deaf children in public schools: Placement, context, and consequences* (Washington, DC: Gallaudet University Press, 1997), 3.

157. ATA (2000), 64–65.

158. Stedt (1992), 83–86.

159. Jemina Napier, "Sign language interpreter training, testing, and accreditation: An international comparison," *AAD* 149:4 (2004), 350–356, 355.

160. Robert J. Hoffmeister, "Cross-cultural misinformation: What does special education say about Deaf people?" *Disability & Society* 11:2 (1996), 171–189, 186–188.

161. Stedt (1992), 86.

162. Youdelman & Messerly (1996).

163. Stedt (1992), 83–84.

164. James W. Carey, "A cultural approach to communication," *Communication* 2:2 (1975), reprinted in James W. Carey, *Communication as culture: Essays on media and society* (1988; New York: Routledge, 1992), 13–36.

165. Foster (1989), 52.

166. Catuogno (1999), 56.

167. Jim Reisler, "Technology: Improving sound, easing fury," *Newsweek* (24 Feb. 2003), 16.

168. Author interview (2004).

169. California Deaf and Hard-of-Hearing Education Advisory Task Force, *Communication access and quality education for Deaf and hard-of-hearing children* (1999), 3.

170. Robert E. Stepp, Jr., "A technological metamorphosis in the education of deaf students," *AAD* 139 (1994), 14–17, 17.

171. Alice M. Moell, "Instant transcript: A courtroom first," *NSR* 43:8 (1982), 18–20.

172. Sandra Klink, "Helping a judge hear with his eyes," *NSR* (Jul. 1984), 36.

173. Cassie Lee Bichy, "Realtime success: Aptitude and attitude," *JCR* (Mar. 2002), 60–64, 61.

174. Hutchins (1989), 139.

175. Randall A. Czerenda, "Man vs. machine," *Transcript* (Jan. 1984), 16–20, 18–19.

176. Michael Brentano, "A history of CAT: From the CIA to the CIC," *NSR* (Jul. 1990), 24–26, 26; H. Paul Haynes, ed., *An assessment of the 'Courtroom of the future' pilot installations* (Washington, DC: EMT Associates, 1988), 4.

177. Elmer deWitt, "The courtroom of the future," *Time* (4 Aug. 1986), 60; Roger Strand, "The courtroom of the future," *Judges' Journal* (spring 1989), 8–11, 47–48, 10; Haynes (1988), 27.

178. Lynn A. Brooks, "Is your CAT a dinosaur?" *NSR* (Jul. 1988), 34–36, 34.

179. Robert A. Lowe and David C. Steelman, *A literature search and analysis of evaluations of alternative court reporting technologies* (North Andover, MA: NCSC, 1988), 20; Darla Lynn Mileni and Jill Berman Levy, "Comparing CAT systems," *NSR* (Dec. 1990), 55–57, 55.

180. Ric Williams and Jean Gonzalez, *Advanced court reporting technology: Computer concepts* (Huntington Beach, CA: Middleton/Wasley, 1991).

181. Frank Andrews, "Computer-integrated courtrooms: Enhancing advocacy," *Trial* (Sep. 1992), 37–39, 38; Williams & Gonzalez (1991), 6–1; Urs von Burg, *The triumph of Ethernet: Technological communities and the battle for the LAN standard* (Stanford, CA: Stanford University Press, 2001).

182. Haynes (1988), 5.

183. Larry G. Johnson, "Using the court reporter's 'electronic transcript,'" *ABA Journal* 71 (1985), 142.

184. Heywood Waga, "New court reporting systems," *Trial* (Sep. 1992), 40–42, 40.

185. Haynes (1988), 7; William E. Hewitt and Jill Berman Levy, *Computer aided transcription: Current technology and court applications* (Williamsburg, VA: NCSC, 1994), 55.

186. Andrews (1992), 38.

187. Strand (1989), 10.

188. Andrews (1992), 38.

189. Quoted in Wacht (1994), 53.

190. Lane, Hoffmeister & Bahan (1996), 353–354.

191. Amy Dockser Marcus, "Visual translation help the deaf in court," *WSJ* (21 Feb. 1990).

192. Peter Wacht, "Tips about realtime and hearing-impaired jurors," *JCR* (Jul. 1995), 42–45, 43; Catuogno (1999), 58.

193. Bauer (2003).

194. Brentano (1992), 25.

195. Andrews (1992), 38; Gregg Shapiro, "Uncertified notes: A rose or a thorn?" *JCR* (Aug. 1995), 29–31, 29.

196. Hewitt & Levy (1994), 10.

197. Scott Dean, "Teaching attorneys and the public about our technologies," *JCR* (Apr. 1992), 26–30, 28; Waga (1992), 40; Herb Landman, "CIC debuts at 60 Centre Street," *Transcript* (summer 1996), 12–13.

198. Jennifer L. C. Pridmore, "Realtiming in teams," *JCR* (Jun. 2001), 60–62, 61.

199. Shirley Barrett with Jim Woitalla, "Remote scoping," *JCR* (Jun. 1998), 62–63.

200. Nancy J. Hopp, "Realtime fear and loathing myths debunked," *JCR* (Feb. 2000), 66–69, 66.

201. Haynes (1988), 18; Morphy (1980), 18.

202. Kevin William Daniel and Mary Cox Daniel, *Writing naked: Principles of writing for realtime and captioning* (Vienna, VA: NCRA, 1998), ii.

203. Lewis Friedman, "Bringing technology to 60 Centre: CIC offers a glimpse of the future," *Transcript* (fall 1994), 13–15, 13.

204. Carol M. Neal, "Adding new value to official reporting," *JCR* (Jan. 1995), 30–33, 30.

205. Hewitt & Levy (1994), 44.

206. Debra Cassens Moss, "Courtroom of future is here," *ABA Journal* (Feb. 1989), 26.

207. Hewitt & Levy (1994), 55.

208. Haynes (1988), 13.

209. Herb Landman, "Courtroom 2000—with realtime—a success in NYC," *Transcript* (winter 1998), 13–16, 14.

210. Kathleen M. Shapiro, "Captioned TV moves to the courtroom," *New Jersey Law Journal* (13 Jun. 1994).

211. Brentano, "Hearing—and seeing—testimony" (1993); Judith H. Brentano, "Computer-integrated courtrooms invite the disabled to participate," *New York Law Journal* (20 Sep. 1994); Hewitt & Levy (1994), 63; Paul Haynes, "The new millennium reporter," *JCR* (Jul. 1997), 28–31, 29.

212. Friedman (1994), 15.

213. Patricia Nealon, "Trial spotlights flaws in court transcript technology," *Boston Globe* (5 Nov. 1997).

214. Peter Wacht, "The future is now: Computer-integrated courtrooms," *JCR* (Mar. 1996), 29–31, 30.

215. Martin H. Block, "Computer integrated courtrooms: Moving the judicial system into the twenty-first century," *Trial* (Sep. 1990), 51–52; Waga (1992), 41.

216. Donald R. DePew, "Court reporting goes real-time at Southern District," *Transcript* (winter 1995), 13–14.

217. Oliver, "Real-time captioning" (1989), 143; Karen Finkelstein and Tammie Shedd, *Real-time writing: The court reporter's guide for mastering real-time skills* (Falls Church, VA: NCI, 1991), 3.

218. Block (1992), 27.

219. Illinois Occupational Skill Standards and Credentialing Council, *Court reporter/captioner* (Springfield: Illinois State Board of Education, 2000), 1–3.

220. Daniel & Daniel (1998), 31.

221. Marianne Lindley, "The real world of realtime," *JCR* (Jul. 1992), 49.

222. Hewitt & Levy (1994), 45.

223. David H. Buswell, "Court reporter plays key role at world-famous facility," *JCR* (Nov. 1998), 60–63, 61.

224. Hewitt & Levy (1994), 9, 44–46, 91, 135 n.12.

225. Hopp (2000), 69.

226. Peter Wacht, "Service is the name of the game," *JCR* (May 1995), 34–39.

227. H. Pauletta Morse and Diane C. Davis, "How realtime reporters spend their time," *JCR* (Mar. 1998), 35–37, 37.

228. Haynes (1988), 7, 22.

229. "Trends in freelance reporting," *JCR* (Nov. 1997), 54–56, 56.

230. Jill Berman Levy, "Planning for a computer integrated courtroom," in *Technologies in court reporting* (Denver: NCSC; Vienna, VA: NCRA, 1991), 5–42.

231. Bichy (2002), 60.

232. Rebecca A. Askew, "Technology in the courtroom," *JCR* (Feb. 2003), 70–72, 70.

233. Wacht, "The future is now" (1996), 29.

234. Haynes (1988), 22; Jesse T. Cantrill and William H. Hoffman, *Analysis of the work of the official court reporter* (Washington, DC: Hay Management Consultants, 1992), 9.

235. Sherri Day, "Electronic Order in the Court," *NYT* (29 May 2003).

236. William Weber, "Real-time technology saves time, eases research," *National Law Journal* (13 Aug. 2001).

237. Wacht, "The future is now" (1996), 29; Levy (1991), 5–24; Brentano (1994); ICR (2001), 6.

238. J. Edward Varallo, "Reporters as managers? I think not," *JCR* (Dec. 1997), 38–39.

239. Hewitt & Levy (1994), 91.

240. Eugene A. Sattler, "The case against videotape recording," *Transcript* (Mar. 1973), 2–4, 2.

241. Raymond F. De Simone, "On the use of videotape," *Transcript* (Mar. 1974), 9–11.

242. David C. Steelman and Samuel D. Conti, *An evaluation of Kentucky's innovative approach to making a videotape record of trial court proceedings* (North Andover, MA: NCSC, 1985), 11–12.

243. NSRA, *Celebrating our heritage* (Arlington, VA: NSRA, 1976), 67.

244. Edmund J. Lenahan, "The tyranny of the technologist," *Transcript* (Jun. 1979), 31–33, 32; William D. Mohr, "1994," *NSR* (May 1976), 29–31, 30.

245. Conference of Court Administrators, Committee to Examine Court Reporting Services, *Court reporting practices among the state court systems: Survey results* (Willliamsburg, VA: NCSC, 1984).

246. Steelman & Conti (1985), 21–22.

247. Karen Klages, "Court reporters on way out?" *ABA Journal* (Feb. 1989), 28–29.

248. Steelman & Conti (1985), 3, 5, 17.

249. Klages (1989), 28.

250. Klages (1989), 28.

251. Don J. DeBenedictis, "Excuse me, did you get all that?" *ABA Journal* (May 1993), 84.

252. NSRA, Committee on Computer Aided Transcription, *CAT needs assessment: Buying the right CAT for you* (Vienna, VA: NSRA, 1985), 46.

253. DeBenedictis (1993).

254. Williams & Gonzalez (1991), 9–2.

255. Shapiro (1994).

256. Joseph R. Karlovits, "TAC and its effect on our profession," *JCR* (Jul. 1992), 24–25.

257. Weber (2001).

258. Andrews (1992), 38.

259. DeBenedictis (1993).

260. Stanley Rizman, "Total Access Courtroom unveiled in New Jersey," *JCR* (Dec. 1992), 36–39, 37.

261. Charles Toutant, "Court reporters becoming 'real-timers'" *New Jersey Law Journal* (19 May 2003).

262. Rizman (1992), 39.

263. Toutant (2003).

264. Vicki Y. Johnson, "How to save & make reporting jobs," *JCR* (Aug. 1994), 47.

265. Mary H. Knapp and Robert W. McCormick, *The complete court reporter's handbook*, 3rd ed. (Upper Saddle River, NJ: Prentice-Hall, 1999), 11, 237.

266. Toutant (2003).

267. Day (2003).

268. Administrative Office of the US Courts, *Courtroom technology manual* (1999).

269. Jerry Kelley, "Court reporting in the year 2000," *JCR* (Dec. 1993), 32–33.

270. Hutchins (1989), 142.

271. Sheldon Altfeld, "How to start your own cable TV network," www .cablemaven.com (visited 28 Sep. 2005).

272. Jones (1984), 50.

273. Brennan (1985).

274. Rogers & Stambler (1995).

275. Block & Okrand (1983), 640.

276. Amy Bowlen and Kathy DiLorenzo, *Realtime captioning . . . The VITAC way* (VITAC, 2003).

277. Al Franken, *Lies (and the lying liars who tell them): A fair and balanced look at the right* (New York: Dutton, 2003), 187–188.

## Chapter 5. Public Interest, Market Failure, and Captioning Regulation

*Epigraph:* Robert J. Rickelman, William A. Henk, and Kent Layton, "Closed-captioned television: A viable technology for the reading teacher," *Reading Teacher* (Apr. 1991), 598–599.

1. Peter J. Boyer, " 'Cosby' captioning sparks dispute," *NYT* (6 Jan. 1986).

2. Norman Black, "Future of group providing 'closed captioning' threatened," *AP* (21 Feb. 1982).

3. Norman Black, "NBC postpones withdrawal from closed-captioning," *AP* (8 Mar. 1982).

4. "NBC drops service for the deaf," *AP* (26 Aug. 1982); Boyer (1986).

5. *Developing technologies for television captioning: benefits for the hearing impaired: hearing before the Subcommittee on Science, Research, and Technology of the Committee on Science and Technology, U.S. House of Representatives, Ninety-eighth Congress, first session, November 9, 1983* (Washington, DC: GPO, 1984), 40–41.

6. Leonard R. Graziplene, *Teletext: Its promise and demise* (Bethlehem, PA: Lehigh University Press, 2000), 68–71.

7. Commission on Education of the Deaf (COED), *Toward equality: Education of the deaf* (Washington, DC: GPO, 1988), 116.

8. Charles M. Firestone, "Rehabilitation Act of 1973: *Gottfried v. Community Television of Southern California* (1983)," in John V. Van Cleve, ed., *The Gallaudet encyclopedia of deaf people and deafness* (New York: McGraw-Hill, 1987), 2:416–420, 417.

9. Louis Schwartz and Robert A. Woods, "Public television and the hearing impaired," *Journal of College and University Law* 9:1 (1982), 1–25, 3.

10. Sara Geer and Mary-Jean Sweeny, "Rehabilitation Act of 1973," in Van Cleve (1987), 2:407–412, 407–410.

11. Thomas W. Fothergill, "*Community Television of Southern California v. Gottfried:* Defining the role of the television industry in serving the needs of the hearing impaired," *New England Law Review* 19:4 (1984) 899–916, 907.

12. Ann P. Michalik, "The public interest standard in the Communications Act and the hearing impaired: *Community Television of Southern California v. Gottfried,*" *Boston College Law Review* (Jul. 1984), 893–918, 904 n.128.

13. Michalik (1984), 893.

14. Michalik (1984), 902.

15. Firestone (1987), 417; Melissa N. Widdifield, "Access of the hearing-impaired to television programming," *Loyola Entertainment Law Journal* (1985) 188–197, 189.

16. Schwartz & Woods (1982), 11; Firestone (1987), 417; Widdifield (1985), 190.

17. Firestone (1987), 417.

18. Widdifield (1985), 190; Firestone (1987), 418.

19. *Community Television of Southern California v. Gottfried,* 459 US 498, 5008 (1983).

20. Michalik (1984), 906, Michalik's words.

21. Firestone (1987), 418.

22. Rowland L. Young, "Public TV station wins in Rehabilitation Act case," *ABA Journal* (May 1983).

23. Firestone (1987), 419.

24. Widdifield (1985), 191.

25. Schwartz & Woods (1982), 4–5.

26. Firestone (1987), 419.

27. Firestone (1987), 419; Schwartz & Woods (1982), 17.

28. Schwartz & Woods (1982), 5.

29. Firestone (1987), 419.

30. Schwartz & Woods (1982), 19; Firestone (1987), 419.

31. Schwartz & Woods (1982), 19.

32. Firestone (1987), 419; James H. Rubin, "High court rejects appeal in California TV-captioning case," *AP* (18 Jun. 1984).

33. Firestone (1987), 420.

34. Michalik (1984), 917.

35. Schwartz & Woods (1982), 10.

36. Michalik (1984), 913.

37. "John Ball: TV's wordsmith," *Broadcasting* (27 Aug. 1990), 87.

38. House Committee on Appropriations, "Departments of Labor, Health and Human Services, Education, and Related Agencies Appropriations for 1984," part 8, CIS-NO: 83-H181-75 (5, 9–12, 16 May 1983), 98th Cong., 1st sess., 365; COED, *First set of draft recommendations* (28 Aug. 1987); Carl Jensema, "A demographic profile of the closed-caption television audience," *AAD* 132:5 (1987), 389–391.

39. *Developing technologies* (1984), 4.

40. Norman Black, "Captioning institute survives with ABC contract," *AP* (7 Sep. 1983); "Public awareness of closed-captioning services," *NCI Research Bulletin* (Apr. 1983).

41. Jones (1984), 32; Sherman & Sherman (1989), vii; House Committee on Appropriations, "Departments of Labor, Health and Human Services, Education,

and Related Agencies Appropriations for 1985," part 9, CIS-NO: 84-H181-87 (1–3, 7 May 1984), 801.

42. House Committee on Appropriations, "Departments of Labor, Health and Human Services, Education, and Related Agencies Appropriations for 1986," part 10, CIS-NO: 85-H181-94 (22, 29, 30 May 1985), 99th Cong., 1st sess., 244; Patricia Brennan, "National Captioning Institute," *Washington Post* (29 Sep. 1985).

43. House Committee on Appropriations, "Departments of Labor, Health and Human Services, Education, and Related Agencies Appropriations for 1985," part 9, CIS-NO: 84-H181-87 (1–3, 7 May 1984), 98th Cong., 2d sess., 799801.

44. "The hearing impaired community & cable television," *NCI Research Bulletin* (Feb. 1983).

45. NCI, *Characteristics of the audience for closed captioned television on Dec. 31, 1984,* report 85-1 (Falls Church, VA: NCI, 1985).

46. US Census Bureau, *Decennial census of population, vol. 1: General social and economic characteristics, part 1: United States summary* (Washington, DC: GPO, 1980), Table 230; NCI (1985).

47. NCI, *The hard of hearing market for closed captioned television,* report 83-4 (Falls Church, VA: NCI, 1983), 2–5.

48. Bruce A. Austin with John W. Myers, "Hearing-impaired viewers of prime-time television," *JOC* 34:4 (1984), 60–71, 65, 69.

49. House Committee on Appropriations (1983), 375.

50. Donald V. Torr, *Captioning project evaluation: Final report* (Washington, DC: Gallaudet College, 1974), 12–15.

51. House Committee on Appropriations (1983), 364–371.

52. Jill Lai, "Closed captioning lets hard of hearing read TV," *UPI* (12 Oct. 1987).

53. Nan Decker, *The caption workbook* (Silver Spring, MD: NAD, 1984).

54. WGBH Caption Center, "The caption kit," [advertisement] in Episcopal Awareness Center on Handicaps, *Resource kit on captioning* (Washington, DC: EACH, 1986).

55. Joe Ahlgren, *Connecting your closed captioning decoder with a video tape recorder and cable TV* (Bethesda, MD: SHHH, 1985).

56. SHHH, *Telecaptioning: New horizons for hearing impaired people* (Bethesda: SHHH, 1987).

57. House Committee on Appropriations, "Departments of Labor, Health and Human Services, Education, and Related Agencies Appropriations for 1987," part 10, CIS-NO: 86-H181-57 (6, 7 May 1986), 99th Cong., 2d sess., 85, 92.

58. House Committee on Appropriations (1984), 805–806; SHHH (1987).

59. Sherman & Sherman (1989), 41, 45.

60. COED, *First set of draft recommendations* (1987).

61. John Carmody, "The TV column," *Washington Post* (6 Jun. 1988); Sherman & Sherman (1989), vii, 42.

62. Sherman & Sherman (1989), vi.

63. Sy Dubow, "The Television Decoder Circuitry Act—TV for all," *Temple Law Review* (summer 1991) 609–618, 614.

64. Eileen R. Smith, "National Captioning Institute, Inc. telecaption decoders: Expanding the horizons of television, 1989" (1989), NPBA, Bill Reed papers, b.32, f.9.

65. House Committee on Appropriations (1983), 364.

66. *Developing technologies* (1984), 194.

67. House Committee on Appropriations (1984), 799.

68. House Committee on Appropriations (1985), 242.

69. House Committee on Appropriations (1984), 813.

70. Lai (1987).

71. Dubow (1991), 613; Steven Weinstein, "Close-captioning can be fast, furious," *St. Petersburg* (FL) *Times* (19 Feb. 1988).

72. *Caption* (Fall 1980).

73. Hoynes (1994), 4; "NBC drops service for the deaf" (1982); Black, "Captioning institute survives" (1983).

74. Black, "Captioning institute survives" (1983); Lai (1987); Jun. Farrell, "Reliability and resourcefulness: Key to NCI's commitment to captioning," *BE* (May 1983), 30.

75. House Committee on Appropriations (1984), 803.

76. House Committee on Appropriations (1983), 363.

77. House Committee on Appropriations (1984), 800.

78. House Committee on Appropriations (1988), 112.

79. John E. D. Ball, "National Captioning Institute," in Van Cleve (1987), 2:221–223, 222.

80. Ball (1987), 222; SHHH (1987).

81. Caption (1992).

82. House Committee on Appropriations (1984), 799.

83. "The caption race: Who's ahead and what's ahead," *Deaf Life* (Feb. 1990), 9–14, 13.

84. Lai (1987); House Committee on Appropriations (1987), 433.

85. "Advertising interest in captioned TV shows grows," *Advertising Age* (10 Apr. 1978), 59.

86. Larry Chase, "Captioning TV programs: competition and controversy," *Disabled USA* (winter 1982), 25–29, 25.

87. "Advertising interest in captioned TV shows grows" (1978).

88. "Sponsors rush to caption TV commercials," *Business Week* (2 Jun. 1980), 104.

89. Rita A. Gesue, "Captioned television: Viewing patterns and opinions of adults in an urban deaf community" (Ph.D. thesis, University of Pittsburgh, 1986); Austin & Myers (1984).

90. NCI, *The attitudes of hearing-impaired viewers toward closed-captioning television commercials,* report 81-15 (Falls Church, VA: NCI, 1981).

91. "Sponsors rush to caption TV commercials" (1980).

92. Philip H. Dougherty, "Advertising," *NYT* (9 Sep. 1980); "NCI announces additional captioned programming," *DA* (Apr. 1980), 3.

93. NCI, report 81-15 (1981).

94. Randall Bloomquist, "Advertisers reach out to deaf and hearing-impaired," *Adweek* (18 Jan. 1988).

95. Dan Cray, "Listen up: Advertisers move to close-captioning," *Adweek* (2 Jul. 1990).

96. Bloomquist (1988); Brennan (1985).

97. "Closed-caption viewers endorse captioned commercials," *NCI Research Bulletin* (Sep. 1990).

98. Colin McIntyre, "Teletext in the United Kingdom" in Efrem Sigel, ed., *The future of videotext: Worldwide prospects for home/office electronic information services* (White Plains, NY: Knowledge Industry Publications, 1983), 2, 115.

99. Richard Veith, *Televisions' teletext* (New York: North-Holland, 1983), 14.

100. Kenneth Edwards, "Information without limit electronically," in Michael Emery and Ted Curtis Smythe, eds., *Readings in mass communication: Concepts and issues in the mass media,* 4th ed. (Dubuque, IA: Wm. C. Brown, 1980), 202–216, 209; McIntyre (1983), 115.

101. Edwards (1980), 209; Veith (1983), 14.

102. McIntyre (1983), 115; Veith (1983), 15; Sigel (1983), 2; Edwards (1980), 205.

103. Ivarsson & Carroll (1998), 24; Edwards (1980), 205.

104. Graziplene (2000), 22.

105. Edwards (1980), 209.

106. Graziplene (2000), 22–23; McIntyre (1983), 114.

107. Graziplene (2000), 30.

108. Andrew Pollack, "Teletext is ready for debut," *NYT* (18 Feb. 1983).

109. Edwards (1980), 202.

110. Les Brown, "New device calls up printed matter on TV," *NYT* (5 Feb. 1980); N. R. Kleinfield, "CBS plans to test teletext on coast," *NYT* (14 Nov. 1980); Peter J. Schuyten, "CBS backs French bid on teletext," *NYT* (31 Jul. 1980).

111. Graziplene (2000), 30.

112. Philippe Durand, "The public service potential of videotex and teletext," *Telecommunications Policy* 7:2 (1983), 153.

113. McIntyre (1983), 116.

114. Ernest Holsendolph, "British system favored in teletext model vote," *NYT* (7 Aug. 1980).

115. Pollack, "Teletext is ready for debut" (1983).

116. Pablo J. Boczkowski, *Digitizing the news: Innovation in online newspapers* (Cambridge, MA: MIT Press, 2004).

117. Paul Baran, *Potential market demand for two-way information services to the home, 1970–1990* (Menlo Park, CA: Institute for the Future, 1971), 6.

118. William J. Broad, "Upstart television: Postponing a threat," *Science* (7 Nov. 1980), 611–615, 611.

119. Kleinfield (1980); CBS Television Network, Engineering & Development Dept., "Teletext field tests, phase I" (Sep. 1979), GUA, Donald V. Torr papers (MSS 28), b.1, f.9; Jeffrey Silverstein, "Videotext in the United States" in Sigel (1983), 53.

120. CBS Broadcast Group, "Teletext and the hearing-impaired," and "Teletext background and information," from Gallaudet demonstration of teletext (20 Nov. 1980), GUA, Donald V. Torr papers (MSS 28), b.1, f.7; Broad (1980), 612.

121. Schuyten (1980).

122. Brown (1980).

123. CBS News, press release on CBS Reports, "How much for the handicapped?" (5 Mar. 1979), GUA, Donald V. Torr papers (MSS 28), b.1, f.6.

124. Brown (1980).

125. Letter from John E. D. Ball to Gene F. Jankowski (31 Jul. 1980), GUA, Donald V. Torr papers (MSS 28), b.1, f.6.

126. E. Marshall Wick, "Reply comments of a concerned consumer before the Federal Communications Commission in the matter of Amendment of Part 73, Subpart E of the Rules Governing Television Broadcast Stations to Authorize Teletext" (17 Oct. 1980), GUA, Donald V. Torr papers (MSS 28), b.2, f.8.

127. Letter from Don Torr to Heina Temple, Jan Richardson, and Rosie Patterson "CBS Demonstration" (7 Nov. 1980), GUA, Donald V. Torr papers (MSS 28), b.2, f.7.

128. CBS Broadcast Group (20 Nov. 1980), GUA, Donald V. Torr papers (MSS 28), b.1, f.7.

129. Carl Jensema, "The battle of captioning vs. teletext," *SHHH Journal* (Mar.–Apr. 1981), 11.

130. Robert Baker, *Subtitling television for deaf children* (Southampton, UK: University of Southampton, Dept. of Teaching Media, 1985), 1; McIntyre (1983), 122; Veith (1983), 34.

131. E. Marshall Wick, "Teletext and deaf people: A look at the issues and concerns" (8 Jan. 1982), GUA, Donald V. Torr papers (MSS 28), b.1, f.8, 3; Kleinfield (1980).

132. "NCI's non-growing pains with captioning," *Broadcasting* (16 Mar. 1981), 202–211, 204.

133. Joseph Blatt, "Teletext: A new television service for home information and captioning," *VR* 84:4 (1982), 209–217, 212.

134. "CBS captioning urged by the deaf," *NYT* (20 May 1982).

135. CBS Broadcast Group (20 Nov. 1980), GUA, Donald V. Torr papers (MSS 28), b.1, f.7; "CBS decides to force issue on teletext," *Broadcasting* (4 Aug. 1980), 62–63.

136. Holsendolph (1980).

137. Wick (1982), GUA, Donald V. Torr papers (MSS 28), b.1, f.8, 4.

138. "Proposed authorization of transmission teletext by TV stations," *FR* (14 Dec. 1981), 60851–60858, 60852.

139. Ernest Holsendolph, "Teletext authorized by F.C.C.," *NYT* (1 Apr. 1983); Ball (1987), 222.

140. "CBS starts its teletext service," *NYT* (5 Apr. 1983).

141. *Developing technologies* (1984), 27.

142. "Deaf picket CBS stations over captioning," *UPI* (28 Sep. 1983).

143. House Committee on Appropriations (1983), 364.

144. *Developing technologies* (1984), 5, 6, 23, 26, 30, 181.

145. Norman Black, "CBS announces closed-captioned broadcasting for the deaf," *AP* (5 Mar. 1984); Ball (1987), 222.

146. Marc Okrand, "Telecommunications: Captioned television," in Van Cleve (1987), 3:264–267, 266; Joseph R. Karlovits, "Captioning 2000: A glimpse of the future," *JCR* (May 1999), 60–62, 60; "The Caption Center turns 20," *Deaf Life* (Mar. 1992), 28.

147. "Teletext still better," *Communications Daily* (6 Mar. 1984).

148. "Teletext still better" (1984).

149. Graziplene (2000), 73.

150. Andrew Pollack, "Time Inc. drops teletext experiment," *NYT* (22 Nov. 1983).

151. "Time Inc. may drop teletext," *NYT* (16 Nov. 1983).

152. Richard H. Veith, "Videotex and teletext," *ARIST* 18 (1983), 3–28; Donald O. Case, "The social shaping of videotex: How information services for the public have evolved," *Journal of the American Society for Information Science* 45:7 (1994), 483–497.

153. "Proposed authorization of transmission teletext by TV stations" (1981), 60852.

154. Michael Tyler, "Videotex, Prestel and Teletext: The economics and politics of some electronic publishing media," *Telecommunications Policy* (Mar. 1979), 37–45, 38.

155. Larry Goldberg and Geoff Freed, "Making multimedia accessible on the World Wide Web," *Technology and Disability* 8 (1998), 127–132, 131; John Burgess, "Web firms seek a bigger slice of TV channels," *Washington Post* (3 Apr. 1998).

156. Pollack, "Teletext is ready for debut" (1983).

157. Martin Elton and John Carey, "Computerizing information: Consumer reactions to teletext," *JOC* (winter 1983), 162–173, 165–166.

158. Elton & Carey (1983), 168.

159. Broad (1980), 611.

160. Pollack, "Teletext is ready for debut" (1983).

161. *Developing technologies* (1984), 27.

162. Graziplene (2000), 74–77.

163. Sherman & Sherman (1989), v.

164. Perkins (1972), 11.

165. George Punch Shaw, "Television viewing habits of hearing-impaired high school students in a residential school setting," in Propp (1980), 180.

166. Robert R. Davila, "Effect of changes in visual information patterns on student achievement using a captioned film and specially adapted still pictures" (Ph.D. thesis, Syracuse University, 1972), 5.

167. George Propp, "An experimental study on the encoding of verbal information for visual transmission to the hearing impaired learner" (Ph.D. thesis, University of Nebraska, 1972), 20.

168. Lawrence O. Reiner and Dale L. Rockwell, *A study of instructional film captioning for deaf students* (Rochester, NY: NTID, Division of Research & Training, 1971), 12; Davila (1972), 75; Propp (1972), 89; E. Ross Stuckless, *A review of research at NTID, 1967–1976* (Rochester, NY: NTID, 1978), 16–17.

169. Brenda W. Rawlings and Frank S. Rubin, *A survey of media equipment available in special education programs for hearing imparied students, 1977–78* (Washington, DC: Gallaudet College, Office of Demographic Studies, 1978).

170. Jill Shulman and Kirk Wilson, "A multi-level linguistic approach to captioning television for hearing-impaired children," in C. R. Vest, ed., *Learning technology for the deaf* (Warrenton, VA: Society for Applied Learning Technology, 1978), 73–80, 75; Jill Shulman and Nan Decker, "Multi-level captioning: A system for preparing reading materials for the hearing impaired," *AAD* 124:5 (1979), 559–567, 565.

171. John E. D. Ball, "Reaching millions of viewers through closed captioning," *PTR* (Nov.–Dec. 1980), 30–32, 38, 31; *Caption* (fall 1980).

172. NCI, *The closed captioned television viewing preferences of hearing-impaired children,* report 81-13 (Falls Church, VA: NCI, 1981), 9.

173. NCI, *School use of closed captioned television,* report 84-2 (Falls Church, VA: NCI, 1984).

174. NCI, *Using closed-captioned television in the teaching of reading to deaf students,* report 85-2 (Falls Church, VA: NCI, 1985).

175. Patricia S. Koskinen and Robert M. Wilson, *Have you read any good TV lately? A guide for using captioned television in the teaching of reading* (Falls Church, VA: NCI, 1987), 3.

176. Ball (1987), 221.

177. House Committee on Appropriations (1983), 363.

178. House Committee on Appropriations (1984), 807.

179. NCI, *A report on two pilot projects which used captioned television in schoolrooms to teach reading to hearing children,* report 83-6 (Falls Church, VA: NCI, 1983); "Captions for deaf help some kids too," *AP* (23 Sep. 1983); NCI, *Closed-captioned television: A new technology for enchancing reading skills of learning disabled students,* report 86-1 (Falls Church, VA: NCI, 1986).

180. Janet Gottlieb, letter to the editor, *AAD* 119 (1974), 11.

181. Stephen Dee Reese, "Multi-channel redundancy effects on television news learning" (Ph.D. thesis, University of Wisconsin-Madison, 1982).

182. Richard M. Ruggerio, "Impact of television captioning on hearing audiences" (Ed.D. thesis, University of California Los Angeles, 1987).

183. House Committee on Appropriations (1986), 96.

184. Jeffrey A. Tannenbaum, "TV decoders find a wider audience," *WSJ* (19 Oct. 1988), 1.

185. Smith (1989), NPBA, Bill Reed papers, b.32, f.9, citing *American Demographics* (Mar. 1989).

186. Senate Committee on Commerce, Science, and Transportation, *Television decoder circuitry act of 1990,* 101st Cong., 2d sess., 1990, S. Rep. 101-393, 2.

187. Susan B. Neuman, "Using captioned television to improve the reading proficiency of language minority students" (1990), NPBA, Bill Reed papers, b.32, f.10.

188. Dennis Kelly, "TV closed-captions fight illiteracy," *USA Today* (11 Jul. 1990).

189. COED (1988), viii, xxi, 111.

190. Ruggerio (1987).

191. J. Daniel Gifford, "Build this closed-caption decoder," *Radio Electronics* (Nov.–Dec. 1986); Linda Carson, "More questions and answers on real-time captioning," in Harkins & Virvan (1989), 157–160, 157; Jinan Lin and Maximillian Erbar, "A digital high-performance multi-standard video data slicer," *IEEE Transactions on Consumer Electronics* 44:3 (1998), 1103–1106, 1103.

192. House Committee on Appropriations (1988), 104.

193. House Committee on Energy and Commerce, *Television decoder circuitry act of 1990,* H. Rep. 101-767, 101st Cong., 2d sess. (27 Sep. 1990), 9.

194. Carson (1989), 157.

195. *Television decoder circuitry act of 1990,* HR 4267, 101st Cong., 2d sess. (1990), http://thomas.loc.gov/cgi-bin/query/D?c101:1:./temp/~c101wb1uzy:: (visited May 10, 2007), 9; Nancy Benac, "Decoder-equipped TVs seen benefiting more than hearing-impaired," *AP* (18 Oct. 1989).

196. Dubow (1991), 617.

197. *Television decoder circuitry act of 1990* (1990), 9.

198. Rodney Ferguson, "Law may soon ease access to captioned TV for deaf," *WSJ* (22 Aug. 1990).

199. Senate Committee on Rules and Administration, *Amending Senate Resolution 28 to implement closed caption broadcasting for hearing-impaired individuals of floor proceedings of the Senate,* S. Rept. 101-54, 101st Cong., 1st sess. (1989), 2–5; Don Phillips, "House shows signs it will start using closed-captioning," *Washington Post* (25 Sep. 1988).

200. *Americans with disabilities act of 1989,* HR 2273, 101st Cong., 1st sess. (1989), http://thomas.loc.gov/cgi-bin/query/D?c101:1:./temp/~c1010EpKgp:: (visited May 10, 2007).

201. Dubow (1991), 615.

202. Dubow (1991), 616. See also Hugh R. Slotten, *Radio and television regulation: Broadcast technology in the United States, 1920–1960* (Baltimore: Johns Hopkins University Press, 2000).

203. "The Caption Center turns 20" (1992), 29; *Television decoder circuitry act of 1990,* SR 1974, 101st Congress, 1st sess. (1990), 6.

204. *Television decoder circuitry act of 1989,* HR 4163, 101st Cong., 1st sess. (1990), http://thomas.loc.gov/cgi-bin/query/z?c101:H.R.4163: (visited May 10, 2007).

205. Kim McAvoy, "Captioning bill on move in House," *Broadcasting* (7 May 1990), 57.

206. Ferguson (1990); "New law will expand TV captions for the deaf," *NYT* (16 Oct. 1990); House, *Television decoder circuitry act of 1990* (1990), 8.

207. Dubow (1991), 613.

208. Senate, *Television decoder circuitry act of 1990* (1990), 3, words of committee report.

209. Ferguson (1990).

210. "John Ball: TV's wordsmith" (1990), 87.

211. Harry Jessell, "HDTV bill tabled, closed captioning law likely," *Broadcasting* (1 Oct. 1990), 59; Senate, *Television decoder circuitry act of 1990* (1990), 1.

212. Ferguson (1990).

213. Frank G. Bowe, letter to the editor, *Fortune* (24 May 1999), 32.

214. Dubow (1991), 616; Senate, *Television decoder circuitry act of 1990* (1990), 5–6.

215. Ferguson (1990).

216. McAvoy (1990), 57.

217. House, *Television decoder circuitry act of 1990* (1990), 2, 5.

218. Senate, *Television decoder circuitry act of 1990* (1990), 7.

219. Dubow (1991), 609.

220. *Caption Center News* (summer 1991).

221. Harry Jessell, "Captioning capability for TV sets becomes law," *Broadcasting* (22 Oct. 1990), 44.

222. David J. Jefferson, "Television caption providers are gearing up for 1993," *WSJ* (21 Oct. 1991); Jessell, "Captioning capability for TV sets becomes law" (1990), 44.

223. Jeff Hutchins, "The captioning technology of the future is being developed today," *SHHH Journal* (Sep.–Oct. 1993), 22–24, 23.

224. Stuart N. Brotman, "Closed-caption television finally comes into its own," *Christian Science Monitor* (23 Dec. 1991).

225. *Caption* (Fall 1989).

226. Brotman (1991).

227. Howard Rosenberg, "Caption services get a workout," *San Francisco Chronicle* (26 Feb. 1991).

228. Brotman (1991).

229. Tim Berners-Lee, with Mark Fischetti, *Weaving the Web: The original design and ultimate destiny of the World Wide Web by its inventor* (San Francisco: HarperBusiness, 2000).

230. W. Russell Neuman, Lee McKnight, and Richard Jay Solomon, *The gordian knot: Political gridlock on the information highway* (Cambridge, MA: MIT Press, 1998).

231. *National communications competition and information infrastructure act of 1994*, H. Rep. 103-560, 103rd Cong., 2d sess. (24 Jun. 1994).

232. Chris McConnell, "FCC launches closed-captioning rulemaking," *B&C* (11 Dec. 1995), 22.

233. *Developments in aging, 1996, vol. 3: A report of the special committee on aging, United States Senate*, S. Rep. 105-036, 105th Cong., 2d sess. (30 Apr. 1998).

234. "FCC 47 CFR Parts 73 and 76, In the matter of closed captioning and video description of video programming: Notice of inquiry," *FR* (18 Dec. 1995), 65052–65054.

235. *Telecommunications act of 1996: Concerence report (to accompany S.652)*, S.Rep. 104-230 (Washington, DC: GPO, 1996).

236. McConnell, "FCC launches closed-captioning rulemaking" (1995), 22; *Developments in aging* (1998); "FCC, closed captioning and video description of video programming: Notice; report to Congress," *FR* (14 Aug. 1996), 42249–42250.

237. James H. Snider, "Local TV news archives as a public good," *Harvard International Journal of Press Politics* 5:2 (2000), 111–117, 112.

238. Carl Jensema, Ralph McCann, and Scott Ramsey, "Closed captioned television presentation speed and vocabulary," *AAD* 141:4 (1996), 284–292, 285.

239. Chris McConnell, "FCC proposes closed captioning by 2005," *B&C* (13 Jan. 1997), 6; "FCC 47 CFR Parts 25, 26, 73, 76 and 100, Closed Captioning of Video Programming: Proposed rule," *FR* (3 Feb. 1997), 4959–4965.

240. Chris McConnell, "FCC asked to go easy on captioning," *B&C* (25 Mar. 1996), 22.

241. "NCI tells FCC opponents are overstating closed-captioning costs," *Communications Daily* (2 Apr. 1997); "NCI tells FCC opponents are overstating closed-captioning costs," *PBR* (4 Apr. 1997).

242. Chris McConnell, "Programmers seek exemptions from closed caption rules," *B&C* (10 Mar. 1997), 23.

243. "TV industry supports captioning rules, with many exemptions," *PBR* (7 Mar. 1997).

244. "TV industry supports captioning rules" (1997); Chris McConnell, "No Hammersteins here, cable insists," *B&C* (28 Jul. 1997), 22.

245. David Hatch, "Everybody likes captions, but read between the lines," *Electronic Media* (10 Mar. 1997).

246. Chris McConnell, "New shows must be captioned by 2006, FCC mandates," *B&C* (11 Aug. 1997), 12; "FCC 47 CFR part 79, closed captioning of video programming: Final rule," *FR* (16 Sep. 1997), 48487–48496.

247. "FCC begins move toward closed-captioning, provides exemptions," *PBR* (22 Aug. 1997).

248. Bill McConnell, "FCC requires more captioning," *B&C* (21 Sep. 1998), 11.

249. "Group says captioning rules will leave 'huge gaps' for hearing-imparied," *PBR* (19 Sep. 1997).

250. "FCC 47 CFR part 79, closed captioning of video programming: Proposed rule," *FR* (21 Jan. 1998), 3070–3075; "FCC 47 CFR part 79, closed captioning of video programming: Final rule," *FR* (20 Oct. 1998), 55959–55963.

251. McConnell, "FCC requires more captioning" (1998), 11.

252. Michael R. Brentano and Jefferson C. McConnaughey, "TV access for the deaf, blind reaches new heights," *National Law Journal* (30 Oct. 2000).

253. Lawrie Mifflin, "Closed-captioning opposed for 'Springer' " *NYT* (5 Mar. 1998); "Congress opposed to captioning Springer," *Weekend All Things Considered* (8 Mar. 1998).

254. Dan Trigoboff, "Senators fight 'Springer' captions," *B&C* (9 Mar. 1998), 14.

255. "Closed captioning controversy," *Good Morning America* (6 Mar. 1998).

256. Trigoboff (1998), 14.

257. NAD, "Education Secretary Riley responds to senators on Springer caption funding" [reprint of Education Secretary Richard W. Riley to Sen. Joseph Lieberman, 30 Mar. 1998], www.nad.org/site/pp.asp?c=foINKQMBF&b=178021 (visited 21 Feb. 2004).

258. "Congress opposed to captioning Springer" (1998).

259. Joseph Pitts and Roy Blunt, "Send dollars to the class or fund 'The Jerry Springer Show': It's your choice," *Congressional Press Releases* (9 Apr. 1998).

260. Oklahoma Council of Public Affairs, www.ocpathink.org/ (visited 23 Jul. 2003); Susie Dutcher, testimony, *Federal Document Clearing House Congressional Testimony* (22 Apr. 1998).

261. Pete Hoekstra, testimony, *Federal Document Clearing House Congressional Testimony* (17 Jul. 1998).

262. *Individuals with disabilities education act amendments of 1996*, S. Rep. 104-275, 104th Cong., 2d sess. (20 May 1996).

263. "Department of Education, Office of Special Education and Rehabilitative Services; Special Education—technology and media services for individuals with disabilities program: Notice inviting public comments," *FR* (17 Dec. 1999), 70981–70983.

264. "Department of Education, Office of Special Education and Rehabilitative Services: Notice inviting applications for new awards for fiscal year 2003," *FR* (28 Jul. 2003), 44317–44328.

265. Zay N. Smith "Advocates for deaf charge censorship in closed captioning," *Chicago Sun-Times* (20 Feb. 2004).

266. NAD, "Television captioning censorship hurts family values," press release, 2 Oct. 2003, www.nad.org/site/pp.asp?c=foINKQMBF&b=179707 (visited 7 May 2007).

267. Smith (2004).

268. NAD, "Television captioning censorship hurts family values" (2003).

269. "Captioned TV demonstrated at Tennessee conference," *DA* (Jan. 1972), 23.

270. *Caption* (fall 1985); *Caption* (fall 1989).

271. Bill Carter, "Julius Barnathan, 70, innovator in television technology at ABC," *NYT* (5 Dec. 1997); John Eggerton, "Julius Barnathan, 1927–1997," *B&C* (8 Dec. 1997), 10.

272. *Late Night with Conan O'Brien* (29 Aug. 1995; 6 Nov. 1995; 26 Dec. 1997).

273. "Eric Daggens discusses CNN Headline News' new look and its intended audience," *All Things Considered* (6 Aug. 2001).

274. Caryn James, "Splitting. Screens. For minds. Divided," *NYT* (9 Jan. 2004).

275. TV Shield, www.tvtape.com (visited 3 Mar. 2004).

276. Norman Black, "Deaf will soon be able to follow prime time television," *AP* (6 Jan. 1980).

277. Jerome D. Schein and Ronald Hamilton, *IMPACT 1980: Telecommunications and deafness* (Silver Spring: NAD, 1980), 99; Karen M. Bellefleur and Philip A. Bellefleur, "An experimental RF subcarrier system to provide a mass

communication system for the deaf," in Vest (1978), 31–34; K. M. Bellefleur and P. A. Bellefleur, "Radio-TTY: A community mass media system for the deaf," *VR* 81 (1979), 35–39.

278. M. Goldfarb, testimony, *Federal Document Clearing House Congressional Testimony* (24 May 1994).

279. Matt Dellinger, "Meet Oscar's transcriber," *New Yorker* (27 Mar. 2000), 39.

280. "Resolution of the ad hoc group to promote closed-captioned television," *DA* 33:7 (1981), 2–3; Harlan Lane, Robert Hoffmeister, and Ben Bahan, *A journey into the Deaf-World* (San Diego: DawnSign Press, 1996), 437; Edward Dolnick, "Deafness as culture," *Atlantic Monthly* (Sep. 1993), 37–53, 53.

281. R. Greg Emerton, "Marginality, biculturalism, and social identity of deaf people," in Ila Parasnis, ed., *Cultural and language diversity and the deaf experience* (New York: Cambridge University Press, 1996), 136–145, 142.

282. Schein & Hamilton (1980), 6.

283. Robert Angus, "I want my CCTV: TV goes to the masses," *Omni* (Sep. 1993), 16.

284. Cynthia M. King and Carol J. LaSasso, "Your crucial role in the future of television captioning," *SHHH Journal* (Sep.-Oct. 1992), 14–16, 15.

## Chapter 6. Privatized Geographies of Captioning and Court Reporting

*Epigraph:* Mary H. Knapp and Robert W. McCormick, *The complete court reporter's handbook,* 3rd ed. (Upper Saddle River, NJ: Prentice-Hall, 1999), 19.

1. David Robb, "Captioners reject contract," *BPI Entertainment News Wire* (16 Aug. 1993); David Robb, "NCI vote authorizes strike," *Hollywood Reporter* (26 Aug. 1993); David Robb, "NCI reading picket signs," *Hollywood Reporter* (9 Sep. 1993); David Robb, "Captioners, NCI come to terms on 3-year pact," *Hollywood Reporter* (1 Oct. 1993).

2. Robb, "NCI vote authorizes strike" (1993); Daniel Cox, "NABET gets OK to strike H'wood org," *Daily Variety* (26 Aug. 1993), 5.

3. "Closed captioning calls for writing between the lines," *Washington Post* (30 Jan. 1995); Ann Brown, "A stroke of success," *Black Enterprise* (Aug. 1997), 30.

4. Peter Wacht, "Captioning: Opportunities, challenges and the future," *JCR* (Jul.–Aug. 2000), 52–59, 55.

5. WGBH Caption Center, *Closed-captioned local news: Getting started in your town* (Boston: The Center, 1990); Larry Goldberg, "Article left out other captioning agencies," *Electronic Media* (26 Mar. 1990); "The Caption Center turns 20," *Deaf Life* (Mar. 1992), 29.

6. Al Spoler, "Captioning industry poised for expansion," *Electronic Media* (21 Jun. 1993).

7. Teresa Gubbins, "TV captions fail spelling test," *Dallas Morning News* (6 Jan. 1999).

8. Commission on Education of the Deaf (COED), *Toward equality: Education of the deaf* (Washington, DC: GPO, 1988), 111, 117.

9. David J. Jefferson, "Television caption providers are gearing up for 1993," *WSJ* (21 Oct. 1991.

10. Jo Ann McCann, "Who pays for all this? Federal funding for closed-captioned television," *SHHH Journal* (Sep.–Oct. 1993), 35.

11. "Educational Media Research, Production, Distribution, and Training Programs; Notice of funding priorities," *FR* (7 Nov. 1994).

12. "Closed captioning calls for writing between the lines" (1995).

13. "The Caption Center turns 20" (1992), 20–22.

14. "BellSouth provides closed captioning for NBC coverage of the 1996 Olympic Games," *PR Newswire* (1 Jul. 1996).

15. Jefferson (1991).

16. Jefferson (1991).

17. Peter M. Nichols, "Home video," *NYT* (10 Nov. 1995).

18. Stuart Gopen, *Gopen's guide to closed captioned video* (Framingham, MA: Caption Database, 1993).

19. Computer Prompting & Captioning Co., advertisement, *AAD* 138:5 (1993), back cover.

20. Trudi Miller Rosenblum, "Closed-captioning now widespread," *Billboard* (23 Jul. 1994), 79–80.

21. "A fair hearing for closed captioning," *Washington Post* (30 Jan. 1994).

22. Michael Erard, "The king of closed captions," *Atlantic Monthly* (Sep. 2001), 24–25.

23. Linda D. Miller, "What is real-time captioning and how can I use it?" *SHHH Journal* 10:1 (1989), 7–10, 10.

24. Jefferson (1991).

25. Heather Draper, "Company has FCC regulations to thank for new, plentiful markets," *AP* (5 Jan. 2002).

26. "Closed captioning calls for writing between the lines" (1995).

27. *Caption* (fall 1988).

28. *Caption Center News* (Mar.–Apr. 1990).

29. Author interview (27 Jan. 2004), Madison, WI.

30. US Congress, House of Representatives, *Depts. of Labor, Health and Human Services, and Education . . . Appropriation Bill 1998* (25 Jul. 1997).

31. Dick Youngblood, "With thanks to government," *Star Tribune* (Minneapolis, MN) (1 Oct. 2000).

32. "Department of Education, Office of Special Education and Rehabilitative Services: Notice inviting applications for new awards for fiscal year (FY) 2002," *FR* (8 Jul. 2002), 45277–45281.

33. Peggy Belflower, in Dan Heath, "The thrill of it all" (interview with Belflower), *JCR* (Jul. 1991), 28–29.

34. Charlotte Eby, "Court reporters needed as only one Iowa school offers courses," *AP* (8 Oct. 2001).

35. Carol Ann Riha, "Bipartisan bill seeks $75 million for captioning," *AP* (16 May 2002); John Nolan, "Broadcast captioning luring court reporters," *AP* (17 Jan. 2004).

36. Joseph R. Karlovits, "Captioning 2000: A glimpse of the future," *JCR* (May 1999), 60–62, 60.

37. Wacht (2000), 55.

38. "The Caption Center turns 20" (1992), 28; Wacht (2000), 55.

39. William A. McNutt, "Closed captioning—and translating—the Pope," *NSR* (Feb. 1988), 44.

40. Wacht (2000), 53.

41. "XScribe helps bring Republican convention to hearing impaired," *PR Newswire* (17 Aug. 1988); Marty Block and Jeff Hutchins, "The politics of captioning," *NSR* (May 1989), 20–21.

42. Joseph R. Karlovits, "Captioning *The Today Show,*" *NSR* (Aug. 1989), 20–21; Claudia Coates, "Closed-caption writers tackle *The Tonight Show,*" *AP* (1 Jun. 1992).

43. Jefferson (1991); "The caption race: Who's ahead and what's ahead," *Deaf Life* (Feb. 1990), 9–14, 14.

44. "The caption race" (1990), 10, 12–13.

45. Matt Dellinger, "Meet Oscar's transcriber," *New Yorker* (27 Mar. 2000), 39; Joseph R. Karlovits, "Has reporter education hit the skids?" *JCR* (Apr. 1996), 41.

46. Karlovits (1999), 60.

47. Tim Carvell, "Where do those TV closed captions come from?" *Fortune* (26 Apr. 1999), 57; Wacht (2000), 55; Dan Weissman, "Court reporters find niche in closed captioning," *Newhouse News Service* (28 Jan. 2000).

48. Karlovits (1999), 62.

49. Wacht (2000), 56.

50. WordWave, www.wordwave.com (visited 5 Oct. 2003).

51. Dellinger (2000), 39.

52. Grace Rishell, "One of 'world's best' wrote closed captions for Oscars," *Pittsburgh Post-Gazette* (29 Mar. 2000).

53. Karlovits (1996), 41.

54. Thomas W. Holcomb Jr., "The art of reading television," *NYT* (31 Jan. 2000).

55. Amy Bowlen and Kathy DiLorenzo, *Realtime Captioning . . . The VITAC Way* (VITAC, 2003).

56. B. J. Quinn, "Diary of a captioning newbie," *JCR* (Sep. 2002), 68–70, 69.

57. B. J. Quinn, "Diary of a captioning newbie (part 2)," *JCR* (Nov.–Dec. 2002), 72–73.

58. WordWave.

59. Judith H. Brentano, "Hearing—and seeing—testimony," *National Law Journal* (21 Jun. 1993); Wacht (2000), 52; Judith H. Brentano, "Access through real-time reporting," *SHHH Journal* (Sep.–Oct. 1993), 34–36, 36; Judy Brentano, "Let the realtime games begin!" *JCR* (Feb. 1997), 38–40, 39; Jenny Price, "Demand increasing for court reporters," *AP* (22 Apr. 2001).

60. WordWave.

61. Hope Viner Samborn, "Court reporter buyouts," *ABA Journal* (Aug. 1998), 32.

62. Thom Weidlich, "Consolidation and the reporting industry," *JCR* (Apr. 1998), 52–55, 54; Thom Weidlich, "An update: Consolidation and the reporting industry," *JCR* (Sep. 1999), 62–64, 63.

63. US Census Bureau, *Business support services: 2002* (Washington, DC: US Dept. of Commerce, 2004).

64. US Bureau of Labor Statistics, *Occupational employment statistics survey* (1999, 2004), www.bls.gov/oes/current/oes232091.htm; www.bls.gov/oes/oes_dl .htm (visited 1 Jun. 2007).

65. Samborn (1998).

66. "WordWave launches ReporterCentral.com," *Business Wire* (14 Aug. 2000); Wacht (2000), 55.

67. Holcomb (2000).

68. WordWave.

69. Nolan (2004).

70. Eby (2001).

71. Nolan (2004).

72. Tim Feran, "What's the word? Caption writers for newscasts must combine speed and accuracy," *Columbus Dispatch* (9 Mar. 2004).

73. Lance W. Steinbeisser, "An interview with captioner Meryl, aka Mary Kay, Webster," *Transcript* (summer 1996), 13–15, 14.

74. Wacht (2000), 54.

75. Steve Alexander, "Minnesota company provides closed-captioning for Olympics," *AP* (17 Feb. 2002); Cassie Lee Bichy and Monette Benoit, "Sharpening the cutting edge: From CAT to the Internet," *JCR* (Aug.–Sep. 1998), 56–61, 60.

76. Jennifer L. C. Pridmore, "The good, the bad, and the late breaking," *JCR* (Sep. 2003), 66–67.

77. Gary D. Robson, "Captioning at home," *JCR* (Mar. 2000), 60–62, 61–62.

78. "The Caption Center turns 20" (1992), 21; Weissman (2000).

79. Wacht (2000), 53.

80. Tara Bahrampour, "Court reporters hear the tap, tap, tap of obsolescence," *NYT* (13 Feb. 2000).

81. Jennifer Jacobs, "Doreen Radin provides TV closed captioning from home," *AP* (2 Feb. 2000).

82. Jennifer Jacobs, "News captioner casts her spell," *Post-Standard* (Syracuse, NY) (26 Jan. 2000); Wacht (2000), 54.

83. Michele Pennington, "Working from home: Is it for you?" *JCR* (Oct. 2001), 62–63.

84. Carl Adrian Riffe, *Verbatim court reporting—there ain't such a thing: A common sense approach to becoming and being a court reporter* (Anniston, AL: Higginbotham, 1986), 38.

85. Kathleen Ganster, "Seeing is believing," *Pittsburgh Post-Gazette* (17 Feb. 1999).

86. Charles Etlinger, "Woman uses court reporting skills to help hearing-impaired students," *AP* (17 May 2000).

87. Shera Dalin, "Watching out loud," *St. Louis Post-Dispatch* (9 Nov. 1998).

88. Jennifer L. C. Pridmore, "Transitioning between reporting and CART/captioning," *JCR* (Sep. 2001), 52–54, 54.

89. Denise L. Doucette, "Feast or famine: What to do?" *JCR* (Jun. 2000), 54–57, 56.

90. Robert W. McCormick, "Realtime for dummies," *JCR* (Jul.–Aug. 2000), 84–86, 85.

91. Knapp & McCormick (1999), 246; Illinois Occupational Skill Standards and Credentialing Council, *Court reporter/captioner* (Springfield: Illinois State Board of Education, 2000), 28.

92. Kay Moody, "New markets for court reporters," *JCR* (Apr. 1995), 34–35; Pat Forbis, "Parallel professions: Court reporting & medical transcription," *JCR* (Apr. 1995), 38–39.

93. Monette Benoit, "Wanted: Reporters for cybercasting careers," *JCR* (May 1998), 60–62, 62.

94. Bichy & Benoit (1998), 57.

95. Thom Weidlich, "Realtime to the net," *JCR* (Sep. 2002), 62–64, 63; "Streampipe announces captioning solution and compliance with new federal IT accessibility initiative," *PR Newswire* (27 Jun. 2001); Jennifer 8. Lee, "Retooling products so all can use them," *NYT* (21 Jun. 2001).

96. Cynthia Waddell, " 'Electronic curbcuts': The ADA in cyberspace," *Human Rights* 27:1 (2000), 22.

97. Lee (2001); "Architectural and Transportation Barriers Compliance Board, 36 CFR Part 1194 RIN 3014-AA25, Electronic and Information Technology Accessibility Standards: Final rule," *FR* (21 Dec. 2000), 80499–80528.

98. Jennifer L. C. Pridmore, "Remote CART: When the provider isn't there," *JCR* (May 2002), 66–68, 67.

99. "Centra Software and Caption Colorado partner to provide deaf and hearing impaired populations with access to real-time, online Web collaboration," *Business Wire* (8 Apr. 2003); "WordWave and e-Media partner to deliver first live Webcast with closed captioning" *Business Wire* (21 May 2001).

100. Pridmore, "Remote CART" (2002), 67.

101. Thom Weidlich, "Reporting and technology: What does the future hold?" *JCR* (Mar. 1999), 57–59; 59.

102. Weidlich (2002), 63.

103. Feran (2004).

104. Brentano, "Access through real-time reporting" (1993), 35.

105. Peter Wacht, "Five hundred new jobs: How you can prepare," *JCR* (May 1996), 40–42, 40.

106. Price (2001).

107. Wacht (2000), 58.

108. "Department of Education, Office of Special Education and Rehabilitative Services; Grant Applications under Part D, Subpart 2 of the Individuals with Disabilities Education Act Amendments of 1997: Notice inviting applications for new awards for fiscal year 1999," *FR* (4 Jan. 1999), 351–362; "Department of Education, Office of Special Education and Rehabilitative Services: Notice inviting applications for new awards for fiscal year (FY) 2002," *FR* (2002); Andrew Mollison, "Caption-makers for live TV are few and far between," *Cox News Service* (24 Mar. 2004).

109. Weissman (2000).

110. Weissman (2000); Donna M. Gaede, "Factors affecting machine shorthand skill development required for registered professional reporter certification" (M.S. thesis, Southern Illinois University, 2000), 6.

111. Wacht (2000), 57.

112. Don J. DeBenedictis, "Excuse me, did you get all that?" *ABA Journal* (May 1993), 84; William E. Hewitt and Jill Berman Levy, *Computer aided transcription: Current technology and court applications* (Williamsburg, VA: NCSC, 1994), 22.

113. Sacha Pfeiffer, "Instant access to Law's testimony," *Boston Globe* (9 May 2002); Weidlich, "Reporting and technology" (1999), 58.

114. Gaede (2000), 20.

115. Hewitt & Levy (1994), 21.

116. US Dept. of Labor, Bureau of Labor Statistics, *Occupational outlook handbook 2002–2003*, "Court Reporters," www.bls.gov/oco/ocos152.htm (visited 25 Apr. 2003); Mark J. Golden, "NCRA gains $5.75 million for training real-timers/captioners," *JCR* (Mar. 2002), 66–68; William M. Bulkeley, "Why stenotypists are trying to keep Paula O'Regan quiet," *WSJ* (3 Jun. 2002); Jacqueline Schmidt, "5 ways to protect your future," *JCR* (Sep. 2000), 68–70, 68.

117. Nolan (2004).

118. Wacht (2000), 56.

119. Wacht (2000), 56.

120. Gaede (2000), 27.

121. Joyce P. Jacobsen, "Results from a national survey of court reporters," *JCR* (Jun. 1993), 42–47, 47.

122. Gaede (2000), 21, 29.

123. James Hebert, "Closed captioning reaches beyond its intended audience," *Copley News Service* (15 Jul. 2002).

124. Bulkeley (2002); Gaede (2000), 1.

125. Dave Wenhold, "A call to action," *JCR* (Jul.–Aug. 2001), 61.

126. Pat Barkley, "Court reporting done right," *Trial* (Jun. 2001), 67–69.

127. Gaede (2000), 19.

128. Weissman (2000).

129. Mollison (2004).

130. Mollison (2004).

131. Mark J. Golden, "NCRA continues its push for federal funds for training realtimers/captioners," *JCR* (Jul.–Aug. 2001), 58–61, 59.

132. US Congress, House of Representatives, *Making omnibus consolidated and emergency supplemental appropriations for fiscal year 2001,* H. Report 1033 (15 Dec. 2000).

133. Mark J. Golden, "NCRA gains federal funds for training realtimers/captioners," *JCR* (Mar. 2001), 62–64, 62–63.

134. Golden, "NCRA continues its push" (2001), 59–60.

135. Claude R. Marx, "Shortage of closed captioners deprives hard-of-hearing viewers," *AP* (3 Jun. 2001); *Training for closed captioners act of 2001,* HR 2527, 107th Cong., 1st sess. (2001), http://thomas.loc.gov/cgi-bin/query/z?c107:H.R.2527: (visited 7 May 2007).

136. *Training for closed captioners act,* HR 2527 (2001).

137. Golden, "NCRA gains $5.75 million" (2002), 66.

138. "Building the profession's future," *JCR* (May 2003), 58–61, 58; Golden, "NCRA gains $5.75 million" (2002), 66–68.

139. Riha (2002).

140. "Members of Congress urge House leadership to enact communication access legislation for people who are deaf or hard-of-hearing," *US Newswire* (22 Oct. 2002).

141. "Building the profession's future" (2003), 61.

142. Anne Marie Borrego, Jeffrey Brainard, Richard Morgan, and Ron Southwick, "Congressional earmarks for higher education, 2002," *Chronicle of Higher Education* (27 Sep. 2002).

143. *Training for realtime writers act of 2003,* HR 970, 108th Cong., 1st sess. (27 Feb. 2003), http://thomas.loc.gov/cgi-bin/query/z?c108:H.R.970: (visited 7 May 2007); *Training for realtime writers act of 2003,* SR 480, 108th Cong., 2d sess. (27 Feb. 2003), http://thomas.loc.gov/cgi-bin/query/D?c108:1:./temp/~c108m MMtsC:: (visited 7 May 2007).

144. "Building the profession's future" (2003), 58–61.

145. *Training for realtime writers act of 2003,* SR 480 (2003).

146. *Departments of Labor, Health and Human Services, and Education, and related agencies appropriation bill, 2004,* 108th Cong., 1st sess., 2003, S. Rep. 108-81, http://thomas.loc.gov/cgi-bin/cpquery/R?cp108:FLD010:@1(sr81): (visited 7 May 2007). Jun.

147. *Training for realtime writers act of 2004,* SR 480 (2004).

148. Mollison (2004).

149. Mary Beth Johnson, "From a distance," *JCR* (Apr. 1998), 67–68.

150. Jacqueline Schmidt, "Remote teaching," *JCR* (Jan. 2002), 66–67, 66.

151. Price (2001).

152. Eby (2001).

153. "Midwest City's Rose State College receives $1M court reporting grant," *Journal Record* (Oklahoma City, OK) (24 Jun. 2002).

154. Patrik Jonsson, "No one to tap the keys of justice," *Christian Science Monitor* (24 Jan. 2003).

155. Sacha Pfeiffer, "Court reporters are precious few," *Boston Globe* (26 Oct. 2000).

156. Jonsson (2003).

157. Eby (2001).

158. Michele Kurtz, "Transcript lag a trial for juries," *Boston Globe* (17 Jan. 2002).

159. Jonathan Sterne, *The audible past: Cultural origins of sound reproduction* (Durham, NC: Duke University Press, 2003).

160. R. Houde, "Prospects for automatic recognition of speech," *AAD* 124:5 (1979), 568–72, 568.

161. D. Raj Reddy, "Speech recognition by machine: A review," *IEEE Proceedings* 64 (1976), 502–531.

162. Allen Newell et al., *Speech understanding systems: Final report of a study group* (New York: American Elsevier, 1973); National Research Council (NRC), *Funding a revolution: Government support for computing research* (Washington, DC: National Academy Press, 1999).

163. NRC (1999); Dennis H. Klatt, "Review of the ARPA Speech Understanding Project," *Journal of the Acoustical Society of America* 62 (1977), 1324–1366.

164. Martha K. Reifschneider, "Voice recognition technology and court reporting: Current research efforts and trends at IBM," *JCR* (Jan. 1994), 26.

165. Hebert (2002).

166. Snyder S, Congressional testimony (12 Feb. 1998); Martha Reifschneider, "The latest on voice recognition technology," *JCR* (Jan. 1996), 38–41, 41.

167. Bulkeley (2002).

168. Marshall Jorpeland, "An interview with a stenomask reporter," *JCR* (Feb. 1995), 34–37, 34.

169. Loretta A. Armstrong, "Stenotype vs. stenomask," *Transcript* (Dec. 1979), 41–44, 41.

170. Jorpeland (1995), 35, quoting Anita Glover.

171. Eugene A. Sattler, "On multi-track voice writing," *Transcript* (Jun. 1974), 7–9, 8.

172. "NCRA Blue Ribbon Commission final report on voice technology," *JCR* (Jun. 2003), 62–66, 63.

173. "NCRA Blue Ribbon Commission" (2003), 66.

174. Bulkeley (2002).

175. "NCRA Blue Ribbon Commission" (2003), 65.

176. Bulkeley (2002).

177. Bulkeley (2002).

178. "NCRA Blue Ribbon Commission" (2003), 63–65.

179. "Members reject voice writer amendment," *JCR* (Oct. 2001), 55–56.

180. *Designating August 3, 2001, as "National Court Reporting and Captioning Day,"* SR 95, 107th Cong., 1st sess., http://thomas.loc.gov/cgi-bin/query/z?c107:S.RES.95.IS: (visited 7 May 2007).

181. Robert Jackson, "Television and the deaf," *DA* 26:8 (1974), 10–11.

182. Robert R. Davila, "Technology and full participation for children and adults who are deaf," *AAD* 139 (1994), 6–9, 8.

183. Aaron Steinfeld, "The benefit of real-time captioning in a mainstream classroom as measured by working memory," *VR* 100:1 (1998), 29–44, 29.

## Conclusion

*Epigraph: Caption* (fall 1980).

1. Clarence J. Blake, "The glossograph," *American Journal of Otology* 4 (1882), 190–193.

2. Author interview (27 Jan. 2004), Madison, WI.

3. Carol M. Neal, "Adding new value to official reporting," *JCR* (Jan. 1995), 30–33, 31.

4. David Perry, "Digital video + Transcript = New market opportunity," *JCR* (Jan. 1998), 33–35, 34.

5. Thom Weidlich, "Online transcript repositories: How they work," *JCR* (May 2001), 60–62, 60.

6. Gary Robson, "Transcript format standards," *JCR* (May 2000), 58–59; Eddie O'Brien, "Legal XML: New standards for legal documents," *JCR* (May 2000), 61–63, 61.

7. Peter J. Howe, "WordWave launches net-search division," *Boston Globe* (2 Oct. 2000).

8. Brad Gilmer, "Media asset-management systems," *BE* (1 Aug. 2003).

9. Steven Vedro, "Why metadata matters: it greases digital wheels," *Current* (10 Sep. 2001).

10. Karen Anderson, "Covering your assets," *B&C* (1 Nov. 1999), 34–37, 34; Jeffrey A. Vilensky, Raymond Lee Katz, and John T. Williams, *Digital media: Maximizing the value of content* (New York: Bear, Stearns, 2001), 5–11; Vedro (2001).

11. W. Klippgen, T. D. C. Little, G. Ahanger, and D. Venkatesh, "The use of metadata for the rendering of personalized video delivery," in Amit Sheth and Wolfgang Klas, eds., *Multimedia data management: Using metadata to integrate and apply digital media* (New York: McGraw-Hill, 1998), 287–318, 310; Wei Qi, Lie Gu, Hao Jiang, Xiang-Rong Chen, and Hong-Jiang Zhang, "Integrating visual, audio and text analysis for news video," *Proceedings, International Conference on Video Processing 2000*, 3 (2000), 520–523, 520.

12. Anderson (1999), 34; Vilensky, Katz & Williams (2001), 17.

13. Charles B. Owen and Fillia Makedon, *Computed synchronization for multimedia applications* (Boston: Kluwer Academic, 1999), 1, 58, 117.

14. Mary Ide and Leah Weisse, "Developing preservation appraisal criteria for a public broadcasting station," *Moving Image* 3:1 (2003), 146–157, 155.

15. Gilmer (2003).

16. Heather Salerno, "New at the top," *Washington Post* (7 Jul. 1997).

17. Paul Placeway and John Lafferty, "Cheating with imperfect transcripts," *Proceedings, Fourth International Conference on Spoken Language* 4 (1996), 2115–2118; A. Bagga, Jianying Hu, Jialin Zhong, and G. Ramesh, "Multi-source combined-media video tracking for summarization," *16th International Conference on Pattern Recognition* 2 (2002), 818–821, 818; N. Babaguchi, Y. Kawai, and T. Kitahashi, "Event based indexing of broadcasted sports video by intermodal collaboration," *IEEE Transactions on Multimedia* 4:1 (2002), 68–75, 69.

18. Geoffrey C. Bowker, *Memory practices in the sciences* (Cambridge, MA: MIT Press, 2005).

19. Barbara P. Semonche, "News library history," in Barbara P. Semonche, ed., *News media libraries: A management handbook* (Westport, CT: Greenwood, 1993), 1–45, 4.

20. Alan Lewis, "A history of television newsgathering formats," in Steven Davidson and Gregory Lukow, eds., *Administration of television newsfilm and videotape collections: A curatorial manual* (Los Angeles: American Film Institute, 1997), 11–30, 18; Ernest J. Dick, "Appraisal of collections," in Davidson & Lukow (1997), 31–48, 42.

21. Lewis (1997), 22.

22. Snider (2000), 111–117, 115.

23. Davidson & Lukow (1997), xiv.

24. Ide & Weisse (2003), 148.

25. Snider (2000), 111.

26. Vilensky, Katz & Williams (2001), 12.

27. Anderson (1999), 34.

28. Vedro (2001).

29. Anderson (1999), 34.

30. Vedro (2001).

31. Ide & Weisse (2003), 149.

32. Saul Hansell, "Google and Yahoo are extending search ability to TV programs," *NYT* (25 Jan. 2005); Google Video, http://video.google.com/ (visited 5 Oct. 2005).

33. Video Monitoring Services of America, description at http://cisweb .lexis-nexis.com.ezproxy.library.wisc.edu/ (visited 6 Oct. 2005).

34. Video Monitoring Services of America, www.vidmon.com/ (visited 6 Oct. 2005).

35. Davidson & Lukow (1997).

36. Snider (2000), 111.

37. Ivarsson (1992), 9.

38. Luyken (1991), 171; Dries (1995), 27.

39. Luyken (1991), 42.

40. Ivarsson (1992), 77.

41. Helene J. B. Reid, "Sub-titling, the intelligent solution," in P. A. Horguelin, ed., *La traduction, une profession* (Ottawa: Canadian Translators and Interpreters Council, 1978), 420–428, 425.

42. Dirk Delabastita, "Translation and the mass media," in Susan Bassnett and André Lefevere, *Translation, history and culture* (London: Pinter, 1990), 99.

43. Ivarsson (1992), 11.

44. Henry Fischbach, "Translation in the United States," *Babel* 7:3 (1961), 119–123.

45. Kurt Gingold, "The financial status of translators and how it can be improved," *Babel* 14:4 (1968), 217–221, 217.

46. Lewis Galantière, "On translation as a profession," *Babel* 16:1 (1970), 30–33, 32.

47. Galantière (1970), 33.

48. Fischbach (1961), 120.

49. Gingold (1968), 217–218, 220; Fischbach (1961), 120–121.

50. Gingold (1968), 219.

51. Herma Briffault, "The plight of the literary translator especially in the USA," *Babel* 10:1 (1964), 12–14, 13.

52. Lawrence Venuti, *The translator's invisibility: A history of translation* (London: Routledge, 1995), 11.

53. Galantière (1970), 30.

54. Venuti (1995), 10.

55. Briffault (1964), 12.

56. Venuti (1995), 2.

57. Lawrence Venuti, ed., *The translation studies reader* (London: Routledge, 2000), 1; Venuti (1995), vii, Bassnett & Lefevere's words.

58. André Lefevere, "Translation: Its genealogy in the West," in Bassnett & Lefevere (1990), 14–28, 15; André Lefevere and Susan Bassnett, "Introduction: Proust's grandmother and the Thousand and One Nights: The 'cultural turn' in translation studies," in Bassnett & Lefevere (1990), 1–13, 8.

59. Venuti (2000), 11.

60. Venuti (2000), 5.

61. James W. Carey, "A cultural approach to communication," *Communication* 2:2 (1975), reprinted in James W. Carey, *Communication as culture: Essays on media and society* (New York: Routledge, 1988; 1992), 13–36.

62. Hans Vöge, "The translation of films: Sub-titling versus dubbing," *Babel* 23:3 (1977), 120–25, 120; Dirk Delabastita, "Translation and mass-communication: Film and TV translation as evidence of cultural dynamics," *Babel* 35:4 (1989), 193–218, 193; Venuti (2000), 335.

63. Delabastita (1990), 97.

64. Basil Hatim and Ian Mason, "Politeness in screen translating," in Venuti (2000), 444; Patrick Zabalbeascoa, "Translating jokes for dubbed television situation comedies," *The Translator* 2:2 (1996), 235–257, 235.

65. *Caption* (1991).

66. Lefevere & Bassnett (1990), 12.

67. Lefevere (1990), 15.

68. John C. Lee, "The nature of fansubs" (22 Aug. 1999), www.fansubs.net/view.php3?page–ature (visited 12 Sep. 2005).

69. Emily Nussbaum, "A DVD face-off between the official and the home-made," *NYT* (21 Dec. 2003).

70. Abe Mark Nornes, "For an abusive subtitling," *Film quarterly* (spring 1999).

71. "Device that filters out foul language from TV shows and movies," *CBS This Morning* (8 Jan. 1999).

72. "TV Guardian designed to protect kids," *Morning Edition* (15 Oct. 1998).

73. "Big Brother is listening," *Popular Electronics* (Jul. 1999), 23–25.

74. Ann Donahue, "A new set-top box tames video and TV," *Video Business* (28 Jan. 2002), 27; Beverly A. Carroll, "Schools will help sell profanity blocker," *Chattanooga Times Free Press* (14 Jan. 2002).

75. Clint Cooper, "See no evil, hear no evil," *Chattanooga Times Free Press* (21 Jan. 2003).

76. On Xerox, see David Owen, *Copies in seconds: Chester Carlson and the birth of the Xerox machine* (New York: Simon & Schuster, 2004).

77. Casey B. Smith, "Electrostatic reproductions," part 1, *NSR* (Dec. 1966), 17.

78. Pengad Companies Inc., "Electrostatic reproduction of transcripts cuts into reporters' shrinking profits," *NSR* (May 1967), 26–27; Charles W. Bartlett, "Judicial comfort in re: Xerox, ER, and Itek," *NSR* (May 1967), 21–22.

79. Smith (1966), 17.

80. Casey B. Smith, "Electrostatic reproductions," part 2, *NSR* (Feb. 1967), 36; Casey B. Smith, "Electrostatic reproductions," part 3, *NSR* (Jul. 1967), 26; Bartlett (1967), 22.

81. Pengad Inc. (1967), 26.

82. Smith, "Electrostatic reproductions," part 3 (1967), 25.

83. Joseph L. Ebersole, *Improving court reporting services* (Washington, DC: FJC, 1972), 33.

84. Pengad Inc. (1967), 26.

85. Scott Dean, "What's happening with transcript copy sales?" *JCR* (Feb. 1994), 20–22, 20.

86. Thom Weidlich, "How should reporters be compensated?" *JCR* (May 2003), 55–57, 57.

87. "NCRA gains victory in FLSA battle," *JCR* (Nov. 1995), 28–29.

88. Susan Leigh Star and J. Griesemer, "Institutional ecology, 'translations' and boundary objects: Amateurs and professionals in Berkeley's museum of vertebrate zoology, 1907–1939," *Social Studies of Science* 19 (1989), 387–420.

89. Etienne Wenger, *Communities of practice: Learning, meaning, identity* (Cambridge: Cambridge University Press, 1998), 108.

90. Susan Leigh Star and Anselm Strauss, "Layers of silence, arenas of voice: The ecology of visible and invisible work," *Computer Supported Cooperative Work* 8 (1999), 9–30, 10.

91. Andrew D. Abbott, *The system of professions: An essay on the division of expert labor* (Chicago: University of Chicago Press, 1988), 215.

92. Wenger (1998), 109.

93. Abbott (1988), 9, 36.

94. Jerry Whitaker, "Closed captioning for DTV," *BE* (Aug. 2001), 82–85; R. N. Blanchard, "EIA-708-B digital closed captioning implementation," *IEEE Transactions on Consumer Electronics* 49:3 (2003), 567–570, 567.

95. Gregory Forbes, "Closed captioning transmission and display in digital television," *Second International Workshop on Digital and Computational Video* (2001), 126–131, 126.

96. Jessica Sandin, "Captioning on the Web," *B&C* (13 May 1996), 64–65.

97. Whitaker (2001).

98. M. Rumreich and M. Zukas, "Closed-captioning for PIP," *IEEE Transactions on Consumer Electronics* 44:3 (1998), 922–925.

99. "DTV closed captioning tested," *BE* (Jan. 1999), 18; Glen Dickson, "WGBH develops DTV captioning," *B&C* (4 Jan. 1999), 69; Jun. Senate Special Committee on Aging, *Developments in aging: 1999 and 2000,* vol. 2, SR 107-158, 107th Cong., 2d sess., 4 Jun. 2002, www.gpo.gov/congress/senate/senate 22lp107.html (visited 7 May 2007).

100. Daniel Altman, "The FCC has voted: A digital tuner for every TV," *NYT* (9 Aug. 2002).

101. Eric A. Taub, "HDTV's acceptance picks up pace," *NYT* (31 Mar. 2003).

102. Michael Rogers, "The end of analog TV," *MSNBC* (24 Apr. 2005).

103. Whitaker (2001).

104. Forbes (2001), 129.

105. "Department of Education, Office of Special Education and Rehabilitative Services: Notice inviting applications for new awards for fiscal year (FY) 2003," *FR* (21 Jul. 2003), 43261–43270.

106. Carl Jensema and Michele Rovins, "Instant reading incentive: Understanding TV captions," *Perspectives in Education and Deafness* 16:1 (1997).

107. Carl Jensema, "Viewer reaction to different television captioning speeds," *AAD* 143:4 (1998), 318–325.

108. Carl J. Jensema, Sameh El Sharkawy, Ramalinga Sarma Danturthi, Robert Burch, and David Hsu, "Eye movement patterns of captioned television viewers," *AAD* 145:3 (2000), 275–285.

109. S. R. Gulliver and G. Ghinea, "Impact of captions on deaf and hearing perception of multimedia video clips," *Proceedings, 2002 IEEE International Conference on Multimedia and Expo* 1 (2002), 753–756, 753, 756.

110. Andrew Mollison, "Caption-makers for live TV are few and far between," *Cox News Service* (24 Mar. 2004).

111. "New analytical study of closed captioning finds audiences think it's important but improvements are needed," *Business Wire* (15 Dec. 2003).

112. Andrew Solomon, "Away from language," *NYT* (28 Aug. 1994).

113. Peter S. Schragle and Gerald C. Bateman, "Impact of captioning," *Deaf American Monograph* (1994), 101–104, 102.

114. Harlan Lane, Robert Hoffmeister, and Ben Bahan, *A journey into the Deaf-World* (San Diego: DawnSign Press, 1996), 135; Carol A. Padden, "From the cultural to the bicultural: The modern deaf community," in Ila Parasnis, ed., *Cultural and language diversity and the deaf experience* (New York: Cambridge University Press, 1996), 79–98, 85.

115. Cecelia Tichi, *Electronic hearth: Creating an American television culture* (New York: Oxford University Press, 1991), 8, 11.

116. Bill Miller, "Deaf movie fans sue theaters," *Washington Post* (21 Apr. 2000); Rebecca Winters, "Dialogue for the deaf," *Time* (1 Jul. 2002), 17.

117. Lane, Hoffmeister & Bahan (1996), 127.

118. Douglas Bullard, *Islay* (Silver Spring, MD: T. J. Publishers, 1986), 257.

119. Katherine A. Jankowski, *Deaf empowerment: Emergence, struggle, and rhetoric* (Washington, DC: Gallaudet University Press, 1997), 166.

# Index